TESTIMONIALS FOR

JOURNEY
TO THE FUTURE

"Guy Dauncey has written an imaginative *tour de force*, blending science, philosophy, and fiction into a delightful story about how we can and must change the world, resulting in a bright green future."

~ David R. Boyd, co-chair of Vancouver's Greenest City initiative and author of *The Optimistic Environmentalist*

"For those of us who have a hard time making sense of the present, much less the future, Guy Dauncey has delivered a most satisfying, visionary, and practical remedy. Woven into story form that engages from the start, his newest work reveals an entirely possible destiny of justice, shared intention, and sustainable everything. As a fusion of Dauncey's vast experience, knowledge, and prodigious energy, this story is a fascinating portrayal of how we can live if we simply do what we already know can be done. Best of all, it's thrilling to read!"

~ Mark Lakeman, founder, The City Repair Project, Portland

"You really have to read *Journey to the Future* twice. Once in paper for the great narrative and once electronically where you can click on every footnote for the amazing amount of research and information on which every detail of the narrative is based. By the time the second read is done you will have all the education that you need to go forth and make it happen."

~ Martin Golder, architect and mediator

"To create a better world we must first be able to imagine what such a world would be like. In *Journey to the Future*, Guy Dauncey embraces that task and presents an ambitious, detailed vision for a greener, fairer and very inviting world."

~ Andrew MacLeod, author of *A Better Place On Earth: The Search for Fairness in Super Unequal British Columbia*

"In making our way from where we are to where we need to be, advice from the future would be helpful. A dystopian future world might warn us what to avoid. A more utopian society could give us hope. Somewhere between the two, from future conflicts, revolution and war to a sustainable society is where Guy Dauncey charts us in his novel *Journey to the Future*. It is at one and the same time a great yarn and a call to action."

> ~ Elizabeth May, MP, Leader of Canada's Green Party

"A marvelous read, a modern-day Ecotopia."

> ~ Thom Hartmann, author and radio host

"This unique book will surely inspire hope and action for a better world."

> ~ Michael Marien, futurist, Global Foresight Books

"*Journey to the Future* is a wonderful way to get motivated by the real possibilities confronting us today. Put away your doubts as you follow the entertaining characters who take you into their better future and come away with loads of ideas for making it happen where you are! Helps us start living the future we dream of right now!"

> ~ Elisabet Sahtouris, evolution biologist and futurist

"This is a practical and fantastic guide to lead us forward to a peaceful future. Truly greenspirational. Jam packed with brilliant ideas that are do-able and attractive. Let's get to building this vision!"

> ~ Angela Bischoff, 100% renewable energy organizer, Toronto

"Guy Dauncey is well known as a futurist and environmental activist but he is also a darn good educator for the possibilities for sustainability on this planet. He has encyclopedic knowledge and the educator's ability to make it accessible. And, despite being frank about the dire environmental issues we face, he frames them within the possibility of hope with real and imagined examples and suggestions for making a difference. As a former director of the Carnegie Community Centre at Main and Hastings in Vancouver, it is fascinating for me to read about what this humanly rich and hugely challenged community might look like if the sustainability practices in Dauncey's future world were implemented. It illustrates what can be done wherever we live, and what is on the horizon for a sustainable world."

> ~ Michael Clague, community educator and social planner

"While technically a work of speculative fiction, Guy Dauncey's *Journey to the Future* is firmly rooted in present realities. His prodigious research and astute analysis stimulates our thinking about global problems and possible solutions. But this is much more than a book of ideas. Readers following Patrick Wu's journey of discovery will experience a rollercoaster of emotions—from outrage, pessimism and doubt to hope and motivation. Whether or not we agree with everything Dauncey proposes, he leads us confidently towards a hopeful future, but with a keener appreciation of our chances for survival. If only our political leaders would read this immensely important book."

~ Marshall Soules, author of *Media, Persuasion and Propaganda*

"Guy Dauncey is an eco-hero whose enthusiasm for climate change solutions is infectious. Through remarkable creativity, Guy provides the reader with a glimpse into a plausible future that is vibrant, positive and joyful, while offering workable solutions to the major crises of our time. We owe Guy a debt of gratitude for his compelling narrative that fills us with hope and inspiration for a better future."

~ Andrew Weaver, MLA, climate scientist and author

They say we can't do it, can't fix it or grow it,

Can't change the world, kiddo, what makes you not know it?

But we've got no ears for your know-nothing blow-it,

We'll change this world ten times before you can throw it.

For this is our now time, not do the fuck-all time

We're done with your moaning and dying delays.

We say live! Live again! For it's change the world now-time

So kiss me and celebrate, show me your ways.

Kiss me and celebrate, change-the-world, elevate

Elevate higher than dirt-streets and mire

Elevate up where the highest hopes relevate

Kiss me and celebrate, change the world now.

ALSO BY GUY DAUNCEY

The Unemployment Handbook

Nice Work If You Can Get It:
How to be Positive about Unemployment

After the Crash:
The Emergence of the Rainbow Economy

Earthfuture:
Stories from a Sustainable World

Stormy Weather:
101 Solutions to Global Climate Change

Enough Blood Shed:
101 Solutions to Violence, Terror and War
(with Dr. Mary-Wynne Ashford)

Cancer: 101 Solutions to a Preventable Epidemic
(with Liz Armstrong and Anne Wordsworth)

Building an Ark:
101 Solutions to Animal Suffering
(with Ethan Smith)

The Climate Challenge:
101 Solutions to Global Warming

The Earth Anthem (We Are One)

A Modern Alphabet (poem)

ABOUT GUY DAUNCEY

Guy Dauncey is a futurist who works to develop a positive vision of a sustainable future and to translate that vision into action. He is founder of the BC Sustainable Energy Association, and the author or co-author of ten books, including the award-winning *Cancer: 101 Solutions to a Preventable Epidemic* and *The Climate Challenge: 101 Solutions to Global Warming*. He is an Honorary Member of the Planning Institute of BC, a Fellow of the Findhorn Foundation in Scotland, and a powerful motivational speaker. His websites are www.earthfuture.com and www.thepracticalutopian.ca. You can also find him on Facebook and Twitter.

SYNOPSIS

In futurist Guy Dauncey's engaging ecotopian novel, young Patrick Wu journeys to a future world brimming with innovation and hope, where the climate crisis is being tackled, the solar revolution is underway, and a new economy is taking shape. Yet enormous dangers still lurk.

Patrick has returned to Vancouver after spending his youth in Sudan, where his parents worked with refugees. After becoming deeply concerned by the state of world and prospects for its future, he finds himself time-shifted to Vancouver in the year 2032, by when it has become one of the world's greenest cities.

For four days he explores the city, interviewing people, asking a host of questions. How does a modern metropolis operate without fossil fuels? How are poverty, homelessness and affordable housing being tackled? What gave people the belief that they could change the world?

He visits a flourishing farm, discovers how healthcare and education have changed, spends time in a low-income part of the city, attends a Friday evening Song of the Universe, and learns how a new cooperative economy is being built while capitalism itself is morphing into something quite different. Politics and democracy are changing too, thanks to a popular uprising known as the OMEGA Days.

The answers he receives are practical and detailed, and give him hope for a future that can actually be achieved. Although Vancouver in 2032 is vibrant and happy, its people are still troubled that the rest of the world is not changing fast enough. In spite of the progress in switching to renewable energy, the world's climate is still in crisis.

Throughout his trip, Patrick hears references to something called syntropy, a universal self-organizing principle that operates through consciousness. Before leaving the future, he attends a dinner party where he learns what syntropy is, and how it has the potential to inspire far greater change.

On his return, Patrick writes a detailed account of his dramatic journey. If you share his concerns about the future, this book may give you hope, reason to believe in a better world, and a roadmap to help us get there.

Guy Dauncey's *Journey to the Future* presents an extensively documented, entirely plausible ecotopian vision. With a few exceptions, all of the innovations and solutions that Patrick Wu encounters in 2032 have already been described in academic and scientific journals and mainstream media. To follow the endnotes and to explore other dimensions of the book, go to www.journeytothefuture.ca.

JOURNEY
TO THE FUTURE
A Better World is Possible

An Ecotopian Novel by
GUY DAUNCEY

Agio
PUBLISHING HOUSE

PUBLISHING HOUSE

151 Howe Street, Victoria BC Canada V8V 4K5
www.agiopublishing.com

We gratefully acknowledge permissions to quote from Kate
Braid's *To This Cedar Fountain* (Caitlin Press, 2012, first
published by Polestar Publishing, 1995) and from Arsenal Pulp
Press for lines from Bud Osborne's poem *Never believe*.
Cover images from BigStockPhoto.

*For rights information, bulk orders, retail orders and promotional
copies, please contact the author through
www.journeytothefuture.ca*

To learn more about the book, and to view the book's over 940
endnotes, most with web-links to original sources,
see *www.journeytothefuture.ca*

Journey To The Future: A Better World is Possible
ISBN 978-1-927755-33-4 trade paperback edition
ISBN 978-1-927755-34-1 Kindle/mobi ebook edition
ISBN 978-1-927755-35-8 epub ebook edition
ISBN 978-1-927755-36-5 PDF ebook edition

Cataloguing information available from
Library and Archives Canada

Agio Publishing House is a socially responsible enterprise,
measuring success on a triple-bottom-line basis.

10   9   8   7   6   5   4   3   2   1 c

To all of Earth's People, young and old

The future belongs to those who believe
in the beauty of their dreams.
– *Eleanor Roosevelt*

The visions we offer our children shape the future.
It matters what those visions are.
– *Carl Sagan*

We are at the very beginning of time for the human race.
It is not unreasonable that we grapple with problems.
But there are tens of thousands of years in the future. Our
responsibility is to do what we can, learn what we can,
improve the solutions, and pass them on.
– *Richard P. Feynman*

The whole future of the Earth seems to depend
on the awakening of our faith in the future.
– *Pierre Teilhard de Chardin*

TABLE OF CONTENTS

It's a race. A race between the expanding reach of our empathy as it stretches across the world, bringing love and intelligent cooperation for the good of all, and the clutching fear of tribal distrust, made more powerful by modern technology. Which will win?

1

Hungry for Hope

MY NAME IS Patrick Wu, and I have just spent four days in the future.

Why? Because I was hungry for hope. Because I found the world confusing and I needed something to light the darkness. And because I could.

How did I do it? I have some theories, but first you need to understand my background.

Although I was born here in Vancouver, in Western Canada, I spent most of my childhood out of the country. My family left Vancouver when I was eight and for twelve years we lived in East Africa and the Middle East, from Lebanon to South Sudan, in all sorts of conditions, while my parents worked for refugee aid organizations. By the time my younger sister Daria and I were teenagers, we had probably seen more poverty and suffering than most people will see in a lifetime. I returned to Vancouver with Daria when I was eighteen, under unhappy circumstances.

East Africa is not all brown and dusty, the way the camps are. Ethiopia has rugged mountains and lush farmland. South Sudan has tranquil villages in a sea of savannah filled with trees. And then there were those nights camped out under the thousand myriad stars of the vast deep African sky, each so mysterious, each saying '*Look at me—see how little you know.*'

Our parents were wonderful. My father declared himself a 'cosmic pantheist', and in the cracks between the refugee camps he loved to discuss the big questions with us. Was there a God? Or was there nothing? Or was 'God' simply a word people used to express the unfathomable vastness of it all? To my father, God was a G.O.D.—a General Omnipresent Diaphany. That's how he liked to put it, and it took some digging to discover that the word diaphany comes from the Greek, meaning 'to appear through.' So the Creator is revealed by an essence that shines through every person, every leaf, every creature.

Can miracles really happen? How can I say they can't, he would answer, when we know so little about how the Universe works and who we really are? "We are not human beings having a spiritual experience," he liked to say, quoting the French scientist and priest Teilhard de Chardin. "We are spiritual beings having a human experience."

My mother was more wrapped up in the lives of the people we lived among— the families fleeing violence and war; the women and children who had been raped and terrorized, their villages burned to the ground. She was sometimes inspired by our father's musings, but more often she was impatient. "If God exists," she said, "he must be a sadist."

"No," my father would answer, "it is we who are the sadists, because we have not yet been moved by the diaphany, by the compassion that can enable us to see the love that infuses everything."

It was heady, inspiring stuff for a fifteen-year-old. I knew how many trillion stars there were and how far their light had travelled, but what did it mean that all this had been created out of a Universe that was fourteen billion years old? All the music, the love, the imagination, the hurt, confusion and suffering—that somehow it had all emerged from the same shared origin? It blew a fuse in my brain, trying to wrap my mind around it.

Why are some humans so cruel and self-centered? Why are some people so warm and kind, while others become bullies, thugs and murderers? How was it possible that a myriad disconnected cosmic particles had organized themselves into atoms, jellyfish, elephants, butterflies, humans, music and science? How could I fathom out the mismatch between the awesome *unity* of the cosmos and the awful *dis*unity of poverty, war and suffering? You may think those to be teenage questions, but they were very real ones, concerning real issues.

It's not as if Daria and I were totally isolated growing up. We were able to keep up with world events and our parents had taught us how to distinguish news from propaganda. By the time I was eighteen I knew that as well as a place of incredible beauty, the world was also a matrix riddled with power, corruption and greed.

Take Somalia's pirates, who kidnapped foreign freighters and their crews, then ransomed them for millions. The Somalis were not there because they woke up one morning and said, "I have an idea—let's play pirates!" They were there because their parents had been fishermen, supporting their families as their ancestors had done for thousands of years until foreign industrial fishing fleets arrived off their coast and began massively plundering the fish, shooting at their tiny fishing boats with water cannons and firearms, making it impossible for them to feed their families. So who were the first pirates—the Somalis or the foreigners?

And then there was that awful day in Jerusalem. I was sixteen. We drove around a corner and suddenly we were in the aftermath of a suicide bombing. There were people with their legs blown off, their bodies flayed by shrapnel. Acrid smoke. People shrieking. Corpses.

A girl was still conscious. Most of one of her arms was gone but I had seen enough of my parents' triage work to know what to do—stop the bleeding. I tore off my T-shirt, tied it tight above the bloody end and pressed her matted scarf firmly into the wound while she gasped and spasmed.

As I knelt over her, she looked up at me, scared and confused. Same age as me, more or less. Wearing a black and white checkered keffiyeh—maybe Palestinian?

Sirens were wailing, people shouting warnings and orders in Hebrew.

The girl was staring at me with the deepest black eyes I'd ever seen and the darkest black hair. Then out of nowhere, *such a deep, powerful connection.* For a moment I'm sure my heart stopped. She tried to say something, but I couldn't understand her.

Then a stretcher appeared beside us, and someone pulled me aside. As she was rushed into an ambulance, I grasped her hand.

"What's your name?" I called out, but her reply was too weak.

One day, I'll go back and try to find her. If she's alive.

———✦———

IT WAS MY father I was closest to. Daria took after our mother, with her red hair and her practical get-it-done approach to life. But then there was that other awful day. Daria and I were in Khartoum. Our parents had flown up to Cairo for some meetings, leaving us to look after ourselves.

We spoke to them every day, but one evening they didn't call. Time ticked by ominously. Eventually, just after midnight, there was that awful phone-call—the one you never want to receive. A bus had hit a hole in the road, veering into a fuel tanker that immediately exploded into flames. The taxi my parents were riding in had been following too close and had plowed into the inferno.

Everything they were to us—their joy, their laughter, their love, their parenting, their wisdom, their work, their nagging, their fussing, their hopes for our future—they ended right there, leaving nothing but memories.

———✦———

A MONTH LATER we were back in Vancouver. I was eighteen, Daria fifteen. We had enough money to rent an apartment and pursue our studies.

As our grieving subsided, we grew to love Vancouver, especially the rain and the greenery, so refreshing after the dusty brown dryness of the refugee camps. The ocean, the forests and the mountains are so stunning. If you've been here, you'll understand. So Vancouver became our city of healing. But my teenage questions took on a new urgency as I learned about the distress that climate change was inflicting on the world. When I read about the climate crisis and all the trouble that was coming, my stomach would feel tight, and I would have difficulty sleeping. Surely, there *had* to be a better way.

I also began to understand how deep was the dysfunction at the heart of the global economy—the stuff they call capitalism. If there was anyone at the helm—and maybe there wasn't, maybe there was just an ideology that people followed—they were driving us at high speed toward the edge of a cliff, humans, elephants and orca whales alike. If we didn't embrace fundamental change, not just locally but everywhere, we were going to be in awful trouble.

But what could I do? I was studying politics and international development at the University of British Columbia and I was active in an environmental group, but I had yet to find my path, my purpose in life.

I enjoyed playing Ultimate Frisbee, but my friends thought I was way too serious. "Chill, dude," they'd say. But how could I, when the stuff I was reading told me we were sleepwalking towards disaster? Where was the vision we so urgently needed, that could offer us a better future?

IN 2010 VANCOUVER, led by its inspirational Mayor Gregor Robertson, announced that it was going to become the greenest city in the world. It was an exciting vision and it was Daria who suggested that we get involved, going to meetings and forums. The *Greenest City Plan* addressed many issues, including transportation, food and the green economy, but it didn't address poverty, homelessness or the larger global problems such as the climate crisis, or the way the plutocrats were looting the planet. How could it? It was a plan for one city, not for the world.[1]

'Where do we come from? What are we? Where are we going?' These were the questions the painter Paul Gauguin asked more than a hundred years ago, in the sensual beauty of Tahiti. Today there's a similar feeling in the air, a sense beneath the skin that a crisis is brewing. Top hedge-fund managers are buying farms with private airstrips in New Zealand so that they can get away when the shit hits the fan. The rich are piling up their goodies, but every year millions more people are becoming trapped in poverty and debt.

We need economic growth to tackle the poverty and the hunger, but the same economic growth is gnawing at Earth's vitals. The tropical rainforests in Borneo are being burnt and torn down to make way for palm-oil plantations. The ocean is being stripped of its tuna and sharks and filled instead with plastic. By burning fossil fuels we are melting the icecaps. The rich are getting richer but so many people are struggling, while nature's ecosystems are being weakened, often to the point of collapse.

Wherever did we get the idea that this world is ours to treat as we like, that we are *entitled* to cut down the rainforests to drill for oil, *entitled* to catch all the fish in the ocean, *entitled* to spray pesticides over the soil, trees and hedgerows? Everything we consume seems to be tearing at the fabric of nature.

And to trump it all, so much of the destructive activity is being done on borrowed money.[2] It cannot be right that governments have to struggle with such massive debts, forcing them to cut back on education and healthcare, while big corporations like Apple, Google, Walmart and Starbucks pay almost nothing in taxes because their accountants have found cunning ways to game the system.[3] It cannot be right that some people have become so wealthy, they have to hide their billions in tax havens.[4]

With all these troubles, there was no shortage of apocalyptic scenarios.[5] But who can live off that stuff? It seeps away at the soul, and makes life dimmer. I craved a future that would reflect the hopes I carried in my heart, a vision that would give me something to work *for,* as well as being against the things that caused grief and pain.

We needed to shift to 100% renewable energy, to replace all fossil fuels—but how was that possible? Our entire system of transportation depended on oil. We needed to transform the world's economy—but what were the alternatives?

Daria and I were talking these things over with some friends one wet February evening, in our Vancouver apartment. The rain was pelting down on the street outside and we were enjoying a cozy home-cooked curry when Daria posed the innocent question that was to change my life: *if the future is so important, what if we could find a way to visit it?*

After the inevitable laughter and the jokes about what she was smoking, she persisted. "We know Vancouver has set a goal to become the greenest city in the world. What if we could somehow travel to that future and see what it looks like twenty years from now, when it has actually *become* the greenest city? And not just the greenest. What if people had also discovered how to end poverty and homelessness? What if they had built a culture that was rich and fulfilling? We could learn how they did it, and use their knowledge to start building that future today."

It was a crazy idea, but once I had thought about it, it got its hooks into me. I found myself dreaming about it. For sure, we need to say 'NO' to the many things that threaten us, but we also need to say 'YES' to a vision so positive that people will yearn for it, and spend their lives working to make it happen. If we want change, the power of our vision must be so much stronger than the power of our fears.

Then late one evening a week or so later, when it was just Daria and me in our apartment together, I felt a strange and sudden sense of lightness.

I sneezed.

Daria said, "Gesundheit!"

I sneezed again.

"Bless you!" Then she laughed.

And again.

I shook my head vigorously to clear my mind.

"Did you know there's an ancient superstition about sneezing, how it opens a crack to the spiritual dimension?" she continued....

To this day, I don't know how it happened. All that I know is that I suddenly found myself in the future, just as if I had been dreaming.

Maybe it was my father who made it possible—who knows what's possible once you are dead? I've thought about it over and over, but I've not found any plausible rational answer. So now I just accept it for whatever it was.

The words you are about to read are my recollection of the four days I spent in the future in June 2032, and the many conversations I enjoyed.

But before you begin, let me say one thing. Even though this book is set in the future, it is not really a work of fiction. It is simply an account of the myriad things I experienced.

So if it's not fiction, what is it? It's about the way the future world is tackling the climate crisis. It's about solutions to poverty, debt and homelessness—and a new economy that is no longer capitalist. The people I met talked to me about food, farming and our health; about transportation and street life; about community organizing and education. They also talked about protest—lots of protest, and about politics and democracy.

It's also about spirituality, philosophy and science—the big ideas my father loved to discuss. Maybe that's why he made it possible, if that's how it happened.

To repeat, I really don't know how it happened: it just did. I found myself sitting on this bench, in the future, with an inner knowing that I had just four days to explore things and learn as much as I could, after which I would return. When I tuned into my father, as I often do, I felt that he was smiling contentedly.

ENDNOTES www.journeytothefuture.ca

Journey to the Future contains over 940 endnotes, almost all of which have web-links referencing the original material on which the novel is based. You can find the endnotes on the book's website, and many other resources. You will also be able to subscribe to Guy Dauncey's newsletter *A Better World Is Possible*, that tracks progress on the many things Patrick saw and learned.

2

June 2032: The Journey Begins

IT WAS GETTING dark, but the late evening air was fragrant and enthralling. I was sitting in a cob-earth shelter on a quiet residential street in what felt like Vancouver, but instead of rows of parked cars I was looking out onto a sea of trees, shrubs and plants.

The road was still there, but it was narrower than it was in my time, and it was meandering instead of straight. On the wall of the shelter I read these words, which someone had painted beautifully:

> *Forget not that the earth delights to feel your bare feet*
> *and the winds long to play with your hair.* – Khalil Gibran

I took a deep breath, to see if I was real. Check. I pinched my thigh. Check. I stood up and looked around. Check. This was too weird. I peered out of the shelter.

And then I panicked. Where was I? What was happening? My heart beat faster and my breath grew short, but then I remembered a trick my father had taught me. *Pause, breathe. Go inside and picture yourself surrounded by a circle of light. Then say quietly, 'I am here to serve. Please help me and guide me.'*

I did as he'd taught me, and a familiar voice inside me said, "Relax. Enjoy."

I took a deep breath, and looked around.

Down the street to my left three women were walking towards me arm-in-arm, laughing and joking as if they'd had a good night out. As I watched, two of the women said goodnight and peeled off into one of the houses. The third woman continued down the street. Black, medium height, well built. Shock of frizzy orange hair.

"Hello—who do we have here?" she asked, looking at me.

I needed to answer, but my brain jammed up.

"What's up—cat got your tongue?" She placed her hand on her hip.

"Er, no. Sorry. I must have been daydreaming," I fumbled. "My name's Patrick. Patrick Wu. I've been living in Sudan, in East Africa, and I'm visiting Vancouver on a research project. I don't suppose you know if anyone around here offers bed and breakfast?" Where did *that* come from? Don't ask me.

"Sudan, you say? You don't look very Sudanese to me."

"I grew up in East Africa. I must have spent too long in the laundromat to get this pale skin of mine."

"The laundromat! Is *that* how you folks got so white? I never knew. After all these years! But you said 'Wu.' Sounds Chinese."

"True - but I'm a true Canadian. My father had Chinese ancestry; my mother's Irish. They died in Egypt six, almost seven years ago."

"I'm sorry to hear that. May the blessings be. So what brings you to our neighborhood on such a lovely night, Patrick Wu?"

Relax. Enjoy. Invent.

"I'm a member of a student society at the University of Khartoum and we're very troubled about the future. So they sent me to Vancouver to learn how you had become the world's greenest city."

"One of the greenest. We don't like to boast."

"I'm sorry. I've got four days to learn what I can before I go on to Portland, and I need a place to stay. That's why I asked about a bed and breakfast."

"Well, you seem like a decent young man. But you're leaving it a bit late, aren't you? It must be your lucky night, since I offer rooms myself. That's my home across the street. You'd best come in and tell me all about it. I'm Dezzy. Dezzy Brooks."

She led me through some shrubs and up the steps to her porch. I recognized the heritage style, no different to houses I knew in Vancouver from my time. The door wasn't locked, and a teenage girl came running down the staircase.

"Hey there, gorgeous!" Dezzy said, giving her a hug. "How's he been? Is he asleep?"

"Yes—and he's been great. We played pick-up soccer in the park and I helped with his school project on gravity waves. He understands so much more than I do! Then we went to watch the sunset to see if we could detect any gravity wave clouds. And we did! At least, we think we did. Jake says that when he grows up he wants to crack the secret of gravity."

"That boy. I expect he'll do it, too. You've been great, Miranda. How much do I owe you?"

"Thirty dollars. Half in cash, half in Deltas, please."

Dezzy reached for her phone, pressed a few buttons and tapped Miranda's phone, giving the babysitter the money. Then she gave her a kiss and said goodnight.

It all felt so normal: that was the weirdest thing. Throughout my journey things like 'Deltas' or the way she paid by tapping her phone would feel normal, as if I'd known them forever. But then I suddenly thought, 'Money! How am I going to cope in this future without money?' I reached into my pocket, and was surprised to find a plastic card. *Phew! Thanks, Dad.* Now I would be able to pay for things I needed, and not worry about being dragged before some strange bankruptcy court in this unknown world of the future.

"She's a great girl," Dezzy was saying, oblivious to my flash of panic. "Daughter of some friends of mine."

"How old is Jake? I take it he's your son."

"Yes. He's eight. I try not to push him in his studies, but there's no stopping him. It must be his dad's genes."

"His father? Is he...."

"We separated three years ago. Teaches physics at the University of British Columbia. Long story. But come on in. I'll show you to your room."

It was neat and tidy with too many cushions on the bed, the way bed and breakfasts often are.

"Do you need a bite to eat? Perhaps a nightcap?"

I wasn't hungry, but I accepted a cup of chamomile tea. Two cats appeared and Dezzy picked them up.

"Meet my babies, Chloe and Indigo." The white cat squirmed in protest while the black cat nuzzled her chin, purring loudly.

"It's just me and Jake who live here, plus a young couple in the laneway house. So tell me again—what brings you here? I'm originally from South Africa. I was adopted as a baby by my parents, who are white. They were anti-apartheid activists who emigrated to escape imprisonment."

I told Dezzy about my parents' work with refugees in East Africa, and I invented a story about how my fellow students had crowd-funded the money to send me on this trip.

Dezzy stared at me, and her expression became intense.

"It's unreal," she said. "You remind me so much of my brother Derek. Here— come with me."

She led me into her living room, where one of the walls had been turned into a shrine. In the center there was a photo of a man sitting cross-legged, wearing a dark red robe. He had short black straight hair and his eyes were closed in meditation. It was true; he did look a bit like me. Other photos showed him surrounded by friends and family, speaking in public, kayaking on the ocean, dancing, and, at the bottom of the shrine, his body draped in white, surrounded by people meditating.

"What happened?" I felt shocked.

"He was assassinated during one of our big demonstrations during the OMEGA Days. That was fifteen years ago, but the police have yet to find who did it. My friends say I should take this shrine down, but it reminds me that his spirit is still with us whenever we work to make a better world."

"But he was assassinated? In Vancouver? Your brother?"

"Yes. And his shrine will remain until the police find his killers. But we succeeded, and they're on the run, whoever *they* are. Our enemies are still powerful, but we'll win. My parents didn't raise me to be a wallflower."

"So it seems I've come to the right place to learn about all the changes Vancouver has made?"

"You could say that," she chuckled. "Where do you want to begin? On second thought, let's leave it till the morning. I've had a good night out and it's a bit late to be talking about changing the world."

"Who's this?" I asked, pointing to a photo of Dezzy arm-in-arm with a tall, handsome black man.

"That's Thaba, my husband. Jake's father. He's South African, same as me. He is rather good-looking, isn't he?"

"How'd you meet?"

"We met at a seminar at the Perimeter Institute in Waterloo, Ontario. It's one of North America's leading centers for advanced physics and cosmology. Thaba had been invited to speak about his work on dark matter and its relationship to gravitational waves. In college he was really good at math so he was recruited to study at the African Institute for Mathematical Sciences founded by Neil Turok from the Perimeter Institute. Neil dreams that the next Einstein will be an African. If our Jake carries on the way he is progressing, maybe it'll be him. Does Afro-Canadian count, I wonder?"[6]

"How did you come to be at the Institute? Are you a physicist too?"

"No. I'm just a humble computer scientist. I work with Embryo. We're competing with Google to build the world's fastest quantum computer. I spent a year at the Institute trying to understand quantum physics." She smiled proudly. "You're probably wondering what a black woman like me was doing at the world's top cosmology center? I thought the same until I met the other Africans. My god, they're smart! And so good at physics and math, both men and women. It was like a big dark cloud lifted. All my self-doubts, the feeling that I didn't deserve to be doing what I was doing—they vanished, just like that."

Cosmology? Quantum computers?

"We were young, and head-over-heels in love. Thaba Mabaleka. That's his full name. Our son's name is Jake Azisa Mabaleka Brooks. I married Thaba because we loved each other, and so that he could immigrate to Canada. He got a job in the physics department at UBC and we'd go for long evening walks on the seawall around Stanley Park, and up Grouse Mountain on the weekend. When Jake arrived, we took him with us everywhere. But Thaba found me increasingly difficult to live with. He didn't know how to handle an independent woman, and I can be pretty stubborn at times. He felt entitled to dominate, believing I should 'fit in'. So we argued relentlessly, and in the end I left him. I still love Thaba, and we're the best of friends, and he's a great dad for Jake. We just don't live together anymore."

"Can you explain how a quantum computer works?" It seemed a big question, and the hour was getting rather late, but I had to ask. How often do you meet a quantum computer scientist from the future?

"Well, you're a sucker for punishment. How long do you have?" she laughed. "I've been studying them for fifteen years and even I don't understand them totally. But let me try to explain. In classical physics, a thing was always a thing. Space and time were absolutes, and matter had a comfortable solidity. And ever since the earliest days scientists have used math to unravel the secrets of the Universe. Okay? But math is digital: zero, one, two, three, four. We use calculus to study the flow of change between the units, but calculus is still essentially digital.[7]

"Quantum physics developed because when you get down to the tiniest level of matter the digital approach fails. It turns out that reality is rather *slippery*. You can't pin it down to a one or a two, on or off. Everything in life keeps oscillating between a digital reality and something else that can't be pinned down. We call it a *probability wave,* and quantum math gives us incredible precision when we incorporate the uncertainty into the math.

"Essentially, reality isn't digital. It's analog. Continuous. Time doesn't flow in bumpy discrete units. Everything in the Universe is a continuum, and in a continuum the space between two digits can have an infinite number of expressions. In a digital computer, a gigabyte of memory gives you a billion bytes to store your games and videos. Today's fastest digital computer can do 5,000 quadrillion floating point operations a second. Are you following me?"[8]

I didn't know if I was following, but I was fascinated.

"In the quantum world, we dive into the continuum where there are no points until an entity of some kind chooses to express itself as a discrete particle. The digital world emerges out of the continuum. Now imagine that you can take the space between two digital units, containing an infinite number of possible quasi-positions, and build a computer that uses quantum math to calculate in the world of the continuum. Within every byte of digital information a quantum computer can find infinite possibilities, making it vastly faster. The trick is to pin the possibilities down to quantum bits, using the quantum properties of something fundamental like a photon. Our quantum computing is being used in the human neurome project to map the neurons of the human brain.[9]

"We are within striking distance of creating a computer that will have the same processing capacity as the human brain," she continued, "and mimic the brain in the way it operates. One of our partner companies is working on a project to create a BioBot with a brain that will self-organize and learn at the same pace as the brain first of a human embryo, then of a baby, a child, and finally an adult. They are deliberately making it *not* seem human, to avoid triggering robophobia—the fear some humans have of robots that resemble humans, but its mental capacity will be the same as an adult human."

"Isn't that rather scary?" I asked. "And what about the ethics—is it okay to do that?"

"It depends whether the BioBot is conscious, the way you and I are. There are some who say that's impossible, that it will still be just a machine. But there are others who argue that if the core of all reality is conscious there's no way to prevent a BioBot from being conscious too. It's a very lively debate. Moore's Law has been delivering incredible capacity, which makes it theoretically possible to build a BioBot with a human-size brain."

"What's Moore's Law?" I felt like an endlessly curious five-year-old.

"It's the prediction made way back in 1965 by Gordon Moore, the co-founder of Intel, that computing capacity would double every year.[10] The doubling has slowed to every five years, but the implications are enormous, especially when you consider the deep future."

"The deep future...?"

"Yes. If we survive the climate crisis, and barring an unexpected massive meteorite strike, humans should be able to inhabit the planet for more than a billion years. After that, and counting, the Sun will grow too hot and we'll need to migrate to a cooler planet. That's what I mean by 'deep.' A billion years is a hundred thousand times longer than the history of civilization so far, and every year, our science, technology and understanding will improve. Hopefully our wisdom will too, and our relationship with nature. It's going to happen. No-one has ever suggested that it's possible to stop time."

I was floored.

"Cat got your tongue again?"

"That's what I like about cats," I replied cheerfully. "They're good listeners—or they are just ignoring us. Getting back to what you were explaining about time, I've often been amazed at how long the *past* is, but I've never thought about the future that way."

"Don't worry: that happens to all of us. We're not equipped by our primate ancestry to think about the deep future. Even the immediate future is a challenge for most people. As long as consciousness exists in the eternal present, the future will be an eternal mystery."

"Well, I suppose that's reassuring. Is there a connection between your quantum computing and your work to make a better world?"

"Straight for the jugular, eh? Why not start with a *really* difficult question? I can try to answer, but you'll have to put your thinking cap on. It concerns the nature of the continuum, and the fundamental story we tell ourselves to explain why we're here, and what we're doing. In reductionist science, everything was material and measurable in a digital form. Most scientists now accept that time, space and the universe are a continuum, not a digital reality. That's why pi, 3.14159 etcetera, is an irrational number with an infinite number of digits after the decimal point. It's because a circle can only be truly measured using quantum math, which captures the essence of the continuum. The connection is consciousness, which is as real at the core of reality as it is in you and I. In the old reductionist story, science didn't have much to say about the human adventure, apart from how fascinating it was. In the new story, consciousness is an unfolding reality that allows us to reach ever more deeply into the fundamental unity of existence, both scientifically and through social and political change. In other words, our efforts to build a better world and our efforts to build a quantum computer are expressions of the same deep syntropic drive, as units of existence self-organize to achieve greater harmony and wellbeing."

What? My dad would have *loved* this stuff. But could *I* understand it? That was the question. And what was this 'syntropic drive' Dezzy had just mentioned?

"Cat got your tongue again?" she teased.

"No—I'm taking it all in. I think I get it," I lied. I wanted to ask about the 'syntropic drive,' but I decided to leave it. Maybe later. "Do the other people where you work think this way?" I asked.

"Some do, some don't. Most of my fellow quantum workers don't think much about the big picture: they're happy to crunch the numbers and get the work done. We do have one guy who's really into it, however. Rajendra Choudhury. He even named his children after famous quantum scientists—Max, Erwin and Emmy. When he was seeking the perfect woman to marry he took genetic samples from his various lovers and had them analyzed. He was looking for the genes for intelligence and empathy. He wanted his children to be smart, but not at the expense of empathy and social skills.[11] And it worked! He married the woman whose genes showed the best results, and they've been happy together ever since. The irony is that his son Max became a hockey player, his second son Erwin runs a restaurant and his daughter Emmy, the youngest, named after Emmy Noether, the famous German Jewish mathematician, she's a pole dancer! He was most upset when she told them what she was doing. He blames his wife for the fact that they did not all become quantum scientists, but she just smiles and quotes Stephen Hawking, who apparently said that God not only plays dice, but that he sometimes throws them where they can't be seen."

"That's hilarious! It's maybe fate playing a joke on people who choose their partners that way. But can I ask you something else? I'm thinking about your brother Derek. When was he assassinated? And what was going on?"

Dezzy stopped laughing. "It happened the year after the financial crash. Life got pretty crazy then—even crazier than during the OMEGA Days. The media calls that decade the Terrible Twenties, but I prefer the Transformative Twenties."

"I was a young teenager at the time," I said (or rather, I lied). "What was it like?"

"Well, on one level, life continued. People still went to concerts and dances. We still invited friends to dinner and people still had to pay their rent or mortgage. But almost everyone had their challenges, and their tragedies. I had baby Jake to keep me busy. Thaba was struggling to find work, and my grandmother was beginning to go senile. So, yes, it was hard.

"There was a constant feeling of crisis," she continued. "I had friends who were actively engaged, some locally and some on the global level, but many people had a hard time just coping with the food crisis, the carbon rationing, and finding money to pay the bills."

Food crisis? Carbon rationing?

"Did *you* have personal tragedies to deal with, as well as Derek's death?"

Dezzy paused.

"Yes, I did," she finally replied. "Thaba and I had two daughters before we had Jake. Gabriela's in Montreal now, studying anthropology, and then there was Anna. She would have been thirteen this March. She got an ear infection, and there was nothing the doctors could do. It was resistant to every known antibiotic, and it took over her whole body. She was five."

"I'm so sorry—that must have been awful." My mind flashed to a memory from the Melkadida refugee camp in Ethiopia, and the searing grief of the parents of a young Somali boy who had just died from a lung infection. "I saw a lot of children die in the camps where my parents worked. I saw how much parents grieved."

"Yes. It was a difficult time," Dezzy replied. "We really loved Anna. Then a year later my father was killed in a food riot in Capetown. He was walking to work, same as everyday. He'd gone back to South Africa, determined to help. He just happened to be in the wrong place at the wrong time."

She was silent.

"What happened?" I asked.

As Dezzie looked across the room, her whole being filled with sadness. "The riot was happening outside a supermarket that had just been sacked. He was a white man wearing a suit and tie, so I guess they thought he was management. They beat him unconscious, doused his body with gasoline and burned him to death. He never had a chance."

"Oh my god. That's awful." My stomach turned over, and I thought I might vomit. I turned away and put my head between my knees.

"Are you okay?" Dezzy asked.

I made a conscious effort to breathe deeply, as my parents' death flashed before me.

"What just happened?" Dezzy asked, coming over to help. "Did that trigger something for you?"

I told her how my parents had been killed, and I was about to explain my whole true story when something sealed my mouth. She wrapped her arms around me, comforting me.

"It's okay," I said after a while, wiping my eyes and blowing my nose.

"Life can be hard," she said, slowly sitting down. "And it can be very cruel."

"Yes," I replied. "How did *you* cope when your father was killed?"

"I was angry. I was distraught. But what can you do? There were tens of thousands of people who were killed in food riots around the world. Some were crushed to death; some set on and murdered. My father had a PhD. He worked for an engineering company developing safe drinking water for the townships. But so what? There were thousands of black kids whose mothers scratched a living raising chickens, who might have earned a PhD if given half a chance. Plenty enough of them were killed, too."

It was painfully true. I had met many children in the refugee camps who wanted to become a doctor or an engineer—if only life would give them half a chance.

"What about your mother?" I asked.

"She's never been the same since. But there's no point dwelling on the negative. There's a lot we've achieved that she's very proud of. But look now, Patrick Wu from Khartoum, I need to get to bed, since Jake's an early riser and I expect he'll be bouncing in at six in the morning. We can talk more tomorrow. I'll see if I can set you up with some meetings that might be useful for your research project."

"You're very kind."

She gave me another hug, and I took myself off to bed.

As soon as I turned out the light, my mind replayed the hours since I had arrived. Dezzy's brother Derek. The OMEGA Days, whatever they were. The food riots. And so much death: Dezzy's father, Dezzy's daughter, my own parents.

And then the girl in Jerusalem. *Relax, breathe*, I heard inside my head.

So as I so often do, I imagined her happy and healed, with an intricately patterned artificial arm. I held her close. If she was alive, I hoped she would feel it.

3
The Solar Revolution

WHEN I CAME down the next morning Dezzy was sitting at the kitchen table with her son Jake, getting him ready for school.

Jake looked up and gave me a lively grin. "Mom says you're from Africa and you're really interested in physics! Can you help with my project on gravity waves?"

"Well," I replied. "It's true that I come from Africa. But Dezzy's exaggerating if she thinks I know enough physics to help with your project."

But Jake wasn't daunted. "Do you think we'll be able to develop an anti-gravity machine that will, like, enable us to fly through space at the speed of light?"

"Who knows what's possible? But if I remember rightly, Einstein proved that if we were able to travel close to the speed of light, time would slow down. So if you went on a longish trip, when you came back your friends would be decades older, perhaps even dead. If you were gone for a year, when you returned the world would be millions of years older."[12]

"Whoa! That would be really wicked," Jake replied. "Maybe there'd be another kind of machine that would let me, like, travel back in time to get back to my friends! Like, an anti-anti-gravity machine!"

"Maybe you've already been a million years into the future and just come back, but you don't remember."

"Okay you two," Dezzy said. "Maybe you can focus on some time travel right now, Jake, so that you're not late for school."

"I'm going to make a gravitational wave attractor for my bike, so that I can get there like, *Zoom!* and I'm there."

As Jake readied himself for school Dezzy offered me a glass of veggie juice, followed by porridge with honey, hemp seeds and homemade pear sauce. Her kitchen was at the back of the house, overlooking the garden. The windowsills were full of jars of sprouting seeds, and she had window boxes full of herbs and salad greens.

"So tell me again," she asked me when Jake had left, "what brings you here?"

I repeated my invented story about studying at the University of Khartoum, and how I was here to learn how Vancouver had become one of the greenest cities in the world, so that we might have more confidence in the future and not feel so besieged by worries and fears. I had four days to explore Vancouver. Then I had a night ride to Portland on Sunday followed by a trip to San Francisco. I figured that would explain my late-night departure when the time came to leave.

"I want to learn all sorts of things, like, are you still using fossil fuels? How do you get around? Have you made any big social and political changes? And what gave people the confidence to believe that Vancouver could become one of the greenest cities in the world?"

"That's a lot of questions. And you want to do all this in just four days?"

"I was thinking that if I could travel around and meet enough people I could learn a lot."

"That's pretty ambitious. Where do you want to begin?"

"How about here in your home? Where do you get your energy from, for instance?" I knew that the oil and gas used to heat buildings in my time was adding to the climate crisis, so it seemed like a good place to begin.

"Okay. I'll give you a tour before I leave for work. If you take a look around the garden, I'll be with you in ten minutes."

The garden path led to a seat inside an arbor wreathed in honeysuckle, surrounded by vegetables. The sun was warming the soil after the overnight rain and the garden was steaming like a jungle—and buzzing with bees, I was happy to see, considering the trouble they were in back in my time due to the use of neonicotinoid pesticides.

There was a fruit tree in each corner of the garden, and the paths were lined with fruit bushes. Tucked in the back there was a small two-storey cottage, blue and white, pretty as they come, and over the fence I could see that the neighboring gardens were also creating havens for insects, birds and wildlife. In the heart of the urban jungle, the city was being rewilded. Dragonflies were flitting around a pond. I've always been fascinated by the way they spend most of their lives underwater and then, by the miracle of metamorphosis, put on wings and become things of such beauty. There's a sustainability leader in Portland, Oregon, called Darcy Hitchcock who wrote a book titled *Dragonfly's Questions – A Novella on a Positive and Sustainable Future*.[13] To create a positive and sustainable future, she wrote, we must first envision it.

"Where shall we begin?" Dezzy asked when I walked back into the kitchen. "The house was built in 1922. It used to have an oil furnace, but the owner before us converted to gas. We did our upgrade ten years ago, increasing its efficiency, installing the heat pump and adding the solar."

"How much did it cost you?"

"It was $20,000 for the building upgrade, $6,000 for the solar PV and $4,000 for the solar hot water. We got a 100% tax credit for everything we spent, which was handy, and Vancity—our credit union—gave us a low-interest loan, which I'm repaying through my BASE account."[14]

"Your BASE account?"

"It's a separate bank account that I use for all my energy costs. When we took out the loan we used the BASE account, and the savings cover the repayments. If

I was to sell before the loan was paid off the new owner would take over the payments. That's what BASE stands for: Building Attached Sustainable Energy."[15]

"That's pretty smart. How much solar did you install?"

"Four kilowatts. It produces around 4,500 kilowatt hours a year."

Four kilowatts of solar PV for $6,000? Back in my time solar cost $3 to $4 a watt to install, so four kilowatts would have been more like $14,000.[16]

"Only $6,000 for your solar?"

"Yes. It was $1.50 a watt, which comes to around five cents a kilowatt-hour. Regular electricity is far more expensive, at almost twenty cents. Over thirty years it will save me far more than the cost of installing it. No wonder everyone's doing it. The power from the large utility-scale solar plants is even cheaper, at around $1 a watt."[17]

"Did other people do the full building upgrade, the way you did?"

"Absolutely. Everyone was scrambling to reduce their carbon footprints because of the carbon rationing. It may not have affected you if you were living in East Africa, but here in Vancouver it was huge. Every year we had to slice ten percent off our use of fossil fuels or buy additional carbon rations from someone who didn't need them—and they weren't cheap. When I flew to South Africa for my father's funeral I had to pay two hundred dollars a tonne since my rations were maxed out for the year. The flight itself was amazing. The plane had no windows, and instead it projected footage of the passing world onto giant screens. Not that many people fly much these days, since it's so expensive, and most airlines have yet to go renewable."[18]

"When did the carbon rationing start?"

"Soon after the global carbon cap was imposed. There was a panic as people scrambled to reduce their use of fossil fuels. Carsharing, ridesharing, cycling—anything that reduced your footprint. Our house had gas central heating, so we were in a hurry to replace it with a heat pump, but so many people had the same idea that the installers were pushing up the prices. That was when the street organizing took off, since people realized that if they got together they could get a better price, and better service too. We used the Vancouver Renewable Energy Co-op, who took a thermal heat-loss photograph of each building and then did everything from the efficiency upgrades to the solar. We set up a tent where they took their lunch breaks and when they finished we organized a big party."[19]

"What other upgrades did you do?"

"Let me show you around." We walked through her kitchen to the living room, a comfy space filled with the normal clutter of everyday living, except for the shrine.

"We wanted to get as close as we could to zero net energy, generating as much energy as we use. So we covered the outside walls with rigid foam insulation and we sealed up all the cracks, using an infrared wand to show where heat was escaping. We doubled the attic insulation, and we replaced the gas boiler with a solar heat pump."

"A solar heat pump?"

"Yes—it's called an Eco Cute.[20] They used to be called air-source heat pumps, but solar heat pump is a better description, since the air is heated by the sun. They sell more with that name, too.

"Then we changed the windows to triple-glazed and put in a heat recovery ventilator—you can see it up there on the wall. We followed the Enerfit Passive House guidelines for retrofits, which ensured that we got the greatest possible efficiency.[21] We installed timer-thermostats that adjust the heat according to our schedule and we changed all the lights to LEDs, which use very little power.[22] Then we put our appliances on a hard-wired home energy scanner that tells me where I'm using power, and we signed up for a home energy management program that lets me monitor the house from wherever I am. I say 'we,' because it was me and Thaba at the time. Now it's just me. Then we installed the solar PV, using the roof-integrated SunStation system,[23] and we backed it up with a Tesla PowerWall battery.[24] We got all our supplies from IKEA, who have made a specialty in serving the zero carbon market.[25] The graphene solar shingles are more efficient, but by the time mine need replacing who knows what might be on the market. Have you seen the new solar membranes? They've got them on BC Place, Vancouver's big sports stadium. Maybe in twenty years there'll be a solar membrane that will cover the entire south face of a house, and we'll be able to do away with panels altogether.[26] Speaking of solar, there's a huge floating solar installation on the ocean off English Bay. It's a square kilometre in size—they say it produces enough power for 18,000 homes. You can boat out to it and tie up—people use it to swim from. It's a scaled-down version of the absolutely massive installation the Chinese have built on the East China Sea, ten thousand square kilometres in size, which produces enough power to meet a quarter of China's electrical demand, or so they say."[27]

Floating solar? Ten thousand square kilometres? That was crazy. There were so many things I was longing to ask.

"How long does your battery store power for?"

"About nine days. We got it after the earthquake. Most people have them now that they've become so cheap."[28]

"You had an earthquake? A big one?" I knew Vancouver was in an active earthquake zone, but nobody worried about it much back in my time.

"Yes, but not the big one, luckily. Magnitude 6.9. It did a bit of damage, but nothing that couldn't be fixed. I try not to imagine what might happen if the big one struck in one of our really dry summers, when the forest fires spark so easily. It would be absolute mayhem, like something out of the Book of Revelations. As it was, there was an old hospital that took quite a battering and some of the high rises lost their windows, but luckily it was at night when the streets were empty.[29] The amazing thing was that the scientists knew it was coming, and they were able to warn us several days in advance."

"How did they know that? I thought earthquakes were something no-one knew how to predict."

"They've got a new system that combines three methods. The first monitors the behaviour of ants, which stay awake at night and refuse to go into their mounds in the days before an earthquake; the second uses camera traps to monitor the behaviour of small mammals in the forest, which stop moving around in the days preceding a quake; and the third monitors the airborne electric charge caused by the subterranean grinding of rocks in the days before an earthquake. It ionizes molecules in the air, which can be picked up by a very-low-frequency receiving station up to two weeks before a quake hits. When you combine the methods they give a very high rate of successful prediction. We had a week's notice of a possible quake and three day's notice of a probable quake. Most people chose to leave the city, which got pretty wild, but almost everyone knew what to do. It was a very crazy time."[30]

"I can imagine. It must have been pretty distressing. But getting back to your home, when you were renovating, did you worry about the paints and resins off-gassing toxic fumes?"

"No. You can't buy that kind of paint any more. I used a smart-smell app to double-check, just to be sure."

An app to test for fumes? Interesting.[31]

"So anyway, for the house, the installers finally gave me a switch that lets me turn everything off when I go out, except the fridge and freezer. The solar panels will be good for 50 years, with a gradual decline in performance, and when the loan is paid off my energy costs will be zero. And I mean that—zero. My home insurance has gone up because of all the climate disasters, but my utility bills will be zero."[32]

"Zero?"

"Yes. It's all laid out here." She picked up her screen, tapped a few keys and showed me a chart.[33]

"I've reduced my electricity use to 4,000 kilowatt hours a year, which is about how much the solar generates. Having an electric car would increase my demand, but why bother when there are so many shared vehicles in the neighborhood?"[34]

"So the solar revolution has finally arrived?"

"Totally. It's rare to see a roof that's not solar these days.[35] I have a friend, Li Wei-Ping, who was telling me how fast things are moving in North Africa and the Middle East. In Egypt, he says, almost every car is electric and they're getting the extra electricity they need for the cars from a solar farm in the desert outside Cairo that's eighty-five square kilometres in size—just a tiny fraction of Egypt's total area. He says they're building similar solar farms elsewhere, with the goal of generating all the power they need for the entire country. He says the same thing is happening throughout the Middle East, and in many other countries."[36]

"What about storage? What do they do when it's dark?"

"Batteries. They're also building solar thermal generation plants which capture the heat of the sun and store it in vast tanks of molten salt. That way, they can keep the lights on twenty-four hours a day."[37]

This was really positive news, which made me feel hopeful.

"How much power does solar provide here in British Columbia?"

"About ten percent. The rest comes from a mixture of hydropower from the dams, wind, tidal energy and the new geothermal plant. Our tidal plant is small, nothing like the big new one that's been built on the Bay of Fundy, which doubles up as a marine sports centre. It's quite amazing.[38] And it's all zero-carbon. America's the same for solar—they're getting ten percent of their power from it.[39] They say that by 2050 solar PV could cost as little as two cents a kilowatt-hour—can you imagine?[40] The price of wind energy has been falling steadily too. There's a big project going on to electrify all of Canada's railways, with solar along the tracks. Lots of surface car parks are getting solar roofs, and every commercial rooftop is being covered—it can save the owners up to $10,000 a year."[41]

"Is the solar revolution proceeding fast enough to make a difference to the climate crisis?"

"Is anything? It's like a cage match, Solar Revolution versus the Climate Monster, each trying to take the other out. The Climate Monster has the laws of physics on its side, alas, thanks to all the carbon that's accumulated in the atmosphere, and it keeps hitting us. It's like BAM! A megadrought here. BAM! A massive forest fire there. BAM! BAM! BAM! A super-typhoon, a monster deluge, an unprecedented heat wave. BAM! Another chunk falls off Antarctica. BAM! Another freezing Arctic Apocalypse, like the one they had on the east coast two winters ago.

"On Solar Revolution's side, using the term very broadly, we have to achieve five major victories. We've got to knock out every carbon producing fuel source on the planet. They've all got to be driven down to zero and substituted with renewable energy. We've got to stop every last bit of deforestation, both here in North America and in Africa, Asia and the Amazon. We've got to transform farming and forestry so that they start sucking carbon out of the atmosphere, weakening the Climate Monster's power. We've got to put a bung in the methane, which includes persuading people to eat far less meat because of all the methane that's being burped by a gazillion cattle. And we've got to pour billions of dollars into climate aid to mend the damage the Climate Monster is inflicting, so that people in the developing world don't feel abandoned and lose the will to persist. So who's winning? It's hard to say.

"There's a huge offshore wind farm being built in the Gulf of Mexico, along with a massive floating solar installation. That's the kind of thing we need. As well as producing loads of renewable energy they are protecting the Gulf Coast, taking the bite out of the hurricanes and reducing their impact.[42] China's doing the same, combining offshore wind and floating solar to create a buffer against typhoons. And they're doing a great job on the Salton Sea in California, with a massive floating solar installation that's reducing the water temperature and slowing evaporation. It's win-win, since the high water temperature was causing algae blooms that killed the fish and evaporation that increased the salinity, which was bad for fish and wildlife. It's things like that that make me feel hopeful. It might be fifty years before we get a clear result, however, because we can't undo the impact of three hundred years of fossil fuel craziness overnight. It's going to be a long drawn-out battle."

Wow. This was a lot to take in.

"That's really impressive," I said. "So I'll need to come back in fifty years if I want to find how the story ends?"

"Something like that. Maybe longer. It will depend how successful nations are at sucking the carbon out of the atmosphere, and how fast they can make it happen. It's unlikely that we'll be able to stop the temperature from rising by two degrees Celsius, and that will move us into territory the climate scientists say will be highly dangerous. But who's to say? Twenty years ago almost nobody thought we'd make such rapid progress. There's a reason why people call it The Great Turning."[43]

The Great Turning? Maybe better left till later.

"Changing the subject," I said, "do you know everyone on your street? I get the feeling that you do."

"We certainly do now," Dezzy replied, "but we didn't when we started. You'd best talk to Betska about that—she lives down the street. I'll send her a message to see if she's free. Maybe she'll have time to meet you."[44]

Dezzy picked up her screen and touched a button. "I love my Streetlife connection.[45] Here's Jonathon, for instance, one of my neighbors, asking if anyone wants a ride to Victoria. And here's someone telling me that my weekly food box will be late because of a hold-up in the depot. And here's a neighbor who's got a load of composted leaf-mulch to spare."

She touched a few buttons and typed a message. "If Betska's in she'll get back to me shortly. I don't want to call in case she's with a client."

"Are there other things you use that for?"

"Yes. I can click here to update my shopping list, here for anything to do with the house, here for the street, and here for concerts or events I'm interested in. It also gives me a daily reminder of bills due, street activities, public events and so on. I don't know how I'd cope without it."

"Does it use Wi-Fi?" I asked.

"No—I'm all done with that. Most people have switched to Li-Fi these days."

"Li-Fi?"

"Yes. You see this lamp? It transmits information directly to my screen, eliminating any risk of EMF pollution. I can use a laser pointer to send the information wherever I want."[46]

"Does it work for your cellphone too?" I asked. "And is electromagnetic radiation still a concern?"

"Yes. It's good wherever there's a transmitting light source. And as to the cellphone danger—yes, it's a huge problem. They tried to cover it up, but you can't suppress the science forever. I've a cousin who used her cellphone all the time, ever since she was a teenager. She used to carry it in her bra. She even slept with it switched on under her pillow. Five years ago she was diagnosed with a breast tumor in the exact spot where she carried the phone. Direct cause, the surgeon said. So now she's had surgery to remove her breast and she's on a host of drugs to stop the cancer from spreading. She says the side effects are really unpleasant. But she's lucky. She could have had a brain tumor, like a friend of mine. He says the surgery has totally wrecked his libido, and it's ruining his marriage."

"It's that bad?"

"There's been a big increase in brain tumors among people who used their phones a lot. So why take the risk? At Jake's school we banded together to get the Wi-Fi removed and go hard-wired. It can also reduce male sperm count, so you'd best be careful. I know we need population control, but that seems a bit extreme."[47]

It was a lot to absorb. Back in my time, cellphones were everywhere. Maybe one person in a hundred was aware of the danger. I wanted to learn more about Derek, however.

"Last night you showed me Derek's shrine. Is that something you're okay talking about?"

"It's kind of you to ask," she said, turning to look out of the window. "And yes, it's fine. I know I'm adopted, but he was my brother. My parents were committed activists. They named me after Desmond Tutu, the South African priest who was a close friend of Madiba, Nelson Mandela. All we were doing was walking down the street, demanding a better world. That's no reason to kill someone."

"How come the police have been unable to find his killer, after all this time?"

"That's what we ask ourselves every day. They were very persistent for the first year or so, but then the police chief changed and they seemed to lose interest. They say they had very few leads. The shooter used a high-powered long-distance precision-guided sniper rifle that combined facial recognition with a laser-lock, so Derek never had a chance.[48] It was a ghost-gun too, so there was no trace of it in any system."[49]

"What's a ghost-gun?"

"It's a home-made gun that's printed on a special machine. Not one of those cheap crappy 3-D printed guns, but a really sophisticated affair. The bullets were purchased across the border in Washington State. I received a hand-written note shortly after he was shot with words about God, Justice and the Judgment Day, but it was almost certainly from a crank. Derek didn't have any personal enemies—just people who were opposed to what he was trying to achieve."

I said nothing, just listening.

"He was really involved with the OMEGA Days. So was Lucas, who lives in the laneway house. He's a good person to talk to. But Derek was one of the leaders. He wrote the manifesto, *The New Conscience*, which was one of the inspirations for the OMEGA Days. He wanted the bailed-out banks to be taken over and turned into public banks. He wanted mandatory pay ceilings for the rich, the closure of the tax havens, and prison sentences for the big tax evaders, including the bosses at Goldman Sachs, who he said were pillaging the world. He wanted a Citizen's Income for all citizens. He was a vocal supporter of a woman's right to choose and physician-assisted suicide, and he said that only the police and bona-fide hunters should be allowed to own a firearm. So he upset a *lot* of people, from the fundamentalists to the gun owners, the bankers, and anyone who was hiding money in a tax haven. There were probably a *lot* of people who were happy to see him dead. But this is Canada, for goodness sake. We don't do political assassinations. They aren't part of our culture.

"We know someone was after him," she continued. "Derek used to ride an electric motorbike and in the week before the rally there were three occasions when he was buzzed by a drone. He was sure they were trying to force him off the road. But he had resisted buying a helmet-cam and the street-cams had been hacked, turned off at the very time Derek was being buzzed, so it was just his word."[50]

Dezzy sat silently, cradling her coffee. Indigo jumped up onto the table and pushed under her arms, nuzzling her face.

"It wasn't just Derek they were after. It was all of us. It was a crazy time. The tax authorities were picking on every activist who was self-employed, subjecting them to punishing audits. And they had no scruples about playing dirty. Someone went out of his way to pick on me, knowing I was Derek's sister and was active in my own right. We were renting a unit on the 25th floor of a condo building, Thaba and me, and we thought we had privacy. But one morning I had just come out of the shower, so I was completely naked, and I pulled up the blinds and there was this drone, hovering there. It was filming me, and within an hour the photos were on the Internet. They were trying to shame me, make me shut up. Close your big black mouth, they said, and we'll take the photos down.

"Was his assassin a gun nut?" she continued. "Was it a right-to-life fundamentalist? Was it someone who wanted to make sure their tax-cheating was never exposed? Was it someone hired by the big banks that were fighting the campaign to create a public bank that would threaten their monopoly on the creation of money? There was a minister in the government who had a survivalist brother who had made a crazy video about the OMEGA movement's leaders, naming names and telling everyone where we lived. He got a year for uttering death threats, but they never found any evidence that he was connected to the shooting. None of the leads went anywhere. I still receive comments on his Facebook page from people expressing love, but there has been no progress on the file for years."

Dezzy looked at me.

"It's so weird. You're so like him. Are you sure you aren't adopted? Maybe you're his twin brother and none of us ever knew."

"No, I'm pretty sure my parents are who they said they were. It's just me and my sister Daria now."

"That's how it was with me and Derek. He wanted to take care of me."

Dezzy was silent, and then she said, "Look, I've got to slip down to The Hive to meet a friend who works there. It's a very popular shared workspace."[51]

Her screen went 'Ping!' and she looked at it.

"Betska says she's free at 10:30 and she'd be happy to meet you. After that, why don't you wander around a bit and take in the local streets? I'll be back for lunch and if Lucas is home maybe he'll come over and tell you about the OMEGA Days. They're a critical part of the story, if you want to understand how we were able to make so many changes. So, Patrick Wu from Khartoum, make yourself at home, and have a great time with Betska."

4

Building a Neighborhood

THE SUN WAS shining, and I had time before my meeting, so I went back into Dezzy's garden. Looking up, I saw all the solar panels on the neighboring houses. Some had a smaller system where a tree shaded part of a roof and some were covered with solar shingles instead of panels. Glancing at my watch, I realized that it had been less than twelve hours since I'd arrived.

I opened Dezzy's door and looked out onto the street. Last night it had been getting dark, but this morning I was able to take a better look.

Instead of three lanes on the road there was just one with passing spaces. The rest of the street looked more like a garden. The sidewalks were lined with borders filled with fuscias, hollyhocks, marigolds and young tomato plants, a creek ran down one side, and there was a play area next to the shelter.

I crossed the road and sat down in the shelter. It was built from cob and its rounded walls were decorated with the shapes of birds and animals—a snake, a frog, a heron. Some bicycles passed, and the occasional car did so slowly, negotiating the speed bumps. The creek seemingly collected storm water from people's roofs, allowing it to trickle into the ground. It felt both wild and homely.

At the end of the street there was a large house carrying a carved wooden sign that said *The Marigolds: A Radical Rest Home*. A brass plate listed several names, including *Anna Betskaya Yureneva, Therapist and Healer*. I rang the bell, and a woman's voice invited me to come on up. At the top of the stairs a pleasant elderly woman with black and white striped frizzy hair smiled at me.

"So you must be the famous Mr. Patrick, from Africa!" she said.

I blushed, and said something incomprehensible.

"Well, come on in!" she chirped in a singsong voice. "I'm sure we'll have lots to talk about." Her living room had deep brown velvet curtains, beautiful antique furniture, a richly colored Kazak carpet, oil paintings, Russian icons, and two deep, luscious armchairs.

"What do you say to a coffee, with a drop of the good stuff?"

I was speechless, but I nodded at her suggestion and made myself comfortable in the planet-swallowing armchair she waved me into. When Betska returned with the queen of all coffees she plumped herself down in the other chair.

"Do you believe in synchronicity?" she asked. "I do. I had a client who was due at 10:30. But just before Dezzy sent me the note about you, he called to cancel. The Universe works in mysterious ways. So what's this research Dezzy says you're doing?"

I explained that I wanted to learn how people in Vancouver had been able to turn their city into one of the greenest in the world—and what was the story they carried in their hearts when they decided to take it on?

"You want to study our souls! Now there's a smart young man. If only the quantum physicists had thought that way a hundred years ago, when they first understood that reality was inseparable from consciousness. All those years wasted on dualism. We're all one—that's what I've always felt, ever since I was a child in Russia. You can't separate the physical from the spiritual. Our inner wealth is as real as our outer wealth. That's what happens to some of my clients. They lose their inner wealth. But as soon as they realize that losing is part of the larger journey, they begin to reconnect. You have to be willing to lose your way if you want to find it. Isn't that so? But you didn't come all this way to listen to me chattering on. You wanted to ask about our neighborhood, right?"

"Er, yes," I replied, feeling overwhelmed by her wisdom and warmth. She seemed like the kind of woman I could open my heart to. I wished I could tell her what I was *really* doing, and about my parents, Daria, the girl in Jerusalem, and the bottomless pit of grief that sometimes overwhelmed me. I had transported myself to this future, but I had brought my present with me. There were times back in my own time when I had to catch myself because the desire to abandon myself and curl up in the arms of an imaginary mother would appear out of nowhere and threaten to drown me with a longing to love and be loved.

"But I *am* interested in what you are saying," I continued, as though these thoughts had never happened. "It's part of our research to figure out what went on in people's minds as they worked to make Vancouver so green. What was the story people told themselves? What motivated them not to give up when there were difficulties?"

"Well now you're onto something," Betska replied. "I'm beginning to take a shine to you. I'm in my late seventies now, so I've lived through a lot. When we started out on this path many people were either in despair or apathetic, living their lives in quiet awareness of the disasters that were unfolding and feeling generally worried about the future. I've lived here on Bunchberry Street for twenty years—fifteen in my home and five in The Marigolds. What a great decision that was, enabling old people like me to live together while keeping our independence.

"I hope you don't mind me chatting away. It's one of the pleasures of old age. I moved here with my mother when I retired from my job as a psychiatrist in Montreal. In those days, almost none of our neighbors spoke to each other. Mama said it was like living in Russia in the old Soviet days. She often went out walking alone. One day, she told me, she saw a furniture van across the street with lots of coming and going so she went over and said, 'Welcome to the neighborhood!' And you know what they replied? They said, 'That's very kind of you, but we've been living here for seven years and we're just moving out!' We've been chuckling over that for years."

I laughed, knowing how true it was.

"But seriously. On her bad days my mother used to mutter that life here was worse than it was in the gulag. At least people there spoke to each other, she said, and the prisoners looked after each other. Admittedly, she didn't make much effort, but that's what she said."

"The gulag? Your mother was in one of Russia's gulags?"

"We both were. I was born in one. I spent the first year of my life there before we were released. Not that I remember anything of it. My mother spent seven years there, from her arrest in 1949 to her release. That's where she met my father."

"What was she imprisoned for?" And there was me, thinking my problems were significant.

"What was anyone imprisoned for? Someone must have filed a complaint, maybe at the hospital where she worked. She was always independently minded, which was not very wise in those days, under Stalin. She was made of iron, my mother. She lived through the 900-day siege of St. Petersburg when she was a teenager, nursing the wounded, the starving and the dying while Hitler was trying to bomb the crap out of the city so that he could raze it to the ground. She lost all her family, every single one of them. She was very bitter about being arrested when the war was over, after sacrificing so much for Mother Russia.

"She was very nervous when the OMEGA Days began, because she had seen what happened to Russia's revolution. But when she saw all the good that came out of it she changed her views, and became one of its biggest supporters."

The OMEGA Days... there it was again.

"And your father? Was he in the gulag too?"

"Yes. He served in a tank regiment in the Russian army and was captured by the Nazis. He was held in a German prison camp for three years, where he almost died. He was lucky. If they had discovered that he was Jewish he wouldn't have lasted long. When the Red Army invaded Poland and liberated the camps, the Russian prisoners were treated with suspicion in case they had been traitors. Most were released, but because my grandfather had been arrested for counter-revolutionary activity at the start of the Russian revolution, they sent my father to a gulag. He died when I was six, a few years after our release."

"I'm so sorry," I said. "Don't you sometimes despair of the human condition?"

"I probably should, but I've come to understand that despair is not a natural part of the human condition. It's a choice that people only make when their primary choice—to be purposeful and optimistic—is taken away, torn away, or knocked out of them. As soon as people find something to believe in they generally grab it with both hands. The instinct to optimism is incredibly deep, irrationally so. Here, have an oatmeal and blackcurrant cookie. One of my neighbors made them."

I did so with delight, for the cookie was as delicious as the coffee.

"How did you start working together on the street?" I asked, changing the subject. "And how did it become the wonderful place it is today?"

"You like our street? That's always nice to hear. It began in the Terrible Twenties, as they call them, soon after I moved here. That was a difficult time. I had worked all my life as a doctor in Montreal and I thought I had a good pension. Little did I know that most of my investments were in the worst possible places: oil and banks. My oil industry shares fell to a fraction of what they had been before the global carbon cap and carbon rationing, and my bank shares—well, we all know what happened to the banks. So almost overnight I found myself relying on my basic old age pension and my Canada pension, plus a few thousand in solar RRSPs, which didn't even cover the mortgage, and there was my grandson Leo struggling to get through college.

"Anyway, you don't get yourself a fancy brass plate on the door without having learned a thing or two, so I took one of my learnings and I applied it to myself. 'A trouble shared is a trouble halved,' I told myself, so I girded up my loins, as they say, and went out to meet my neighbors. Well, *that* was an interesting experience. Troubles? I thought *I* had troubles? I had neighbors whose gas had been cut off, neighbors whose rent was months overdue, neighbors who were eating day-old bread and baked beans. There was a middle-aged mother, Galena, who had two teenage girls she was raising on her own. She was lucky—she still had a job with an insurance company. But her ex had lost his job and he could no longer pay child support or his share of the mortgage. Both of the girls were staying out late, bringing home boys, getting into drugs and all sorts of trouble, and nothing she said was having any effect. When I listened to her she burst into tears, she was so relieved to have someone she could talk to.

"That was nothing compared to some places, mind you. I've a sister in Atlanta, Georgia, and you wouldn't believe what they've been living through. Last summer the temperature hit forty degrees Celsius and the roads started to melt. The vehicles got stuck in the tar, causing the entire highway system to grind to a halt. It was an unbelievable mess. They had to use helicopters to evacuate people who were trapped in their cars in the appalling heat. Back in the Twenties she told me the suicide rate in Atlanta was going through the roof as people struggled with debt, foreclosures and evictions. They're doing a lot more carsharing now, she says. Whatever our difficulties are here, we've had it easier than most places."[52]

"That's pretty incredible," I replied. "But maybe they enjoy carsharing?"

"Maybe! So anyway, there was a group called Village Vancouver—part of a global movement called Transition Towns. They were encouraging neighbors to get together and start a thing called Transition Streets. I liked what they were proposing, so I talked Dezzy into helping me organize a street party. I do wish she'd not broken up with Thaba, her husband. He's a good man. A bit controlling, but a good man.[53]

"So we knocked on everyone's doors and invited them all to a potluck. We closed the street, put up balloons and got the beer out, and afterwards we talked about how we could help each other, and make our street more friendly. Social permaculture, someone called it.

"The city had just launched a contest with a prize for the street with the best plan to become more sustainable, so we entered. The contest was fun, because it brought people together to share ideas. We didn't win, but we worked on our ideas anyway. The cob shelter—have you seen it? That was the first thing we built. There's this group in Portland called City Repair that encourages neighbors to reclaim their streets.[54] They inspired us to build the shelter, and to paint the mandala across the intersection. Have you seen it? Then we organized our first White Dinner and had our first street wedding."

"What's a White Dinner?"

"Oh, my! You haven't lived, young man, if you've not been to a White Dinner. We hold one every August during the annual Block Party weekend. We close the street and everyone dresses in white and brings a table, a white tablecloth, candles, and food to share. It's *so* romantic! And since nobody has to go home we can sing and dance the night away. Talk about a great way to get people friendly with each other."[55]

"I really like the way you help each other," I said. "Dezzy showed me her Streetlife page. Does every street have one?"

"I expect so. It's got so many benefits. I do so prefer it to the phone. I remember when young people of your age used to walk around with their noses stuck to their devices, not even seeing each other. NoFace-book, I used to call it. Anyway, I'm rambling."

"What kind of things did you do to help with people's difficulties?"

"Well, at first we simply created a space where people could talk. Every Tuesday night, in my living room, people would drop in and know there'd be someone who would listen. It was mostly women, but a few men, too. That's how we started the knitting circle—and the babysitting circle, which is a way for old folks like me to help the younger families. Some grandparents live far away and the young parents have no-one they can turn to.

"Then in the spring a group formed to grow as much food as possible. Out of that came the Community Tool Library, with its power tools, ladders and equipment, and the Bunchberry Urban Farmers. They organize workshops on seed saving, canning, raising chickens, things like that.[56] Last winter they did a course on how to skin and cook a raccoon. Ugh! Not to my taste. And last summer they got together and built a community root cellar in an old swimming pool the owners were no longer using. And every Saturday morning in summer we have our community produce stand. I'm getting too old to garden seriously, but we keep our yard full of food. I bake cookies, which I sell at the produce stand. They earn me some useful Dandelion Dollars."

"Is that a local currency of some kind?"

"Yes. I keep forgetting that you're visiting from far away. We have Dandelion Dollars for the street, Diva Dollars for the neighborhood and Delta Dollars for the region as a whole. They were a godsend after the financial crash, when all the credit seized up. They were like a backup economy that kept things going when the mainframe went down. Resilience—that's what it's about. Not being held hostage to a system that requires so much central control and coordination. You'd have thought they'd have learned from the collapse of the old Soviet empire. But no, the banks had to persist with their ever-more-complicated schemes, their derivatives and their high-risk insecurities, continually expanding the debt bubble until they hit the perfect storm and the whole thing came crashing down...."

I made a mental note to learn about the financial crisis—which might be due soon back in my time. "It's great!" I said. "You have so much going on."

"Everything is so much more advanced today," Betska replied. "In those days we had no Village Councils, no Citizen's Income, no free college education. We had municipal elections where most people didn't bother to vote and a healthcare system that was falling apart at the seams. Not like today—most things are so much better. And I forgot—there's also Freda's Friday StartUp group for teenagers who want to start their own business. That's really important, helping young people learn the skills they need to run a business and look after themselves. Leo—he's my grandson—I tried to persuade him to join, but he says it's not his cup of tea. He's happier with his books and his big ideas.[57]

"Then let me see, we have our monthly potluck, our annual street party, and a big get-together after Christmas when we celebrate our successes and consider what else is needed. We celebrate our failures too, for you can't expect success unless you also expect a number of failures.[58] And I nearly forgot! Last year the young people organized a community work bee to repaint Mrs. Wilson's house. She's getting old and couldn't afford it, so they got it done over a weekend."

"Who's behind all this? Do you have a paid organizer?"

"My goodness, no. We're just neighbors, doing what we can. But I will admit, some of us do have a secret weapon. I'm one of several people on the street who are members of the Sustainable Living Co-op. It came out of Transition Streets. We each pay $10 or $20 a month into a central kitty, depending on what we can afford, and we use the money to hire a coordinator who helps us share our skills and support each other. It's such a simple idea, but it's been responsible for so many new initiatives, including the community currencies, and the expansion of the carsharing and bike-sharing co-ops. The Carbon Reduction Circles were one of the Co-op's initiatives. Most people were very confused when the carbon rationing started, especially since the rations shrank each year. That really freaked some people out. The Circles helped people analyze their use of fossil fuels in a rational manner, and find ways to reduce them. Paying $10 or $20 a month actually saves us money, since the Co-op helps us participate in the sharing economy. And twice a year we have a Neighborhood Swap, when people put out things they're happy to part with."[59]

The sharing economy—it was something I wanted to learn more about. It sounded so much more fun than the private economy.[60]

"How many people belong to the Co-op?"

"When it started there were just thirty. They each put a thousand dollars in to get it going. By the time I joined there were several thousand, and today I believe there are forty thousand members in the region as a whole. That's less than two percent of the population, but it's a powerful force for change, with a budget of three million dollars a year. That pays for fifty part-time staff, based in the OMEGA Centre for Sustainable Living. When we wanted to convert three of our houses into The Marigolds we could never have done it without the Co-op's help. There were so many legal, financial and regulatory barriers to overcome, but they're trained in that kind of thing."

"How do you find time for all these activities?"

"Well, I'm retired, so what else would I be doing? But for other people, the four-day week helps. And I think many people are less stressed than they used to be thanks to the Citizen's Income, the sharing economy, the tax and benefit changes, and all the other community changes. The Citizen's Income takes the edge off the worry, knowing you have a guaranteed income you can depend on each month."

Whoa! So many changes! But this one I had to explore.

"Is that like welfare?"

"Yes and no. The important thing about it is that it's not conditional on anything apart from being a resident Canadian citizen who has lived here for ten years or more. Everyone gets it, rich or poor, no other conditions attached."

"How much do you get? And how is it financed?"

"Every adult gets $700 a month, on which you pay tax if you're earning at a taxable level. It's not enough to live on, but it's enough to take the precariousness out of life, and to end the constant fear that poverty used to bring. It's financed by a general increase in taxes, including raising the tax level for the super-rich to 65%. There was a lot of change that happened after the OMEGA Days, and the Citizen's Income was part of it. You'll have to ask someone else if you want the details."[61]

A guaranteed basic income for everyone... yes, that would make a difference.

"Going back to when you reclaimed the street—was it easy to get everyone on board?"

"Well, there's always someone," Betska replied, topping up my coffee. "There was one man, Jan, from Slovakia. He was convinced that the climate scientists were in league with big business and that carbon rationing was part of a conspiracy to establish world government. There was no persuading him on rational grounds. He'd been listening to those crazy late night talk-radio shows and all his thoughts were scrambled. But he started to change when his wife joined the knitting circle. Through her, he got involved in the community tool shed, and now he helps the kids with their carpentry. It seems to have mellowed him. He's a nice man, really."

"This has been fantastic, Mrs. Yureneva. Can I ask what it cost you to redesign the street?"

"Call me Betska, please. As far as I remember, the mandala and the cob bench cost us nothing, apart from a few headaches with the city planners until the people at City Repair in Portland came up and spoke to them, which changed their attitude. You wouldn't believe the things they've got going on down there in Portland. They have an annual Sustainable Streets Contest in which streets compete with each other to see who can achieve the highest overall score for things like sustainability, growing food and local happiness."[62]

"With a prize for the winner?"

"Yes. Everyone on the winning street gets a reduction in their property taxes for a year and they get to represent Portland at the West Coast Sustainable Streets Championship."

"What's the prize for the street that becomes the West Coast Champion?"

"They get an even bigger tax-break and everyone gets a ticket to the Sustainable Planet Expo at Disneyland. But you're distracting me. You were asking about the cost of the street reclaiming. After the mandala, we wanted speed bumps. The engineers said they'd cost $1,500 each, but we persuaded them to let us do it ourselves using a kit we found on the Internet. They ended up costing $1,250 for the two.[63]

"Then someone passed a video around by an American called Jason Roberts on how to build a better block, and people started talking about a much larger retrofit, which led to what you see today. That cost $40,000. It would have been $200,000 if we had used the city's workforce or hired contractors, but we offered to do the landscaping ourselves, and after a bit of negotiating with the union, the city agreed. Shared among forty homes, it adds just $60 a year to our taxes. But here's the thing! Not long after we'd done it the city published a study on the cost of upgrading the storm drains to carry the increased volume of run-off from the monster rainstorms we've been having, thanks to climate change. Tremendous deluges. A million dollars per street! That's what it came to. But on our street we don't need new storm drains because of the swales. And all for $40,000. So the city got a real bargain. We were one of the first streets to do it. People liked it so much that whenever one of our houses came on the market it was snapped up immediately."[64]

I was impressed. It was such a change from the car-dominated streets of my time.

"It's been invaluable for building our sense of community," Betska continued. "When there's busy traffic it's hard to make friendships with people on the other side of the street. Having a stronger sense of community also builds a stronger sense of ethics, to tell you what's right and what's wrong. I've seen what happens when people lose that, and I can assure you, it's not pretty."[65]

"Who maintains the boulevards? You've got a lot of food growing here."

"We all do, under the city's Green Streets program. I used to do more when

I was a bit more nimble.[66] I've got this nifty new electronic walking stick, mind you. It even tells me where to go if I get lost!"[67]

"What exactly is The Marigolds? The sign says it's a Radical Rest Home. What does that mean?"

"It means we live together and support each other in our old age, while retaining our independence. You have to be over sixty to live here, and we've a spare room where the grandchildren stay when they come to visit. There's a whole network of Radical Rest Homes across Canada, with at least ten more here in Vancouver. I used to own the whole of this house, but I sold it to the Rest Home. I now own shares in it and I live in these two rooms. There are twenty-four of us who live here. We joined the three houses together and added some extensions. We have a common living space and a kitchen where we sometimes eat together and where we can have meetings, film nights and parties. It's so much more fun than living alone, and far, far better than going into a home."[68]

"What happens if someone gets sick, or begins to go senile?"

"We have an arrangement with the Community Health Centre, and a nurse comes in once a week. She's like our family doctor. Most of us like to go for a morning walk, and we spend time in the Seniors Playground doing our balance and strength exercises.[69] We also make a big effort to get our Omega-3 fatty acids and our greens, to keep our brains agile. Our nurse has drilled into us the foods we need to eat and what to avoid if we are to stave off the dreaded Alzheimer's, like reducing our consumption of meat and dairy. We have a daily Scrabble and crossword bee after breakfast, and in the school holidays we run a weekly story-telling morning for the children, which keeps our brains active. We must be saving the healthcare system millions. There's a new system in which people who volunteer to help people with dementia get paid in Time Dollars, which they can trade for other kinds of service. It was developed in Japan, and now it's working here."[70]

"How did you afford to buy the three homes and renovate them? It must have been very expensive."

"It was, but we had some serious help from the Sustainable Living Co-op who helped us get grants from the city and the Ministry of Health in lieu of all the costs we're saving them. They own half the building with non-voting shares and we're free to run our lives the way we want. The main condition is that a quarter of the residents must be low-income. They didn't want us being exclusionary."

"This is really impressive, Betska. There's one more thing I'd like to ask. All these changes seem to have happened quite quickly. What happened to make them possible?"

"Well, that's a much longer story. To understand that you need to go back twenty years, when not so many people had a positive vision of the future. All the news was bad, and everything seemed to be getting worse. I'm talking the big issues now, things like climate change, financial collapse, and the collapse of ocean fishing. Most people felt powerless to make a difference. Even many

JOURNEY TO THE FUTURE | 35

greens were cocooning, growing their vegetables and riding their bikes but feeling powerless about the larger crisis that was looming closer every day.

"But then some people started stepping up and making things happen, like the street reclaiming and the local food markets. And Vancouver made its commitment to become the greenest city in the world, and to get all its energy from 100% renewable energy.[71] It was an exciting vision, and people put a lot of energy into it. And then of course, we had the crash, followed by the OMEGA Days with all their excitement, and the Terrible Twenties, or the Transformative Twenties as most of us like to call them. What a time! My, those were the days! After that, everything was different, and it became normal to do what we're doing today. There were leaders stepping up all over the place. It was like people slowly fell in love with the future. Slowly and beautifully."

"What do you mean?"

"Ah, you're so young—all thunderbolts or nothing at all. As more people took up cycling, growing food and building the sharing economy, people liked what they saw. So while their heads were telling them things were hopeless, their hearts had them engaged. I remember being invited to a wedding where everyone came by foot or bicycle and there were no physical gifts at all—only gifts of time. Best wedding I've ever been to. It took away the families fretting, the mountains of stress and the strained bank accounts. People gradually fell in love with the changes. They ceased being so negative. Does that make sense?"

"Yes, it does. But don't the global difficulties still make people feel hopeless?"

"What's in your coffee, young man? Did I give you the wrong mix? You seem stuck in the gloomeries. For sure, there's a world of change still needed, but once you're on the journey it doesn't worry you so much. Think back to World War Two. Did the British feel hopeless when they had been thrown out of France by Herr Hitler? Did the Russians feel hopeless when they were being besieged by the Nazis in St. Petersburg and Volgograd? Far from it. It made them all the more bloody-minded and determined.

"Once you're determined, everything feels different. Now that we've achieved so much locally, people are putting more effort into global change. There's a group two blocks over working with a village in Costa Rica, helping rebuild their school after the terrible mudslide they had last summer. I read that there are three hundred groups here working on one project or another, all with a global connection. They're the Changers, these young people, born in the early decades of the century. Me, I'm an old-fashioned Boomer. Dezzy—she's a Millennial. But my grandson Leo, and the young folks like Lucas and Aliya who live in Dezzy's laneway-house—have you met them? They're all Changers. That's what the world needs: Changers.

"That makes me remember," Betska continued. "Leo—my grandson—he's been telling me about a scientist called Elisabet Sahtouris. Must be almost ninety by now. She's a biologist at the forefront of the new science of syntropy, and I've been reading her latest book."[72]

Syntropy? This was the second time I had heard the word. Little did I know how significant it would become.

"We're all part of a single living system, she says, from the furthest galaxies to the tiniest microbes, even the atoms and particles. At every level, the system is conscious and self-organizing, right down to the atoms and below. That's how bacteria evolved into multi-cellular organisms, and then into humans. Within every unit, they both cooperate and compete as they juggle the pressures around them and seek to evolve to a higher level of order that gives them more freedom, more ability to express their potential, and more cooperation instead of conflict.

"Sahtouris believes that the Earth Harmony Movement, as she calls it, is evolving in the same syntropic way that the early bacteria did, with each cell of the movement finding its own function, then combining with other cells to form a larger organism. In the beginning, she says, groups formed around particular tasks like community gardening, or campaigning to stop a pipeline. They knew of each other's existence, but they didn't really cooperate."

The Earth Harmony Movement? Atoms being conscious? These were things I had to file away and hope to learn more about later.

"But then they started linking up, helping each other and developing a shared vision. They used their consciousness to feel out their surroundings and they created the new intentions that would shape a different future. As the cells of the new society emerged they began to envision a new future and to work together to make it happen. That influenced the surrounding cells, the traditional functions of the old order—things like banking, law and accounting—and they formed change-cells too until the whole function supported the emerging new order. People changed their investments and their banks. They changed the things they bought and the places they shopped. They started buying fair trade goods. Sometimes they left their jobs and developed new careers.

"Eventually, Sahtouris says, the process will extend to Earth as a whole. That's how the E-70 Group of Nations emerged, as nations started working together."

The E-70? Something else to learn about.

"Symgaiagenesis, she calls it. Isn't that a wonderful word? It combines the Greek words *sym*, meaning working together, with *gaia* for Earth and *genesis* for evolution. Symgaiagenesis."

"Symgaiagenesis," I repeated, "The Irish poet James Joyce would have loved it." And then the Irish coffee must have kicked in:

> *"Symgaiagenesis, some guy with a Guinnessis,*
> *brown lagered froth of bacterial genesis,*
> *brewing new life in the bowels of our synthesis,*
> *New worlds-a-simmering, shining and shimmering."*

Where the hell did *that* come from?

"Well, young man," Betska responded. "You are full of surprises! Are you a poet as well as a traveler?"

"No, not at all. Must be my grandfather's influence, on my mother's side. He loved Joyce. It would be a good word for Scrabble. Too many letters, though."

"This has been delightful," Betska said, getting up. "I hope it's been some use. You must come again if you're staying with Dezzy for any amount of time."

I thanked her, and she saw me out onto the street, tucking a bag of cookies into my pocket. My mind was spinning, and I hadn't even been here a full day.

5
Exploring a New Economy

I HAD AN hour before Dezzy returned, so I decided to explore the neighborhood. There was a creek that ran along the footpath, and halfway down the street a pond was covered with lilies. Eat your heart out, Monet.

As I took in the scene I spotted a pair of ducks hiding behind a cluster of reeds. A small willow tree had sunk its roots into the bank and was shading half the pond. There were water boatmen, pond skaters, and brilliant blue dragonflies. A pair of violet green swallows swooped over the water, looking for insects.

For most of human history we have lived in closely-knit villages where everyone knew each other. Even when we started building cities, the streets remained social places where people would talk, and then talk some more. It was only after 1920 that we surrendered them to traffic and some bureaucrat invented the crime of 'jay-walking.' But here was a future where the residents had reversed the flow of history and reclaimed their street.

A mother arrived with two small girls to play at the water's edge. On one side of the pond there was a pale-colored rock onto which someone had painted these words by the Chinese poet Lao Tzu:

> *Stand before it—there is no beginning.*
> *Follow it and there is no end.*
> *Stay with the Tao, move with the present.*
> *Knowing the ancient beginning is the essence of Tao.*[73]

A heron arrived with a noisy flapping, taking up watch on a stone in the pond. The mother hushed her children. Far out in space, the universe circled. A cyclist passed, calling a friendly hello. I got up, nodded to the mother and continued walking down the street.

Several houses had hand-made mosaic signs that showed their number, or said 'Hilda lives here,' and the intersection had been transformed into a vivid mandala painted red and yellow, as Betska had said. It was empty, but I could easily imagine the neighbors meeting here for a party.[74]

I turned down the adjoining street, with its single winding lane. Along with the shrubs and vegetable beds the liberated space had been converted into a play area, with a basketball hoop and goals for street-hockey. Only one of the houses had a fence, and most were growing food in their front yards.

Halfway down the street I turned into the alley that ran behind Dezzy's street. The asphalt had been removed and most of the fences had been taken down, creating a rambling footpath, a ribbon of greenery. On a bench under a tree two young fathers sat chatting as their children played, greeting me as I passed. Several new homes had been built from converted garages, but since they had no car access the alley remained quiet, almost rural.

At the end of the alley I passed under an arch and found myself in a new development with a semicircle of five townhouses on either side of the street. The road was closed to through traffic, and the area between the houses had a play area and barbecue pit. But when I came to the end of the street I was shocked to discover that the next street over had a large expanse of asphalt with a few parked cars, the same as the streets were back in my time. The houses had solar roofs and many were growing food, but it felt as if the street had become trapped in the past.

"What happened?" I asked Dezzie, when she returned for lunch. "Why did that street miss out?"

"It's sad," she replied. "Most of the residents wanted to follow our example, but they needed the signatures of seventy-five percent of the residents to do so. Several houses were being rented from owners who lived abroad, and the tenants couldn't vote. Unknown to the organizers, some of the residents contacted the absentee owners and persuaded them to vote against the change. They said it was the increase on their taxes, but in reality it was to protect what they saw as their freedom to drive. Without the votes of the absentee owners, there was nothing they could do. And here's the crazy thing: the property values on that street have fallen. Who'd want to live like that, when you could live the way we do?"

Dezzy picked up a remote and some soul music came out of the wall. She started dancing.

"Life is good!" she called out. "Come and join me!"

I stumbled to my feet, and she took my hand. My feet tripped from self-consciousness, but she said, "No worries. Enjoy! Life is good!"

Later, collapsing into an armchair, she asked, "So, are you single? Or do you have someone special tucked away?"

I blushed. "No. There's no-one. Maybe I'm still waiting." It wasn't true, but I didn't feel able to say so. In truth, there was no space in my heart because I had already given it to the girl in Jerusalem—the one who might not even be alive.

Dezzy got up to make lunch, tussling my hair. "You won't be short for company, a good-looking young man like you. Don't you worry."

"This is a great kitchen," I said, eager to change the subject. "What's the screen above the stove for?"

"It gives me recipes for the food I have available and tells me what's good for me and what will help me keep my weight down. Here—I'll show you how it works."

She pulled up a menu that said 'salad' and touched the word for each ingredient she was using: red leaf lettuce, arugula, spinach, kale, chopped almonds, mushrooms, flax seed, peas, goat's cheese, grated apple and chopped ginger, topped with a ginger honey dressing with pepper and garlic.

"It lists the nutrients and phytonutrients for each ingredient and tells me how many calories I'm getting. If I use it every day it gives me a running total for the week. If I press this button, it tells me what nutrients I'm missing and suggests recipes to fill the gap. This salad will turn you into a superman: it's loaded with everything you need."

"That's really cool. Does everyone have one?"

"They're becoming quite common, especially with the new healthcare system. Can you shell these peas for me and then bag them? I've picked more than we need so we can freeze the rest. You see the screen on the fridge door? If you tap the letter P you'll see 'peas' and you can enter how many bags we're doing."

"What do you want me to do with the shells?"

"They go in the composting chute, which empties into the bin outside."

"What about rats and raccoons? Don't they go for the compost? And do you get a lot of wildlife here?"

"With all the food people are growing? We've enough wildlife to fill a zoo! The compost bin is tightly sealed and I keep a spray bottle laced with cayenne pepper and chopped habanero chilies, which usually puts the raccoons and skunks off for a while. We've also got crows, herons, frogs, grass snakes, bald eagles, squirrels, bats, and the occasional coyote."

We sat down at the kitchen table overlooking the garden and enjoyed the salad. The early afternoon sun was pouring in, and her black cat Indigo was sitting on the path in the garden gazing at something.

"Can you tell me more about Derek?" I asked, when Dezzy had poured the coffee.

"How far back do you want to go? He was my big brother. He always took care of me. Even when he was at school he was trying to save the world. When he left school he took off travelling for a couple of years, volunteering on organic farms in Europe and the Middle East and then in Australia.[75] When he came back he worked for a while to earn some money and then spent three years getting his economics degree at UBC. He was in his last year when the first financial meltdown happened in 2008, so you could say he had a front-row seat.

"When he finished he could have gotten a job with any big company, but instead he worked at a soup kitchen in the Downtown Eastside, Vancouver's poorest neighborhood. Then he got a job as a neighborhood outreach worker, completed the certificate program in Community Economic Development at Simon Fraser University and got involved with the Occupy Movement, which started out so hopeful."

The Occupy movement: something from my time.

"We were fairly depressed when so little came of it. People didn't know what they wanted. They knew what they were opposed to, but not what to replace it with. Not just here, but everywhere: London, Wall Street, Toronto. That was when

Derek realized that if we were going to make a difference in the world we needed a much clearer vision of the kind of future we wanted.

"I remember Derek coming round one night. He was upset that Occupy had so little to show for itself, apart from some great memories. It was winter, but he took off to stay with a friend in Tofino on Vancouver Island, where he spent three weeks walking the beaches and taking in the storms. When he came back he announced that he was going to tour the world to visit places where they were building a cooperative economy. The financial meltdown had shown what a disaster the current system was, and there had to be a better way."[76]

"Where did he go?"

"It's been a while, but it's fresh in my mind, since I helped write his biography. He had been impressed by a book titled *The Public Bank Solution*, by the American author Ellen Brown, so he took a Greyhound bus across the prairies in midwinter to learn about the Bank of North Dakota, the only publicly-owned bank in America. There are countries all over the world that practice public banking, including many European nations, but here in North America there has been a taboo against it. The bank was created in 1919, following a political upheaval when the local farmers organized successfully against the Wall Street banks and got themselves elected into the state government. The bank has been creating new money and supporting community banks in their lending ever since, making a huge success of it: so much so that North Dakota is still the only state in the US without any debt. Derek took the lesson to heart, and from then on he put public banking at the heart of all the changes he saw to be necessary. Whenever a bank creates money, he realized, the interest goes to the bank's owners. When a bank is public, the interest goes to the government where it can be used to pay for education and healthcare, things like that. It seems so straightforward when you put it like that. Little did Derek know how determined the banks would be not to let go of their monopoly on the creation of money, and all the profits it brought.[77]

"Next he travelled to Minneapolis, where he visited the Institute for Local Self-Reliance, one of the first places in North America to focus on the local economy and the value that can be obtained when a city puts its mind to generating its own energy, recycling its wastes and supporting local stores instead of big-box retail stores, things like that.[78]

"He continued on to Cleveland, Ohio, where he saw how they had developed worker-owned co-operatives by tapping into the purchasing power of local hospitals and universities. He visited coops that were providing solar services, laundry services and food, and he saw the motivational energy that worker-owned enterprises released.[79]

"From there he took a flight to Brazil, and the coastal city of Fortaleza in Ceara State, where a man called Joaquim Melo had set up a system of community banking called Banco Palmas and a thriving co-operative currency called the Palmas, which inspired the formation of fifty community banks. Their goal was to democratize credit, so that ordinary people would no longer depend on the big banks to provide them with loans and microloans.[80] Whenever Derek flew, he offset his carbon by donating to the Solar Electric Light Fund, which helps rural villagers around the world replace the use of kerosene with solar energy."[81]

This was a lot to take in, but I had the feeling that it would all come together, so I encouraged her to continue.

"From Brazil he flew to Lisbon, in Portugal, and he took the bus to a place called Mondragon in the Basque region of northern Spain. In the 1940s, after the misery and destruction of the Spanish civil war, a Jesuit priest, Father José María Arizmendi, had been searching for ways to put the church's social doctrine into practice. He asked the question, 'What is the Jesuit way to develop an economy?' and that led him to the work of the 19th century British socialist Robert Owen, who pioneered the world's first workers' co-operatives. Father Arizmendi went on to found the world's most successful cooperative economy, which employs some 100,000 people in 300 co-operative businesses. They have their own co-operatively owned bank, their own university and their own welfare system. Almost every worker is a member-owner and their wage differential is only 6.5 to 1, compared to as much as 1,000 to 1 in North America, as it used to be. Mondragon survived the financial meltdowns with far less damage than other companies in Spain. When the second crash happened, and the market contracted by 20%, instead of firing 20% of the workers everyone reduced their working hours by 20%."[82]

"That seems like an intelligent way to handle a crisis."

"Yes—but it takes a business that is owned co-operatively and managed democratically to come to a decision like that. When decisions are made by the few whose personal incomes depend on what they decide, it's easy to see why the people who are shut out get to suffer."

"Anyway, from Spain Derek travelled overland to the Italian region of Emilia-Romagna, south of Venice, where he spent six months learning Italian and studying their cooperative economy, which has made their region the most successful in Italy. In a population of 4.5 million, two out of every three citizens belongs to a co-operative, and co-ops make up 30% of the economy. But even among the private businesses there is cooperation and self-organization that doesn't happen elsewhere. The businesses and coops belong to various regional networking organizations, to which they pay a portion of their proceeds, and in return they get help with everything from training to product development. There's a clear reason for their success, Derek discovered, and it's to do with cooperative self-organization, instead of the private competitive approach."[83]

"How did Derek even know about these places? I've not heard of any of them."

"He was a member of the New Economy Coalition," Dezzy replied. "Their members know where all the important innovations are happening, and they have their fingers on the pulse of change. They helped him organize the trip and gave him introductions.[84]

"While he was in Emilia-Romagna he bought a bike and from Italy he cycled north into Austria, to a town called Güssing on the border with Hungary where they had turned their local economy around by making it the world's first 100% renewable energy region, using solar and biomass energy to generate all their heat, power and fuel.[85] Everyone's doing it today, but it was new at the time. Then he cycled across Austria to Switzerland, where he learned about the special system of Swiss banking known as WIR—that's German for 'we'—in which 60,000 small and medium sized businesses share a mutual credit network. By providing a parallel source of credit they are able to balance the ups and downs of the business cycle and provide stability for Switzerland's economy. It's one of the secrets of Switzerland's success, along with their local savings banks."[86]

"This was quite the trip," I said. I was envious. What a great opportunity he created for himself!

"Yes. And he was blogging and filming wherever he went, making videos that he'd post on YouTube. He had tens of thousands of followers on Twitter, Foible, Village, YouTube and Facebook, so all these people were learning about new ways of running an economy along cooperative lines.

"From Switzerland he crossed into Germany and spent a few days in Freiburg, Germany's solar city. Then he cycled to the small town of Prien am Chiemsee in southern Bavaria, where Christian Gelleri, a school economics teacher and his students had launched Europe's most successful local currency, the Chiemgauer, named after the region. It's pegged to the euro and can only be used locally. It loses two percent of its face value every three months, which gives people an incentive to use it. We use the same system with the Delta Dollars, our regional currency here in Vancouver.[87] In Bavaria he was invited to a private house party with some of the biggest names in co-operative currencies and banking reform—people like Bernard Lietaer—and all the time he was pondering how the ideas could be applied in Vancouver and the world as a whole.[88]

"From Germany he crossed into Denmark, where he rode around Copenhagen and experienced their cycling revolution first hand. At the time, something like thirty percent of the people rode a bike to work or school. Today, it's more like fifty percent. It's amazing what they've achieved, and all because local people organized to make it so over a period of fifty years.

"Then he crossed the bridge to Sweden and cycled to the town of Skövde, where a co-operatively owned bank known as the JAK Bank (pronounced *yok*) provides mortgages and loans to its members by charging a small fee instead of interest. At the time, they might have been the only bank in Europe not charging interest, but it's quite common now.

"I'd never thought much about interest until Derek started blogging about it," Dezzy continued. "I had no idea how much it increased the cost of everything we buy. Derek used to repeat a question that a German woman called Margrit Kennedy used to ask. She was one of the cooperative economy's founding inspirations. If an employer offers you a choice of two different pay-raises, she would ask, which would you prefer? You can have either an extra $10,000 a week, or you can take a one cent raise the first week, two cents the next week, and double again each week for the rest of the year. What's your choice?"[89]

"I'd take the $10,000 a week. That would be amazing."

"That's what everyone chooses, because we don't understand the power of compound interest. If you took one cent and let it double every week, by the end of the year you'd be earning $45 trillion a week. Not bad, eh?"

"Whoa! I had no idea."

"That was Derek's reaction too. Margrit had calculated that in Germany, if you removed interest from the price of everyday goods and services the cost would fall by forty percent. That's the power of interest when it compounds on itself, which it does whenever you miss a payment. That's one reason why Derek became so interested in co-operative currencies and new ways of banking.[90] The JAK Bank uses a fractional reserve ratio of 5:1, meaning they keep $20 in the bank for every $100 they lend out, whereas some Wall Street and European banks had been lending on ratios as high as 60:1, which is one of the reasons they collapsed."[91]

"While Derek was in Sweden he also went to the city of Växjö, which, like Güssing, was working to become a 100% renewable energy region using biomass, hydropower, geothermal, wind and solar energy. He was always seeking solutions to the climate crisis as well as the economic crisis.[92]

"From Sweden he crossed to Helsinki in Finland where he met a group of people who were promoting the idea of a guaranteed Basic Income, or Citizen's Income. So many people were living precariously, either because they had no work or because they depended on part-time jobs, and they argued that a completely new approach was needed in which every citizen would receive a guaranteed monthly income, no conditions attached, financed out of taxes. That was where he learned about the Basic Income Earth Network, in which people around the world were collaborating to develop the best models, in readiness for change. Canada adopted the Citizen's Income seven years ago."[93]

"That must make quite a difference in people's lives."

"Yes. It has reduced the fear and desperation many people were feeling. Not eliminated, but certainly reduced. Everyone gets it, children too. I've got a good income in my current job, but when I left Jake's dad and became a single mother it was really hard. Now that we have the Citizen's Income, the four-day week and $20-a-day daycare it's a lot easier for parents, both working and single."

"Was daycare another of the changes that happened after the OMEGA Days?" I still didn't know what they were, but it felt like a safe guess.

"Yes—that and a lot of other things. It's one of the things Derek went on about when he got back—how subsidized daycare was so normal in places like Mondragon, Emilia Romagna and Finland, and how it contributed to family stability as well as economic stability. It was in Finland too that Derek became obsessed with saunas. He was convinced they were the secret to Finland's egalitarian culture."

"Saunas? How come?"

"His theory was that the Finns had learned to see each other as equals, since in traditional Finnish culture everyone has a sauna at least once a week, men and women separately, and always in the nude. When you're naked, he said, it's hard to pretend that you're better than someone else. He wanted to form a co-operative sauna when he got back, but he never got round to it.[94]

"From Finland he took a boat to England, where he spent a year doing an internship with the New Economics Foundation in London. That gave him a chance to get to know some of the country's leaders in green, community-based, cooperative economics,[95] and to get his mind around the role of the government in building a cooperative green economy. He was very impressed with two people: Mariana Mazzucato, Professor in Economics at the University of Sussex, author of a book titled *The Entrepreneurial State*, and Ha-Joon Chang, the Cambridge economist, author of a book titled *23 Things They Don't Tell You About Capitalism*. Mariana Mazzucato persuaded him of the need for the state to play an active entrepreneurial role by investing in new ecologically sound developments, and Ha-Joon Chang persuaded him of the state's importance in guiding and controlling trade and capital flows—the complete opposite of what the neo-liberal economists had been preaching.[96]

"He also visited a number of innovative projects, including the birthplace of the Transition Town movement in Totnes, Devon,[97] and the Findhorn Foundation, a spiritual and ecological community in northern Scotland whose members had the lowest carbon footprint in Europe. He helped build a solar-cob house in their ecovillage, and he participated in a global conference on Visions of the Future.[98] Somewhere along the way he fell in love with a girl named Jenny and he moved to the small Yorkshire town of Todmorden where she lived, which had become well-known for a project called Incredible Edible Todmorden, where the townspeople were growing all their own food."[99]

"How did he support himself while he did all this travelling?"

"Before he left he raised some money by crowdsourcing on Indiegogo, and he worked on organic farms. The Indiegogo connection was great, since the people supporting him followed his travels.[100]

"When he came back he brought Jenny with him. They were inseparable, and she supported him while he spent the next year writing his book, *My Love Song to the Planet*, which played a big part in inspiring the OMEGA Days, alongside his manifesto, *The New Conscience*. It was published just before the financial meltdown. She was a lovely woman. Very spiritual. Believed in angels. Jenny said she could sometimes see angels sitting on people's shoulders. She told me once that she saw my father, who wanted to tell me he was always there for me. He used his pet nickname for me, and there's no way she could have known about that.

Other people had the same kind of experience with her. Don't ask me to explain it: I believe it's real, but it's beyond any rational explanation."[101]

"That's fascinating," I replied. "I sometimes have the feeling that my father is looking after me, too. I have so much to learn," I said. "But tell me, what was it like in Vancouver during the meltdown?"

"It was nowhere near as bad here as it was elsewhere. The regions that depended on just a few employers and the big manufacturing centers—they took it really hard. The bottom fell right out of the money market and businesses that depended on credit found it impossible to function. Unemployment hit 25%, and a lot of people lost their homes because they couldn't pay the rent or mortgage. It was far worse in the States, because Canada's bankers were better regulated and had not been allowed to take the big risks they took south of the border. But with Canada being so closely linked to the US, and our banks being so invested in the carbon bubble, everyone was affected."

"What was Derek's response?"

"To organize. His biggest concern was that there would be an old-fashioned social revolt, a repeat of the Occupy movement. It was important what they achieved first time round, he said, but it wasn't enough. This time round we had to be really clear in our vision and strategy. Luckily he wasn't the only one with this concern. There were some small demonstrations, but the global movement as a whole was able to hold its fire until it was ready with the OMEGA Days."

"What happened then? I was nine at the time, living in east Africa." It was a lie, but a convenient one.

"I'll ask Lucas to come over when he gets back from work. He was very involved, so he'll give you a better story. Meanwhile, do you want to see the video Derek made when he got back from his trip?"

And so it was that I found myself sitting in Dezzy's garden watching Derek's video. Six and a half million hits—that was big for a talk. His manner of speaking was confident, and he had a charisma and charm that was quite magnetic. His video was titled *My Love Song to the Planet*, the same as his book, and it was organized in seven chapters, or verses, using Italian for the titles, a language he said he'd fallen in love with. My notes are just the briefest summary.

VERSETTO UNO: MIRACOLO

In the first verse he expressed his deep love and wonder for the Earth, the ocean, and the cosmic, biological and spiritual evolution that has brought us to where we are. Our existence is a miracle that gives us the ability to dream, to choose, and to make a difference. In spite of all the terrible things we have done to each other and to the planet, he still had faith in humanity and in our ability to build a better world.

VERSETTO DUE: EGOISTICO

Over the last five hundred years we have built our economy as if all humans were selfish and opportunistic, and everyone was a dominator. But we are not—we are also cooperators. The global economy was on its last legs, he wrote, and about to collapse. It would be brought down by the weight of selfishness, expressed in the sheer magnitude of investments and debts that lacked any substance. But we must not fear its collapse. We must understand it, and use the opportunity to build a new economy that would reflect the cooperative, caring side of our nature.

VERSETTO TRE: MADRE TERRA

In our self-centeredness we have treated Earth as if she existed purely for our benefit. The earth and the oceans give us so much, yet like a mother, they never send us an invoice. Our entire existence depends on air that is safe to breathe, water that is safe to drink, soil that is safe to grow food in, ocean plankton and forests that produce the oxygen we breathe, and an atmosphere that protects us and regulates Earth's temperature.

Earth gives us eight million species to share the planet with. We can't continue to mine Earth's resources and harvest her creatures as if she were a cookie jar, there for the plunder. We need to join hands with the First Peoples of the Earth and rethink everything we know about the way we treat the Earth and her eco-systems. We need to craft a new way of living that is in harmony with nature, as well as with ourselves.

VERSETTO QUATTRO: EMPATIA

As Martin Luther King said, the moral arc of the Universe is long, and it bends towards justice. But it does not bend on its own. It bends because people put their hands on it and work to bend it in the direction of justice and compassion. We need to continue to bend it, extending our empathy not only to our fellow humans but also to our fellow species, and to Earth as a whole.[102]

VERSETTO CINQUE: NUOVA ECONOMIA

We need to build a new economy that will reflect our ability to be kind and co-operative as well as to compete and be entrepreneurial. An economy based only on competition and opportunism will always create misfortune and unhappiness, just as it does in our personal lives when we behave selfishly. We need to recreate the way we create money, the way we trade, the way we run our businesses, the way we bank, the way we own land and housing, and the way we develop our economies. Capitalism was built block by block over several centuries, he said. We needed to build the new cooperative economy in the same practical way, block by block.[103]

VERSETTO SEI: NUOVA GOVERNO

We need to reflect our caring for the Earth and our need to live cooperatively through new laws and new methods of governance that take bold steps to advance a cooperative, sustainable economy that can flourish while protecting habitat, restoring damaged ecosystems, and paying for nature's many services.

VERSETTO SETTE: SPERANZA E DETERMINAZIONE

We need to have hope and determination to overcome the bastions of power whose supporters delay, obstruct and prevent progress. We need to be passionate and positive about the future we are about to create. He ended by reciting a poem:

Our world will be changed by love, not anger,
By the creativity of hope, not the sadness of defeat.
Come with me now: your birthright calls.
We are ancient, we are proud,
We are as old as existence, as determined as the stars.
Find hope, where once you found despair,
Find determination, where once you feared defeat.
Come with me, let us build a better world;
It is not I, but your own soul that calls.

"It's very inspiring," I said to Dezzy when I had finished watching.

"It all seems so long ago. It's good to know that it still resonates. By the way, there's an event tomorrow night you might be interested in."

"What's that?"

"It's called *Song of the Universe.* It's for people who like to think and dream both spiritually and scientifically, without any religion. Does that interest you?"

I assured her that it did. There was not much in the religious belief systems I had come across that I liked, but neither did I like the cold materialism that so often went hand-in-hand with science. I didn't believe in God but I didn't believe in materialism, either. There had to be more.

"Okay. We'll go together. And I'll see if I can rustle up some meetings to help you learn how Vancouver has been progressing."

 6

The OMEGA Days

AS WE WERE talking there was a rap on the door and a young man with long blond hair came in wearing a cut-off T-shirt and jeans, bearing a big tattoo on one arm.

"Hi Lucas," Dezzy said. "We were just talking about you. Meet Patrick. He's visiting. He wants to learn about the OMEGA Days."

"Happy to meet you," Lucas said, grasping my hand warmly. "Lucas George." Then turning to Dezzy, he asked, "Did you get the tickets?"

"You bet I did," Dezzy replied. "How often do you get to hear Crocus in full holoconcert? But you're such a contradiction, Lucas. One day you're all simplicity, boycotting plastic and living off raccoon roadkill and the next day you're holotechnology's biggest fan."

"Who's Crocus?" I asked hesitantly. "And what's a holoconcert? You must excuse my ignorance. I'm more used to campfire singing."

"Hey—that's cool! I love campfire singing," Lucas said. "I wish we had more of it. Crocus just happens to be a megastar. She's from Brazil. And yea, I know the holotech stuff seems like a contradiction. But holograms use almost no energy. They're pure creative expression. And it's not as if Crocus is flying here, burning scarce biofuel."

Now I was really confused....

"So how is she getting here?" I asked naively.

"Skypogram. High-definition hologram with a great local band. Some of her vocals give me the chills."[104]

"They've lined up a pretty impressive show," Dezzy said. "The Vancouver Ballet's putting on a holographic performance with dancers flying over our heads and they have two poets doing that travelling words thing with lines of poetry that move through the air. Crocus will be using projected holographs to enrich her songs and there's full harmony sing-along for several pieces. We're going to have to learn our parts."

"So what was it brings you here?" Lucas asked. "I'm sorry—I got distracted."

I explained my reasons for coming, and he replied, "Well, if that's your interest, you should have been here fifteen years ago. That's when we really had some fun. I still think of them as the best years of my life."

"I was hoping you could tell Patrick about the OMEGA Days," Dezzy said. "You spent half the year in juvy, if I remember right. Here, have a coffee."

"Thanks, Dezzy. Yes, juvy was where we dreamed up some of our best protests. The joint has great advantages if the keepers let you keep your connections. We had such good face-time, and we didn't even have to pay!"

"What were you in jail for?"

"Oh, blocking the streets, meditating on the sidewalk, singing in the police station. You name it."

"Our Lucas was quite the hero," Dezzy said. "Derek said he could trust him with his life."

"So much for trust," Lucas interjected. "It wasn't much use when it really mattered, was it?"

"You mustn't say that. I doubt there was anyone who could have prevented it. Let's not go there."

"But how can we not? And let them get away with it? There had to be an insider. Someone in the Vancouver Police Department, I think. How else could his assassination have gone unsolved for so long?"

"Let's not go there, Lucas."

"Okay, but he was my buddy as well. I keep thinking about all the people who had a motive to kill him."

"But Chief Constable Liu Cheng was one of the best. I can't believe he'd hide anything from us."

"That's what all the best con artists are like. They're likable. You'd never guess they were up to something bad. Besides, it's a big department. There were plenty of others. Maybe someone sympathetic to that American group, the Sons of Heritage, who hated everything we stood for."

"Lucas, once again, let's not go there. Please."

"What else happened during the OMEGA Days?" I asked, hoping to change the subject.

"Well, we had our Twenty-Five Solutions plastered all over the city," Dezzy replied, turning to face me. "Everything was linked through the new synthesis. In the end, the politicians were tripping over each other to implement the solutions.

"OMEGA," Dezzy continued. "The O stood for Occupy Democracy... M for Meaningful Work... E for a New Economy... G for a Green Future, and A for Affordable Living. Whenever people saw the OMEGA sign they knew what it meant. And we were all such aptivists, as well as activists, spreading the word in a host of different ways."

"And raptivists," Lucas added. "We had some wicked lyrics." He stood up, shook his body and launched in:

> They say we can't do it, can't fix it or grow it,
> Can't change the world, kiddo, what makes you not know it?
> But we've got no ears for your know-nothing blow-it,

We'll change this world ten times before you can throw it.
For this is our now time, not do the fuck-all time
We're done with your moaning and dying delays.
We say live! Live again! For it's change the world now-time
So kiss me and celebrate, show me your ways.
Kiss me and celebrate, change-the-world, elevate
Elevate higher than dirt-streets and mire
Elevate up where the highest hopes relevate
Kiss me and celebrate, change the world now.

We applauded, and Lucas laughed.

"See what I mean?" Dezzy said. "You were one of our heroes—and Derek's too. Give me a minute—I may be able to find an old poster."

My mind was bubbling with questions. What was this new synthesis? And what *were* the OMEGA Days?

"Was it just in Vancouver that the OMEGA Days happened?" I asked cautiously while Dezzy was out of the room.

"For sure it was—not. Where have you been? In a coma?" Lucas stared at me incredulously. "Oh, sorry, I forgot—Timbuktu, wasn't it? No— it was pretty much global. London, New York, Hamburg—almost everywhere. Paris, Rio, San Francisco, Portland, Toronto, Cairo. Even Shanghai and Beijing, before they were suppressed. It was like the Occupy movement, but far more advanced. They didn't all have the success we had. That's why we need another big global effort to get us over the tipping point—and soon."

"I, uh, you'll have to forgive me," I improvised. "I was only nine at the time, and where we lived most people were concerned with finding enough food to eat."

"I'm sorry. I didn't mean to speak so harshly. The OMEGA Days were pretty big out here, and it's easy to forget that it wasn't the same everywhere."

"Here, I found it," Dezzy said, returning with a beautiful poster featuring a large OMEGA sign created by two people kneeling opposite each other with their heads together.

"Each of the five themes had five solutions," she said.

"It's thanks to the A for Affordable Living that we're able to live in Dezzy's laneway house, our Little Palace," Lucas said. "Back then, Vancouver was so expensive your parents had to be multi-millionaires before you could even consider buying a place. Even renting cost an arm, leg and your friggin' firstborn. I had been living in Prince George with my family but I couldn't take the fighting, and my father constantly picking on me. My mom had a good job with the credit union and my father was a logger but he lost his job when the company cut back due to the timber shortages—thanks to the pine beetle, another climate impact. He was using drugs, and I'm pretty sure he was dealing them too. That's what the fighting was about.

"So Julie and me—she was my older sister—we walked out. Lovely sunny day. I was fifteen. Hitchhiked to Vancouver. Didn't know what to expect—bright lights, excitement maybe. But there were homeless people all over the place, living in bus shelters and stairwells. For three years I lived rough and the worst thing is that one day Julie disappeared. Fuck it, I'm still so pissed at myself."

"What happened?"

"She just disappeared. There's no way she'd have gone for more than a few days without telling me. We were so close. I did everything I could, but we couldn't find her. The police response was pathetic. She's still on the list of missing women, but I've given up. I expect they'll find her remains one day, stashed in some horrible place."[105]

"Oh my god. I'm really sorry...."

"My mother came down from Prince George and we spent a month searching, but it's all so long ago now. I've created a Shine-On with all her favorite music, so it's like she's still there. And it's true; it does help."[106]

"What's a Shine-On? I'm sorry, I've not heard of that before."

"It's an online shrine, that's there, like forever. It's got photos and videos and poems she wrote, memories from her family and friends, and her favorite music. So like, my kids will be able to see who their aunt was and know what a wonderful person she was. Here—I'll show you."

We were sitting at the kitchen table, the sun filtering in through the trees, and just by Lucas talking about his sister Julie and showing me her photos I could feel her presence.

"So, anyway," Lucas continued, closing his device, "I was in with a great crowd and we were all making do—couch-surfing, sharing rooms, living on the streets. We were proud of who we were. 'The Love Liberationists,' we called ourselves. I'd have gone to Europe to join the street revolutions if I'd had the cash. That's where we thought the action was. We never thought it could happen right here in Vancouver. I did go to New York for the uprising, but that's another story. Some of us had degrees but couldn't find work. Some had jobs but hated the dreariness. There was one girl who was a great artist but couldn't sell her work. And some were aboriginal, with ancient souls.

"We were a great mix, and no-one judged us. That was when I got the idea to go to college, but there was no way I could afford it, and anyway, I'd quit school at 15 so I didn't have the exams. So I started watching TED lectures and joining MOOCs, those on-line courses in politics, social change, things like that. I was, like, 18, and doing all these courses at Stanford, Harvard, Udemy, Coursera and the Khan Academy.[107] We had these great discussions going late into the night as we tried to piece it together: our personal lives, the story of civilization, why it was all going wrong. And what to do? We read the European Manifestos, like Indignez-Vous!—which was great—and The Coming Insurrection—which was crap. We thought we wanted to create a street revolution, but we weren't getting anywhere. Something was missing, but we didn't know what.[108]

"Then the financial crash happened and things got crazy, with demos being organized by anyone with a Twitter account—big demos organized by the labor unions, small demos organized by the nurses, school kids, and old folks who were angry that their savings were disappearing and they couldn't afford to buy food. And then there were the crazies who would come in from the suburbs looking for windows to smash, dressed in black, wearing old Anonymous masks. They grabbed the media attention and made it difficult for the rest of us. I still think they were infiltrators. It was a mess, and who were we? For all our grand discussions, we weren't getting much done.

"*Be realistic—demand the impossible.* That was one of our slogans. We tried all sorts of things, like occupying a Starbucks to protest their tax avoidance; occupying a Safeway store to protest their sale of junk food; occupying those automatic neighborhood grocery stores that used to have no staff; occupying a bank to protest their profiteering on student loans. And after, when we'd been dragged off to jail, we felt so alive. *Fall in love, not in line.* That was another of our slogans. *Every great dream starts in the darkness of sleep.* The trouble was, we were confused. We didn't know what our dream was. We wanted things like the end of capitalism, the protection of nature, things that would stop climate change and the continued use of fossil fuels. But we didn't want anyone organizing us or telling us what to do. We wanted to do it ourselves. We didn't trust other people's ideologies or political agendas."

"I didn't know all this," Dezzy said. "Was this before the OMEGA Days?"

"Yea, before. But then I met this guy called Jim, from the labor movement. He got to know us and he taught us how to organize in a completely different way. He gave us a name. Said we were part of a new social class, 'the precariat,' and there were people like us all over the world, people whose lives were precarious who were learning how to organize.

"It was important to learn from the Occupy movement, he said. The most important lesson was that you've got to have a positive vision and practical solutions. It wasn't enough to criticize. We had to build the politics of paradise, he said, taking his cue from a Brit, Guy Standing, who did a lot of work on the precariat and the need for a Citizen's Income.[109]

"Jim got us together in a community center one day and gave us a lecture I'll never forget. He told us about the French Revolution in 1789, Europe's street revolutions in 1848, and how they achieved a lot of great things but how they triggered a huge right-wing reaction, out of fear. It was the same with the student protests in 1968, which led to the revenge of the right. It was all very heroic, but for what? If you're lucky, the government might fall and you'll get a change of regime, but then what?

"A lot of people are willing to embrace change, he said, but they hate uncertainty; it makes them fearful. People don't like being taken by surprise. If it's a choice between being pushed into a new dark place and retreating to a familiar place most people will retreat, even when it's against their better interest. That's why fascism can get a grip at times of crisis, and why right wing politicians get support for their calls to bring out the riot police and jail the troublemakers. We had to inspire people with a vision, he said. We had to offer practical solutions that would address the problems while also speaking to people's need for security and control.

"The European model of street revolution was so old-school, Jim said. It can maybe overthrow a government, but it's like the world it wants to replace. It polarizes the imagination down to just two options—win or lose, fight or surrender—and unless you're careful it can become macho-aggressive, by both men and women. It pushes people into an us-versus-them mindset that encourages violence, which is what the state wants because it knows how to respond to violence. Unless you've got the army on your side the state is always going to win. The old dualism of left versus right, workers versus business owners—that was part of the problem, he said. The real duality is between those who want community and harmony with nature, and those who want individualism and consumerism. That's what it boils down to, not right versus left. There were thousands of businesses that supported the new direction, but the old dualism shut them out.

"Jim said we had to do things differently. We couldn't go on repeating the same old ways, even if we did it with tweets and videos. As well as being a labor organizer he was a Buddhist. He taught us how to be mindful, how to meditate, and how to listen from the heart. Deep listening, he called it. He became my real father, the one I looked up to.

"'*Action, not anger*'—that was the motto he drilled into us. 'Anger will destroy whatever you're trying to do, whether it's personal or political,' he said. 'Whatever you do, do it with love. Be the solution, not the complaint. Treat everyone you meet with respect, the way you'd like to be treated.'"

Just then there was a knock on the door. "That was Laszlo," Dezzy said when she returned. "Can you feed his cats on Saturday night, Lucas? He's got a trade show he needs to attend. I'm away and Jake's staying with his dad."

"For sure. How's he doing?"

"Amazingly well." Dezzy turned to look at me. "Laszlo's a new immigrant from Hungary and he's starting a business helping children to write and illustrate their own stories. He's very creative."

"Tell him I'll be happy to do so," Lucas said. "Maybe I can trade it for some great Hungarian goulash."

"You were talking about Jim," I said. "It sounds like he was a pretty special kind of guy."

"He was. He used to teach a weekly class on the Zen of Higher Purpose. 'What is it that you are called to do with your life?' he would ask. 'You've only got one life and then it's back to being dirt, so the sooner you find out what it is the less of it you'll waste.' If we didn't know, he'd send us off on a wilderness retreat way up on the Sunshine Coast. He also introduced us to a friend of his, the Vietnamese Zen Buddhist monk and peace activist Thich Nhat Hanh."

"Holy crap! You've met Thich Nhat Hanh? My parents used to have his photo in our tent."

"They did? That's really cool! His Fourteen Precepts of Engaged Buddhism are my practical guide to everyday living. *'Do not maintain anger or hatred. Learn to penetrate and transform them when they are still seeds in your consciousness.'* Precept #6. I used to have a lot of anger. Still do, I'm afraid. I blamed it on my father, but it became a habit, an indulgence that stopped me from growing. Precept #2: *'Do not think the knowledge you presently possess is changeless, absolute truth. Avoid being narrow-minded and bound to present views.'* I used to be pretty opinionated, too—thought I knew everything. Aliya says I still am—she's my sweetheart. But it was all a sham, a cover for my fear that really I knew nothing, that I would always be an ignorant bum. Thay (Thich Nhat Hanh) cleared all that away, left me free as a bird. Mindfulness, that's what it's about."[110]

"Was he involved in the OMEGA Days too?"

"I've no idea. It was Jim who brought him here. He must have been almost ninety, but he was still very spirited. He never told us what to do. That was up to us, he said. We had to learn to listen, and let our hearts be our guide.

"Anyway, Jim was a good friend of Derek's, which was how I met Derek and the beautiful Dezzy. This was, like, in the middle of the second financial meltdown, when banks all over the world were either being taken over by the government or closing their doors. There was no more appetite to bail them out, and even if they'd wanted to, the money wasn't there. So the credit dried up and the businesses that survived had to lay off half their workers."

"You were saying how confusing it was with all the demonstrations and street protests. But earlier you were explaining how positive the OMEGA Days were. What happened to change the sense of confusion?"

"More tea either of you?" Dezzy said. "This is quite the trip down memory lane."

We offered our mugs. Mine had a design of a Thunderbird on it and Lucas's had a moose design, both painted by Norval Morisseau, a First Nations artist from Ontario.[111]

"They were given me by a friend," Dezzy replied when I asked. "But do continue, Lucas. Tell Patrick how the OMEGA Days emerged out of the confusion."

"We were about six months into the financial crash. There were so many people who shared the frustration that nothing constructive was happening, and there was a feeling that we were in danger of missing the boat. I didn't know it at the time, but Derek and Jim were part of a global network of people who were working hard to piece it all together and come up with a strategy that would deliver a similar message in every city and every country. That's where the idea of the OMEGA Days came from, with its five themes, each with its five solutions. The strength of the idea was that it covered all the bases, but it allowed local groups to come up with their own solutions.

"The challenge was to create a synchronized global launch, but people needed time to research the solutions and build the coalitions that would be so important. It was all done very publicly. It was announced to the world as The OMEGA Quest and we had six months to do everything—assemble the best solutions, build a broad coalition, raise funds, organize the launch. There was a global website where you could read the best solutions and vote on the ones you liked, and connect and form groups. Each of the five themes had an on-line course where you could learn about the solutions. It was like a global university, with millions of people researching the best solutions in a host of different languages. There was a whole series of gatherings here in Vancouver as people worked to come up with solutions for the city, the province, and Canada as a whole."

Just then Dezzy's son Jake burst in with a friend, fresh from school and bubbling with excitement.

"Hi, Mum! Can Ali come and play?"

"Not until you've given me a kiss," Dezzy replied. "How was school? Nice to see you, Ali—you're looking good!"

Jake gave his mother a quick kiss and then asked, "Can we play on Ben's trampoline across the street?"

"As long as Ben's there it's fine by me. Ali, do your parents know you're here?"

"They're at work until five o'clock."

"Okay—but can you text them to tell them where you are? Here—use my device."

Ali texted his parents, and both boys ran tumbling out of the house.

"What a pair!" Dezzy said. "You'd never believe that just a few years ago Jake used to be sluggish and overweight."

"What happened?" I asked.

"We cut most of the sugar out of our diet, and I gave him an Ubooly," Dezzy replied. "I was overweight too, and my doctor had diagnosed me with Type 2 diabetes. That was a pretty big wake-up call. So we got seriously involved with the Community Health Center. They helped us develop Personal Health Plans and I joined a peer support group to help with diet change, exercise and weight loss. It's amazing what cutting the sugar out did. With that, the change of diet and taking up cycling, it totally cured my diabetes. And you've seen how Jake is. But anyway, where were we?"[112]

"No, wait!" I said. "What's an Ubooly?"

"You've not seen one?" Dezzy replied. "Oh, have you got a treat in store." She left the room and came back with a purple cuddly toy, tossing it to me.

"Patrick, meet Ubooly. Say something."

"Hi, Ubooly."

"Hi, Patrick. What shall we play at?" the toy replied.

"What!" I was surprised. Then I asked it, "Can we play chase?"

"Yes! Let's run into the kitchen," it replied.

"It's a smart toy, packed with games and learning adventures," Dezzy said.

"They keep getting better. Jake uses his to chase around the house, learn Spanish, do physics experiments, all sorts of things. I used it to get Jake and Ali running around the neighborhood, trying to break records and generally getting fit. But I'm sorry. Where were we?"[113]

"Lucas was taking part in a global university," I said, "getting ready for the launch of the OMEGA Days."

"Right," Lucas said. "I wasn't in Vancouver for the launch. I went to join it in New York."

"How was that? It must have been amazing."

"If you can call being crushed, kettled, pepper-sprayed, jailed and fucking strip-searched amazing. It depends on your taste, I suppose. The police completely over-reacted, and everything went sideways. The plan was for a million people to parade down Broadway and circle Wall Street. It was intended to be celebratory, but there were too many groups with different agendas and too many extremists and evangelicals intent on blocking our progress. I didn't know how crazy America was until I went there. I was arrested the very first day and spent a week in jail before I was deported. It was *awful*. They tied my hands behind my back with those plastic cuffs, which were really painful. It wasn't at all like my jail experiences in British Columbia. It was just generally nasty.

"So anyway, in the days that followed, things in New York really deteriorated. There were people throwing Molotoff cocktails into police cars, smashing windows and attacking the banks. I saw the videos, and it was brutal. The police used pain-rays and sonic weapons to force people to disperse, and they were spotting people with their drones, fixing markers on them. There were thousands of arrests. It wasn't at all how it was meant to be."

"That's too bad. What happened in other cities, and other parts of the world? And what about here in Vancouver?"

"Vancouver's launch went off like a treat," Dezzy said. "We didn't do any big marches or parades. Instead we held a thousand house parties, engaging friends and neighbors all over the city. Everyone had one of the OMEGA flags, the same as all over the world, and on the day of the launch the flags appeared everywhere. At night everyone had green light bulbs in their front rooms and porches. That really made an impression.

"Every city did its own thing. Some went for a big demonstration, like New York, while others were more creative. In Berlin they organized a week of non-stop concerts celebrating their twenty-five OMEGA solutions. In Paris they did a 24-hour bicycle ride with 100,000 people circling the city center. In Beijing they organized teach-ins all over the city to avoid a showdown with the authorities in Tiananmen Square. They were still crushed in the end, but for a few months it was really promising.

"It wasn't just the groups and meetings," Lucas said. "It was the incredible sense of commitment. OMEGA Shanghai dreamed up a concept called 'Global Acupuncture.' They created a model of the Earth that showed her as a patient in need of care and attention. They showed all these pressure points and they encouraged everyone to choose a pressure point and get stuck in. If we each chose one point, they said, we would have an impact.

"I chose the forests, and I've stuck with them ever since. It's a great feeling, really getting to know what you're doing. You can actually accomplish something and make a difference.

"We could have done it before if we'd had the vision. Before Jim came along people had forgotten that people on lower incomes could organize too. He inspired us to take control of our destinies, and make things happen.

"If it hadn't been for Jim I might still be in jail," he continued, "and not for blocking the streets. I'd been doing drugs, and occasionally stealing to support my habit. It's not something I'm proud of. But Jim cared. He made me feel worth loving. He taught me to believe in myself. He showed me how to reprogram the negative self-images that had wrapped themselves around my brain and convert my anger into action. Negative energy drains the heart, he said."

"So what changed to create the political will?" I asked.

"Everything. There was such a sense of crisis what with the financial meltdown, the government cutbacks, the credit freeze, the unemployment, people losing their homes, and the growing number of people who identified with the precariat. There were so many people who'd had it up to here with the plutocrats and the corporations not paying their taxes while concealing their wealth in tax havens. Friggin' trillions they hid. Trillions. The governments always seemed to care more about the bankers and the one percent than they did about ordinary people like me and Aliya who had to put up with crap jobs, crap housing and huge personal debts, and all the while the climate crisis kept getting worse. We blockaded the routes of the pipelines that were being planned to ship the tar sands bitumen to China and the railways that were shipping coal to Asia. We had a pretty good time of it, but the elections they came and went. The politicians, they made their promises, but nothing changed. And all the while the climate crisis kept dumping more extreme floods and igniting ever more extreme forest fires while the fossil fuel companies were laughing all the way to the bank."

Lucas was in full stream. I could see why people found him inspiring.

"But it was *our* future that was going down the tubes. Me and my friends, we were young. We had our whole future ahead of us but everything was conspiring to take it away, and away from the creatures we shared the Earth with—the forests, the oceans, the salmon, the grizzly bears. I used to go ballistic each time I saw a photo of some hunter gloating over the corpse of a beautiful grizzly bear he'd just shot—or she, for it wasn't just the men. When we say a green future we mean green, not polluted brown or day-glow pink all fucked up and given back to us stuffed on a wall or in a friggin' shopping catalogue."

"You're the man, Lucas! Viva la revolución!" Dezzy laughed. "But Patrick was asking what made the difference. What was it that enabled us to succeed?"

"Oh yeah. Sorry about that. I get carried away. Jim said it was three things.

First, he said, we learned how to organize and build a broad coalition. I don't know how, but we managed to get the unions, the youth organizations, the eco-groups, the social change groups, the First Nations, the seniors, the churches and the green businesses on board as well as the precariat—people like me and thousands of others who never normally had a voice. Jim says it was because we had learned how to listen, which made people respect us, and not feel that we were just pushing our own ideas at them. We also learned from books like *Why It's Still Kicking Off Everywhere,* by the British journalist Paul Mason, who showed how the lack of a broad coalition had been the downfall of previous breakouts.[114]

"We also learned from two other Occupy mistakes. It's not leadership that's the problem—it's non-transparent, non-responsive, non-democratic leadership. We needed to be proud of the impulse to lead, Jim said, not suppress it. If you've got any passion in you, you've got it in you to be a leader. We needed to trust our inner leaders, he said, and bring them out. That was really important in build-ing the coalition. If we had turned away every potential leader as soon as they showed up, as some of our anarchist friends wanted, we'd never have been able to build a coalition. That was the politics of *dis*empowerment, not success. And we decided to scrap trying to make every decision by consensus. It took far too long, it got people tied up in knots over trivial things and it gave disproportional space to people who were disruptive who might actually have been provocateurs, trying to disable us from doing anything. We reverted back to the use of strong majorities, which made decision-making a lot easier.[115]

"So once we had our twenty-five solutions in place we used social media to organize mini-protests at very short notice with rules on how to communicate, following Jim's insistence on action, not anger. And we trained in non-violent protest, learning from people like Gandhi and the civil rights movement.[116]

"At the street protests, the office workers and waitresses would stop work and join in the singing. Then after five minutes we'd disperse and show up somewhere else the next day. We kept it up for months. One day we'd hold a philosophy discussion in the middle of Howe Street. The next day we'd be out in the burbs planting tomatoes along the boulevards. Then we'd be up at the university, burn-ing the documents that symbolized the student debts we'd paid off."[117]

"You were paying off student debts? How on earth did you do that?"

"There was an OMEGA group at the university buying up the debt for pen-nies on the dollar to keep it out of the hands of the debt collectors and the vulture funds, using crowd-funding to gather the cash. They called it a Rolling Jubilee, after the name of the group in New York that dreamed it up. They were an off-shoot of Occupy Wall Street. That's all history, since higher education is now free, in effect."[118]

"Free? So there's no more student debt?"

"Correct. Students pay for their education with a three percent deduction on their income for twenty-five years. If you have a really low income, you pay three percent of your really low income, and if you're a billionaire you pay three percent of your billions, which strikes me as fair. For shorter courses it's less, and students who graduated before the program started have had their interest capped at two percent to drive out the vulture funds."[119]

No more student debt. That would be a huge burden off young people's lives.

"That's really impressive. So the original Occupy movement wasn't just complaining?"

"The Occupy movement never died. It morphed and transformed. It turned from a caterpillar into a butterfly. The activists who stuck with it developed a positive approach and did some really great things."

"That's very cool. What was the second thing that made a difference?"

"Oh—right. The media. We paid a lot of attention to the conventional media, building good relationships, always trying something different so they'd get a good story. Social media was great for organizing, but conventional media still mattered for influencing public opinion, especially among the older people. Once we laid out a fancy picnic complete with tablecloths and wine glasses right across the road—but no food, to highlight all the people who were hungry. Ten minutes later we were gone, quick as we had come. On another occasion we organized fifty jugglers. The crowd was huge, which gave us a chance to talk to people. And then there was the time we got inside the buildings on either side of Hastings Street, slung a cable between them and had acrobats tightrope walking across the middle of the road. We ran a banner across saying WE NEED TO BUILD A BETTER FUTURE. Humans pay more attention to something new, Jim told us, so every day we needed to dream up something new to keep people's attention."

"Is that what you got arrested for? The street protests?"

"That, and other things. They got good at picking off the leaders. We'd given up trying to organize anything in secret; they were monitoring every phone call, tweet and email. You could hardly *think* something without them knowing. If you bought a book about social change from Amazon they had you listed immediately. But the arrests started to work in our favor, because people saw how we were arguing for practical solutions. So, it built and built. Then in July we organized the Festival of Hope in the Downtown East Side. That was when Derek was assassinated. It was our most powerful moment, which became our greatest fucking tragedy."[120]

Lucas fell silent. Dezzy too.

"We never believed it would come to that. Prison, yes, but not that." Dezzy stared down at the table. Then she looked up and smiled.

"Look... Derek would be so proud of what we've achieved. So why don't you continue with your story?"

"It would be easier if we'd nailed whoever did it. It sticks in my fucking craw to know that someone's out there laughing about it with his friends over a gin and fucking tonic."

"Don't go there, Lucas. It'll eat you up."

"Okay, okay. But I'm not giving up. I still think it was someone connected to the Alphas. They were so cocky the way they mocked us."

"Who were the Alphas?" I asked. "But before you answer, what about Jenny, Derek's girlfriend, who came back with him from England?"

"Oh, Jenny. Dezzy, you tell him."

"Jenny was another tragic story," Dezzy said. "She was right beside Derek holding his hand in the front row of the march, radiant as a little bee. She even managed to swat one of the nanobots when it flew too close, bringing it down. What a cheer that produced!"

"Nanobots?"

"Yes, the micro-drones the police used to track us. That one got too low, and Jenny downed it. We were going down Hastings towards the intersection at Main at the center of the Downtown Eastside. There were about twenty of us in the front row; Jenny was right beside Derek holding his hand when he was shot. He collapsed, and there was a second bullet that might have been aimed at Jenny. It hit a girl behind her, Chanandeep Singh. She survived, luckily."

After a pause, Dezzy continued. "There was pandemonium. People tried to scatter, but there was nowhere to run to—no-one knew what was happening. Then something incredible happened. I was down on the ground with Jenny trying to help Derek and Chanandeep and people started forming a circle around us, then others joined in so there was this complete circle of people holding hands around us. What courage that took, since no-one knew where the shots had come from, or if there'd be more. And then someone started singing, not one of our new songs, but that old one from the civil rights movement: *We Shall Overcome*. My god, what a moment that was as the song spread down the street and tens of thousands of people joined in. Soon after, the police arrived, sirens screaming, and they took control, trying to get us to disperse. I've got to give it to them—the Vancouver police were amazing. They were well trained for non-violent protests, and they always carried body-cams.[121] They never used tasers or pain-rays, the way the cops did in Toronto and New York.[122] They tried to make us disperse, but people just stayed there, singing. Some were praying and meditating, down on their knees in the midst of it all. I think Derek died while they were singing, so that would have been his last memory if he was conscious. He never said anything. The paramedics rushed him to hospital, but it was too late."

I was silent, taking in Dezzy's story.

"I think I knew then that we'd win," she said. "Sitting on the ground holding Derek's hand while Jenny was so desperate, all the people singing, I heard a voice inside me that said we were going to win, that it would all be okay. I could really feel that the power of the people was with us."

"And the funeral," Lucas said. "That was stupendous. It was a huge procession, so many people sharing the same vision, the same determination, the same love."

"What happened to Jenny?" I asked.

Dezzy sighed. "She was so deeply traumatized; we couldn't pull her out if it. She tried sharing in our OMEGA Circles, but it just made things worse. I've never seen anyone so deep in grief. Eventually, she left Vancouver and went to Ecuador, where a friend had asked her to help build an ecovillage. I still hear from her occasionally. She has a new partner and she seems happy; I doubt she'll ever return to Vancouver."

"What were the OMEGA Circles?" I asked.

"They were our support groups," Dezzy replied, getting up and staring out of the window. She looked wistful.

"I'm sorry," she said. "I haven't thought about Jenny for a while. I still feel that we should have done more. We were such good friends. I miss her."

"But you were asking about the OMEGA Circles," she said, turning back to face me. "The Circles were our support groups. It's like Lucas was saying about Thich Naht Hanh. We have to practice love at the personal level as well as the political.

"We needed a support network to help with our various issues and fears. We're none of us perfect; we all sometimes act selfish or stupid when we're under pressure. It's the little things—the self-importance, the defensiveness—that can screw up a group's effectiveness and rob it of its joy. How to speak honestly when you know you might hurt someone? That's the difficulty.

"The answer is with love. It's the only way it can work. The OMEGA Circles were our buried gold. They held us together. Every Sunday night we would share a meal, meet in a circle and then listen to music together lying around on cushions, holding hands or cuddling. The things Lucas said about what helped us succeed are all true, but I doubt we could have done it without the Circles. I've seen so many campaigns lose their effectiveness because they leave no time for love and connection. They're dominated by the big talkers and the workaholic activists, and without meaning to they can rob a campaign of its spirit because everything feels like work. It's so important to celebrate the bonds you form when you work on something you believe in. But I'm getting distracted. You were asking about something else...."

"Yes, the Alphas. But it's important, what you're telling me," I said. "I'm taking it all to heart."

Dezzy smiled.

"The Alphas," said Lucas. "We never knew who they were. They had a website, but they had it on encrypted P5, the highest privacy setting, and we were never able to hack it. They were totally opposed to everything we stood for. They used to joke that Omega came at the end of Greek alphabet, so we represented the scum of humanity, while Alpha came at the beginning, so they were the real leaders of the free world, protecting it against the socialist dictatorship we supposedly wanted to impose. That's typical of fascists, accusing you of the very thing they want to do themselves. They were probably a front for the oil industry, or the gun lobby in America. Or it could have been people who were freaked out by Derek's statements about tax havens and his calls for transparency laws that would have revealed their hidden wealth, with talk of jail time for the big evaders.

"Anyway, within a month of Derek's assassination the public response was so overwhelming that the Alphas began to weaken. We organized an OMEGA Roadshow that toured the province, visiting all the small towns in the conservative heartland, winning people over to our solutions-based approach. There was a massive petition to the Premier, urging progress on a package of legislation that supported the changes we wanted. Here in Vancouver we had people going house to house in every neighborhood getting people to sign. So many of the people we contacted signed and gave us their support, and many followed up by phoning the Premier's office. As the weeks went by we built a really strong coalition, including young people, students, labor unions, retired people, the urban poor, small businesses, and some of the province's most prominent scientists, academics, First Nations leaders, musicians and sports stars. With that much support the politicians started lining up to join us, and the government could see which way the wind was blowing.

"In the end they agreed to our request to create five Citizens' Assemblies, one for each of the five OMEGA themes, each with a hundred delegates chosen at random to hold hearings around the province. The Assemblies were to be truly open as they envisioned the future they wanted and researched the best solutions. The organization that ran the Assemblies even gave them a list of common brain traps to steer them away from confrontational thinking.

"When they reported back six months later they supported most of our proposals, with some small changes and improvements. It was such an amazing feeling. We had done a massive job of getting people involved. We had organized a Festival of Solutions, inviting people to post their best ideas to the website, organized into categories so that all the poverty solutions could be seen in the same place, for instance, and people could make comments and give each idea a rating. That's how the best ideas emerged."

"That's fantastic!" I exclaimed. "It's amazing what you achieved." I looked at Dezzy and Lucas with deep appreciation.

"Well, it's interesting to look back on it," said Dezzy. "And yes, we did accomplish a lot. Derek would be proud. But there's so much more that needs to be done before we're out of the woods."

"At first it seemed like we were dragging a huge weight along the ground, a sack of ideas all jumbled together and crashing against each other," Lucas said. "Then it gathered momentum, like an airplane on the runway, and started to fly...."

I nodded. "So going back, Lucas, you said there were three things that made a difference. What was the third? Or the fourth, if you include the OMEGA Circles."

"Oh, the music!" Lucas responded. "We had the best music: singing, dancing, everything. The Belgian climate activists had taken the World War II classic song

Bella Ciao about Italian partisans leaving to fight the fascists and they had set new words to it—'*Do It Now*'—and it became our anthem."

Lucas sang the words quietly, with a determined emphasis on the words *now, now, now.*

> *We need to wake up,*
> *We need to wise up,*
> *We need to open our eyes*
> *And do it now now now.*
> *We need to build a better future,*
> *And we need to start right now.*[123]

"It was amazing to be part of a thousand people singing it together. We developed some great harmonies, and we put all our songs on YouTube so that people could learn them. We took songs like John Lennon's *All we are saying is give peace a chance* and gave them new lyrics like *All we are saying is give us a home.* Even the Vancouver Bach Choir got involved, if you can believe it. One of their singers wrote a new verse for Oh Canada...."

Lucas stood up, cleared his throat and started singing:

> *O Canada! Protect our future now!*
> *Give us the strength to guard these seeds we sow.*
> *With glowing hearts we see them rise,*
> *Our future green and free.*
> *From far and wide, O Canada, we sow these seeds for thee.*
> *Earth keep our land, fair, green and free!*
> *O Canada, we sow these seeds for thee,*
> *O Canada, we sow these seeds for thee.*[124]

"Everyone knew that the seeds represented the work we were doing to build a better future. When the politicians finally voted to approve the package of legislation that made the OMEGA solutions a reality the public in the balcony of the Legislature in Victoria erupted into song, singing the new verse. Then the politicians joined in, led by the Premier himself—the Green Party's first ever Premier—and afterwards everyone sang *Do It Now*, with the Premier himself singing along in full voice, grinning from ear to ear."[125]

"I was in Chicago at the time," Dezzy chipped in. "It was on the evening news. We knew something important was happening. They showed a clip of everyone singing *Do It Now*."

"The Premier of BC came from the Green Party? How did that happen?" That was a *big* surprise. Back in my time the Greens had only just elected their first MLA.

"They won seven seats in the election," Dezzy replied, "and they offered to form a coalition with the New Democrats on condition that voting was made proportional. The New Democrats had no choice, since they needed the Greens to form a government. In the election after that the Greens won 29 seats to the Liberals 26, the New Democrats 23 and the Vancouver Island Party's one seat, enabling them to form a coalition with the New Democrats."[126]

"What happened to all the people who had been trying to block the changes? Where were they in all this?"

"They fought back," Lucas continued. "It took them a while to get organized, because they never imagined anything would come of it. Canada is not America, where the opposition to change was so well funded. When you were in school, did they teach you about the Tea Party, the so-called spontaneous revolt against government corruption and wastefulness?"

"No, but I learned about it."

"Well it wasn't spontaneous at all. It was planned and financed by organizations with direct links to the tobacco industry, and by the Koch brothers, the coal industry billionaires.[127] But no-one expected anything earth-shattering to come out of Vancouver, so the business interests who might have supported a right-wing pushback weren't well organized. Besides, they'd seen the Occupy movement come and go and they assumed we would be the same. But we weren't. We were so much better organized.

"They started to scramble once they realized the scale of what was happening, and took the time to read Derek's book. Then out of nowhere there was a new player on the block—the Canadian Freedom Foundation. Their funding was very secretive, but we think it came from private sources in the US. They claimed to be a grassroots non-profit upholding the values of western civilization, but they were deeply implicated with the Alphas. They started using the normal means of attack, but after a few months things began to turn ugly. I discovered where things had gotten to when the police battered my door down at five o'clock on a Sunday morning with an arrest warrant. There was a huge dawn roundup, *five hundred* of us. They stripped us of our clothes and devices, so we had no idea what was going on. Some said it was the beginning of fascism—they compared it to Krystallnacht in Germany in November 1938, when Nazi storm troopers destroyed Jewish synagogues and businesses and marched 30,000 Jews off to the concentration camps. By breakfast, we learned that there'd been a so-called 'credible threat' of an algattack against the Toronto Stock Exchange, and Tiger News was claiming to have evidence that we were behind it."

"What's an algattack?"

"It hacks into the algorithms that govern critical parts of the economy. It's far more insidious than a straightforward hack-attack. It replaces the governing algorithms with cloned substitutes designed to achieve very different goals. When an algattack hit the financial sector the money-flows went crazy, causing investors to panic. It's a very serious cyberweapon, and it's almost impossible to know who's behind it."[128]

"How long did they keep you in jail for?"

"Four weeks," Lucas replied. "They invoked the *Cyber-Terrorism Control*

Act they had passed when Russia's financial crisis showed what was possible, but when nothing credible emerged they had to release us. But by then we'd had four more weeks to organize. There was a prisoner with us from the Athabasca Chipewyan First Nation in Alberta who taught us how to do Powwow dancing within the confines of a cell. It was a great way to strengthen the spirit. Catch the anger, he said, and channel it constructively. It created a bond that was worth a million when we got out of jail."

"Do you think the Canadian Freedom Foundation was behind the threat of an algattack?" I asked.

"Maybe. The police never did locate the source. The emails came from Russia and pointed to a local connection, but when our lawyer ran a grammatical algorithm he proved that they were very unlikely to have come from us.

"But the threats worked in our favor, because people saw that we were being victimized, and when push came to shove people liked our positive vision and our practical solutions. They weren't being fooled any more by the claims that climate change was a hoax, that genetically engineered crops could feed the world and people were only unemployed because they were lazy. And there were lots of local politicians and business leaders who supported us, who went out of their way to call the Premier's office.

"The attacks came in two waves," Lucas continued. "The first wave came on the heels of the OMEGA Days—the stuff we've been talking about. The second wave came in the 2020s, when the Greens and New Democrats were in government and they were clear about the changes they planned to introduce. That's when the forces opposed to change began to show their teeth.

"Jim used to say that if you want to change the world, you have to know who your opposition is. Then you have to get inside it and transform it. There are three main fortresses of power that maintain the status quo, he said, and each has to be taken over and transformed. They are the government, with its deep state apparatus of police, security forces, intelligence operations and the military; the banks; and the corporations. The corporations don't occupy a single fortress. They have a whole network of fortresses that protect their interests in fossil fuels, forestry, the media, the food industry, farming, chemicals, fisheries, retail shopping and so on—wherever there's money to be made. Each of the fortresses needs to be transformed.

"Behind them, there's a secondary line of defense in the plutocracy—the elites and the super-wealthy who benefit from the status quo and who work to maintain it, sending their plutokids to private pluto-schools where they become pluto-friends with other plutokids, holidaying together on their private pluto-islands and giving each other positions on their pluto-boards. Taken together, it was a very formidable defensive structure, based on the belief that economic growth, the free market and making money are the most important things in life, and all other goals are secondary.

"It was not until the Greens and New Democrats started moving towards public banking that the Fortress of Banking began to flex its muscles, threatening dire economic calamity, the downgrading of the province's credit rating and the collapse of BC's economy if it moved away from the private, neo-liberal model of banking—but that's a whole other story."

"Well," said Dezzy. "Is this useful, Patrick?"

"My mind is reeling. I have so many questions, but they'll have to wait. But what do you do for work personally, Lucas? Are you still involved in the OMEGA movement?"

"I'm a woodworker in the Cascadia Forest Co-op. And no, I'm not involved, except locally. I sometimes go to our neighborhood meetings. And yes, it's still happening, but not with the same intensity. Aliya's more involved. She's my sweetheart. She's also our street rep on the Neighborhood Council. You'll meet her if you stick around.

"I love working with wood," he continued. "We've got so much forest here in British Columbia, most of it publicly owned, and yet for years the private companies with the timber licenses had been stripping it with no proper oversight or control, creating massive clearcuts hundreds of hectares in size with just a few trees left standing to meet the letter of the law. Not everywhere, but in a lot of places. Everything not big enough to sell, they piled it up to burn. Fir, cedar, maple… it was a crime.[129] Now that we've got back control we're able to do so much more with it, and the forests are being managed in ways that respect the ecosystem as well as the profit system. Now the main thing we have to worry about is all the forest fires, which have been so bad of late. We do a lot of work with the Trust for Sustainable Forestry, which buys up privately owned forest lands that are threatened with bad logging and protects them by enabling the development of small forest villages on a tiny portion of the land, protecting the rest. Right now, we're building a hammer-beamed timber ceiling for a village hall, the way it was done in Europe a thousand years ago. It's amazing. Sometimes when I'm in the forest I look around and I say to the trees, 'You're safe now. We're not going to hurt you anymore.' They're so happy that we've finally stopped the destruction."[130]

"I'd love to talk more with you, Lucas. I'd be fascinated to know what the solutions were for affordable housing."

"Why don't you come over to my place, and we can talk some more? Give me five minutes to do a couple of things and I'll be right with you."

After Lucas took off, Dezzy said, "Lucas is amazing. He's a bit gruff at times, but he's a sweetheart under the skin. Jake adores him. Make sure you're back in time for dinner. I've invited Betska and her grandson Leo. And tell Lucas and Aliya they're invited too."

7

A for Affordable Living

WHEN LUCAS RETURNED he invited me over to the laneway house he called The Little Palace at the back of Dezzy's garden, which he shared with his girlfriend Aliya. From the outside, it looked like a tiny two-storey house with a solar roof. Inside, it was one big room with a high ceiling, kitchen area, sleeping platform and large windows that made it seem spacious. He offered me a glass of water and we sat down at the table overlooking the garden.[131]

"You said you built this yourself? It's lovely!"

"We had help from a friend who's a builder, and friends came over for weekend work-parties. We followed the Passive House Code, which ensures that you will use ninety percent less heat energy than you would with a conventional house, and I attended the five-day training course put on by the Canadian Passive House Institute.[132] We couldn't get it certified, since that's really hard with such a small surface to volume ratio, but look at the result! I can go barefoot in here all year round, and it doesn't get too hot in summer either. And when there's a power cut in winter we can keep warm with just three candles.[133] The plumbing was complicated, so we hired a professional. We have a composting toilet, and we get all our potable water from rainwater stored in a big tank under the house, filtered and UV treated. We wanted to build a greywater treatment system, but the city wouldn't allow it. The water thing is really important. There have been several summers recently when Vancouver's reservoirs almost ran dry due to the long periods of drought we've been having."[134]

"What do you do for heat?"

"We hardly need any. It's a passive house, so it needs 90% less heat than a regular house. It's got ten-inch walls with rigid foam insulation and Magnum Board in place of drywall and plywood.[135] There's foam insulation under the slab and the really tight construction seals out all the leaks. As for heat, see that white box up on the wall? That's our heat recovery ventilator, which recovers ninety-five percent of the outgoing heat."[136]

"That's pretty impressive."

"The windows were the most expensive part. They're triple-glazed with a fiberglass frame—they're made locally, and we got them from IKEA. The heat recovery ventilators are made locally too. There's been quite the employment boom, with new businesses and co-ops starting to serve the flourishing green building scene. We used recycled timber from the Re-Store, and our electricity use is minimal—mainly the toaster-oven and drying clothes in winter. Our solar produces an average four and a half kilowatt-hours a day, and we have twenty kilowatt-hours of battery storage."

"Is this way of building common for a home like this?"

"Yes, pretty much. Every new building in Canada has to meet the Near Passive House Code these days. They adopted the rule after seeing how much success they had in Brussels, Belgium, doing the same thing.[137] Why build something you have to pay to heat, when you can reduce your cost to almost nothing? We got the highest green rating, and Dezzy gets a reduction on her municipal taxes because there's no burden on the city water. We should be looking to buy our own place, but we really like it here."

"I gather there's a rule that you can't park a car. Couldn't you just park one street over?"

"The parking spots are all permitted and the laneway homes aren't assigned a permit. We could buy a spot if we wanted, but who needs a car when cycling's such fun, public transit is so efficient and carsharing's so easy? Our goal is to live as simply as possible, buy as little as possible and have a really light footprint on the Earth. And no plastics."

"No plastics at all?"

"Well, as far as possible. Aliya says we should be okay with compostable bioplastic, since it's earth-friendly, but I'm holding out for no plastics at all except where there's no alternative."

"Do you use Li-Fi, the same as Dezzy?"

"No—we're cabled in directly. But she's right to do it. I've read too much about the dangers of Wi-Fi, and Aliya keeps telling me about people who have cellphone related tumors. She's a nurse, so she sees it first-hand. I've become a bit obsessed, I'll admit. Ever since I learned about the Pacific Garbage Patch when I was a teenager I've been campaigning to stop the use of plastic. Those photos of dead albatrosses with their bellies full of cigarette lighters, plastic bottle tops and other plastic crap really grossed me out. The plastic is constantly breaking down, so it's filling the ocean with tiny fragments that are being swallowed by the fish."[138]

"Is anything being done to clean up all the plastic in the ocean?"

"There's a global treaty to reduce and recycle plastics, and a crowd-sourced initiative to name and shame the countries with the least recycling and the worst beaches, and there are about twenty Ocean Mantarrays at work around the world. They use the currents to catch the plastic and remove it."

"Mantarrays?"

"Yes. They were dreamed up by a Dutch teenager, Boyan Slat. He was a student when he started working on the idea. He developed a system of floating booms anchored to the ocean bottom. It shows what you can do when you put your heart into something."[139]

"That's really impressive. He was just a teenager?"

"Yes. So that got me thinking about the oil plastics are made from. Did you know that at the peak of global oil production the world was consuming enough oil to fill five thousand Olympic-sized swimming pools every day?[140] And all of it from the two hundred million years old remains of ancient sea creatures.

"So I try to never to use plastic and we aim to live as simply as possible, using the sharing economy instead of the consuming economy. As Gandhi said, you have to *be* the change you want to see in the world."

"The sharing economy?"

"Yes—giving and sharing without any exchange of money. It's the oldest economy of all. It was only when we started exploring and invading each other's territory that gifting turned into trading, and then into stealing and slaving. I want to recover the ancient ways."[141]

"But you still enjoy going to—what was it—a holoconcert?"

"Yeah—I know it's a contradiction. We can't all go back to living in tribes, hunting and fishing, but we can enjoy the experience of sharing. Have you ever been in a Gift Circle?"

"No—what's that?"

"It's when we get together and share whatever we can offer, but as a gift, not an exchange. Aliya and I are part of a Circle that meets every month. When we built our house we had tons of help from people in the circle. The more you give, the richer you feel. Aliya's much better at talking about this kind of stuff. She's a lot clearer inside. I still carry a lot of crap."[142]

"Join the gang!" I replied. "I've got all sorts of confusions that rattle around inside me. I wonder how long it takes to get this living thing figured out, so that life's not such a roller-coaster."

"Did I tell you I'm First Nations?" Lucas said. "My father is Carrier, from the Saik'uz First Nation in the interior, near Vanderhoof. My mother's family comes from Wales. My father was pretty distant from his people due to some bad things that went down, and after I came to Vancouver I didn't have much to do with him. But then someone told me about a wilderness camp the Carrier organize to re-introduce people to their culture. So I decided to go."

"How was it?"

"It was amazing. There were twenty of us. We spent a month living in the bush learning how to hunt and track, how to build a camp, listening to the elders and their stories. It took me a week to get over my city hang-ups, like fussing over what time it was, and then I began to feel at home, both on the land and in my own skin, for the first time in my life. People had always talked about 'being close with nature,' but I never really understood what they meant until I spent time in the wilderness. It was like I began to feel a true empathy for the forest, the bears, the birds, and all the other critters, like we shared the same soul. I could really feel how we hurt them when we acted so carelessly with our consumer way of living. We spent three days entirely alone, fasting with just water, and we were asked to go in with a question. Mine was 'What is the question that I'm asking in my life?' What I came out with was *'How can I live a life that nourishes everyone around me, both humans and in nature?'*"

"That's a pretty big question!"

"Yes. I guess it's a lifetime adventure. It's also really cool how many great things the Carrier are doing. They've formed a co-op that's building passive housing, solar projects, cohousing clusters, things like that. They gave us a tour of the Saik'uz Village Project. It's completely car-free, centered around the longhouse, the way villages used to be before the white man arrived. They're also learning and speaking their traditional Yinka-Dene language. Their ceremonies and dancing are just so powerful."[143]

"Do you think you might return, like, go and live up there?"

"I've thought about it. But I'm happy here with Aliya, and the woodworking co-op is going really well. The crew gave me a Certificate of Excellence for my work on our last project. So you could say I'm happy in both worlds: my First Nations world and my city world. And it's so cool that the Premier of British Columbia is First Nations. She's from Haida Gwaii. I'm so proud of her."

"How did that come about?"

"Charlene Jack. She became a Green Party MLA when the Greens formed the government in coalition with the New Democrats. She started out as Minister of Family Services, then Minister of Finance, and then she became leader of the party. In the last election she led the Greens to a majority, and she's still only forty-eight. There's talk that she might run for one of Canada's seats on the new Global Assembly. That would really be something. Imagine being one of seven people chosen to represent your country on the global stage."

A Global Assembly? I was itching to know more, but I held off.

Lucas paused, then out of the blue, "Tell me, do you meditate?"

"I tried it a few times, but I never got into the habit," I replied. "My mother used to meditate when she was in the mood. Why do you ask?"

"Aliya meditates. It's quite a big thing with her. She gets up early and I see her sitting there so silently. She does it before her morning prayers. She's Muslim. So I was just wondering. When I was with Jim we used to meditate a lot, but whenever I do now I feel uncomfortable. I get my peace out in the forest. But look, you said you wanted to find out how we tackled the affordable housing crisis?"

"Yes. Is that okay? It can wait if you want."

"No, it's fine. If we had a day I could take you on a tour and show you all the great things that are happening. Maybe you'll get to see some if you go downtown. But I can tell you about some of them. Can I get you a coffee or a cup of tea?"

While Lucas was in the kitchen area I sat at the table and looked around at his space. I could feel Aliya's presence, though I'd yet to meet her. There was a beautiful embroidered wall hanging and a gorgeous blue, green and gold prayer rug decorated with flowers and minarets.

"So," Lucas said, returning with two mugs of tea, "when I first came to Vancouver the housing situation was ridiculous. Even a tiny house like this used to sell for half a million dollars. That would be almost three thousand dollars a month if you could get a mortgage, and it was more than fifteen hundred a month to rent a one-bedroom apartment. They said Vancouver had the second least affordable housing of anywhere in the world, after Hong Kong."[144]

This was something I knew all too well. Daria and I had a small inheritance from our parents but nothing we could buy a house with, and the two-bedroom apartment we rented cost more than two thousand dollars a month.

"The OMEGA team that tackled the problem wasn't starting from scratch, however," Lucas continued. "The city had put a lot of effort into the problem before OMEGA came along. The most important things they were doing were requiring developers to make twenty percent of the units of any new development affordable, giving incentives to developers to build 100% rental buildings, allowing far more secondary suites around the city, and requiring 35% of all new developments to be family-oriented housing.

"Then during the OMEGA Days, because of the crisis, they brought in a new rule that developers building ten units or more had to sell ten percent of the units to the Affordable Housing Agency at cost, which it rents out to people who work in core services such as healthcare, social services and the police, giving the developer increased density as a trade-off.[145] They also worked to prevent demolitions, and to encourage laneway housing like ours. And they established the Vancouver Rent Bank, which helps renters with short-term loans if they're in a crisis.[146]

"So we've got to give them credit, but it was still not enough. The first OMEGA solution was to build a pool of money that could be invested in affordable co-op and rental housing. So there's now an escalating property transfer tax on top-end real estate sales over three million dollars, and a speculation tax on properties that are flipped within a year of being bought. And before the ban was brought in on property-purchase by non-Canadians, the same as in Australia, there was an annual levy on properties bought through offshore companies or registered offshore to avoid taxes.[147] And there's an escalating series of fines for owners who leave their properties empty, culminating in jail-time for persistent offenders."[148]

"Jail-time?" I was shocked.

"Yes. They took the idea from London, England, where they had a similar problem.[149] The money goes to the Affordable Housing Agency, which distributes it to community non-profits and Neighborhood Associations to build affordable housing and housing co-ops. That was the second OMEGA solution.[150] People can now invest their retirement savings with the Agency, which has increased the pool of available money.[151]

"As well as building new housing the Agency is buying out the slumlords who operate the old rooming hotels, so that they can be restored or demolished and rebuilt, still as rooming hotels for single people on really low incomes but clean and safe with good community facilities. Those places were terrible. Rats, lice, bedbugs—you name it. I lived in one for a while, so I know what I'm talking about."

"What was the third solution?"

"The third addressed the problems people were having around community living. As a result, it's now legal for more than five people to share a house, [152] and for the owners of buildings with flat roofs to build rooftop suites with the same no-car rule that there is for laneway housing.[153] Vancouver also has two new zoning bylaws, one that allows single family lots to be subdivided into five units of three-storey townhouses, which has done a lot to increase the supply of housing,[154] and one that allows micro-villages on land that is temporarily vacant. Here—I'll show you some photos."

Lucas reached for his laptop and projected a series of images onto the wall, straight from the device. They showed tiny villages of tiny homes, complete with gardens clustered around a village green.

"I've a friend who lives in one in East Vancouver where a development proposal has been stalled for a year. His home is even smaller than ours. They're pre-assembled by a builders' co-op in the Fraser Valley to a standardized design, and wrapped in ten-inch slabs of foam insulation. They come with a rainwater capture tank, a UV filter and composting toilet, and they share community greywater treatment. There's a rule that if you want to live there you have to help create the gardens, and you're expected to join the weekly village meeting. It creates a strong sense of community, but as soon as the land is ready to be developed the village has to go, at twelve months notice. The Agency keeps a record of all vacant land, however, so it's often possible to move to a new site. It's not a great solution, since it's so temporary, but the people who live in the villages love them so they put up with the inconvenience of having to move."[155]

"If they weren't allowed, the land would just be sitting there empty, right?"

"Yes. So, moving on, the fourth OMEGA solution was for people in apartment buildings, renting from a landlord. They wanted tenants to be able to form Tenants' Stewardship Councils to address the various problems that arise. So they wrote a Tenants' Charter, spelling out their desire to live in buildings that are comfortable and energy efficient, free of fumes and infestations, with space to grow food and to store their bicycles and recyclables. In return, the Charter spells out the tenants' commitment to look after the property, to abide by a code of respectful conduct, and to agree to a set of conditions if someone can't pay the rent. That was the landlords' biggest headache—dealing with tenants who trashed the property and walked away without paying. They wanted a guarantee that in return for cooperating when a landlord upgraded a building, the landlord wouldn't raise the rent. And finally, they wanted the Tenants' Stewardship Councils to have first right of refusal to buy their building if it came on the market, to convert it into a Housing Co-op with the land being owned by a Community Land Trust."

"Did the landlords agree?"

"It took a year, but then some of them helped draft the legislation, and after that the others came on board. When they realized that the tenants weren't being hostile they saw it as a chance to build a new relationship. The thing they liked least about being a landlord, they said, was all the conflicts and complaints. If the Stewardship Councils could improve that, they were all for it."

I had so many questions, but I was eager to learn about the fifth solution.

"The fifth was for farmers, many of whom were struggling," Lucas explained, "and for all the young people who wanted to farm but couldn't afford the land. As a result, any farmer with more than twenty hectares is now allowed to sell one hectare for development as a clustered farm village, with a series of conditions to ensure that the people living there farm the land. The new farm villagers can also lease or buy as much land as they need from the farmer. It's been a huge success, with dozens of new villages being built."

Lucas threw up a series of photos. The villages seemed timeless, apart from the solar panels on the roofs. Some were built from straw bales or cob and some from timber, and they were often clustered around a shared courtyard, with a barn for farm equipment.[156]

"So," Lucas said. "It hasn't solved the whole housing crisis, and there was a lot more that was needed to help the homeless, but it cracked the biggest problem, which was the chronic shortage of affordable housing."[157]

"That's really impressive. You must feel proud."

"I only played a tiny role. The affordable housing stuff was other people's work. Most of the time I think about all the things that still need doing. Vancouver's great, but it's not the world. The climate crisis is by far the biggest problem we face, and it's going to take everyone working together to solve it, in every country."

"What was it like for you during the Terrible Twenties, as Dezzy called them?"

"The Transformative Twenties, you mean? I remember getting my first carbon ration card. When I analyzed my emissions I found that I needed nowhere near the 4.5 tonnes I was allowed. I didn't drive a car, didn't fly places, and I lived in a rented room which had a baseboard heater, using electricity from zero-carbon hydro. So I made $800 bucks selling the four tonnes I didn't need.

"I remember the trucker's strike, however. The truckers were really upset. They said the carbon rationing was putting them out of business so they block-aded the Chevron oil terminal by Burnaby Mountain, parking their rigs at impossible angles across the entry roads. It was only a couple of days before there were line-ups at every gas station. After a week the whole city was grinding to a halt. The police had to arrest fifty drivers and tow their trucks away to get the oil flowing. It was a great lesson in how dependent we were on oil.

"I also remember the protests against the tar sands up at Fort McMurray, in Alberta. They were pretty scary. I went up with a group of people but after a few days some locals discovered where we came from, and why, and we were surrounded by a crowd of angry oil workers who started to beat us up. The police eventually rescued us, but not before I had a smashed face and three broken ribs. They jailed us for our own safety and I spent a week in hospital under police guard before they got us out of town. A week later a man was shot by a sniper on the road into town. It only takes one nut-case to make it all go bad."

"That's crazy. Was everyone who worked in the tar sands angry like that?"

"No, not at all. They were just really worried about losing their jobs. I spoke to a lot of really good people who said they sympathized with our cause, but they needed the money. There were things like that happening all the time. It was a lot crazier south of the border."

"Going back to what you were saying about affordable housing, Dezzy told me that the A in OMEGA stood for Affordable Living. Did it include other things as well as housing?"

"Yes, it also included affordable healthcare, food, transportation and child-care. But look, I need to get a few things done. Are you going to be around for a while? I enjoy talking with you."

I explained that I had until Sunday night, and then retreated to Dezzy's garden to ponder things. I hadn't expected to be plunged into tales of protest and revolutionary fervor. But what *had* I expected? It was naïve to think that Vancouver could have become such a green city without some degree of upheaval.

The garden was rich with the abundance of June's greenery. There were rows of healthy young lettuces and lots of bees on the purple and white flowers of the broad beans. The rhododendrons were in flower, deepest red and purest white, and the first strawberries were ripe. I sat in a garden chair savoring a couple, then closed my eyes and drifted off.

I had a dream in which I was walking in a forest of tall Douglas fir trees, the sun dropping patches of brightness onto the forest floor. Then out of nowhere a dark hole opened up in front of me. Not quite a hole—more a fetid swamp, oozing something dangerous.

8

A New Synthesis

WHEN I WOKE up I pondered my dream. Was there something I was missing? Then I went into Dezzy's kitchen, where I helped by making a salad while she prepared a quiche and a rhubarb pie.

"Is your mother still alive?" I asked after a while.

"Yes. She lives here in Vancouver, where I can keep an eye on her. She has become very attached to her church."

"When your parents adopted you, was there a reason why they chose South Africa?"

"For sure—they're from South Africa! They were activists during the anti-apartheid struggle in the 1980s and they had to leave in a hurry to avoid being arrested. I think I told you they named me Dezzy after Desmond Tutu, the famous South African priest. They were really happy when I married Thaba. And equally unhappy when I left him, but that's another story."

"Hi Dezzy!" A young woman in her twenties appeared in the kitchen. She had pale brown skin and tightly cropped black hair curled into multi-colored spirals. She was wearing a long summer dress embroidered with a traditional Middle Eastern design, but her expression looked confused.

"Hey, Aliya! Good to see you. You look… what's up?"

"I'm all over the place. Can I have a hug?"

"Well, my sweetest honey-bee… of course you can!" Dezzy embraced Aliya, who burst into tears. It took her a while to calm down, and then she blew her nose and sat down at the table.

"So tell me, what happened? This is Patrick, by the way. He's staying with me for a few days while he visits Vancouver."

Aliya nodded in my direction and gave me a timid smile. Then she turned to Dezzy. "I had the most awful day at the hospital, and then on top of it all I discovered that, that…." She choked up and started to cry again.

"Take your time, girl. You discovered that…."

"I discovered that… I'm pregnant!"

"Oh my goodness. That's wonderful! Does Lucas know?"

"Yes. I just told him five minutes ago. He thinks it's wonderful. We both do. But at the hospital today there was this two-year-old girl with an MDR lung infection we'd been nursing. I really loved her, but we just couldn't save her."

Aliya broke down and cried again. After she had pulled herself together, she said, "I just loved her so much. I don't know what it was. I've seen many people die, but this one, she was so sweet, she got to me right here." Aliya gestured to her heart. "I think she reminded me of a child I knew in the refugee camp in Turkey, after we escaped from Syria. But how many more? How many more are we going to lose? I had to break it to her parents. I feel so mixed up. How can I celebrate a baby growing inside me when they have just lost theirs? How can I even think of bringing a child into the world when this is the reality it's going to face? It's too confusing. I almost feel that I want to give them my baby to make up for their loss."

I didn't want to probe, but I guessed—correctly—that Aliya had been living with her parents in Syria at the time of the civil war.

"Oh, Aliya," Dezzy said. "Come here. Have another hug. You're just too loving. You are *such* a gift to the world."

MDR… I racked my brain and then remembered that it stood for multi-drug resistant, caused by the overuse of antibiotics. My thoughts were interrupted by a knock at the door, and Betska came in accompanied by a tall young man with wavy black hair and an intense expression, wearing black pants and a snazzy black shirt with two vivid vertical green stripes down the front.

"Hi there, everyone!" Betska said. "Good to see you, Aliya. You look as if—" Her sentence was interrupted as Aliya rushed over and threw herself into Betska's arms. Leo looked embarrassed, and then came over and introduced himself, shaking my hand. "I'm Leo—Leo Brankovic Lavric."

"That's quite the name," I said.

"Leo's my name. Branko's my father's name and Lavric is my father's family name. He comes from Slovenia. Long story. I live a few blocks over in a shared collective house with five other people. Betska's my grandmother."

"Aliya, can you choose the art to go with our dinner?" Dezzy asked. "Whatever you feel like. And I mean that."

I watched as Aliya picked up a remote and pointed it at a picture hanging on the wall, a summer landscape by Cezanne, all mountains and greenery. She pulled up a menu that seemed to include artwork from all over the world. She flicked through several, and settled on a powerful photo of a young girl's face in a garden staring directly at you, with the slight hint of a smile.

That's crazy! I thought. She seemed to have the entire universe of art at her fingertips. But why not? It was obviously easy to digitalize. So why wasn't it happening back in my time?

"A perfect choice for a lovely summer evening," Dezzy said.

"It's hard when you have such a great collection," Aliya said. "We're so spoiled for choice."

"No more than we have been for music for all these years," Betska said. "I remember how proud my grandfather was when he bought his first record.

Stravinsky's *Rite of Spring*. That's all he had, apart from what was on the radio. Now look at all the choices we have."

Lucas joined us, happy at Aliya's news, and then Dezzy's son Jake came bouncing in. When we were seated, we joined hands and closed our eyes. The room was still, and nobody broke the spell. Finally Dezzy ended it and we tucked into the meal. The conversation ranged through Aliya's pregnancy, Jake's day at school, and a variety of personal and neighborhood happenings.

After a while, Dezzy asked me, "Since you're visiting, is there anything particular you'd like to ask about Vancouver, and all the changes that have been happening here?"

"Well, I have so many questions," I replied. "But yes, there is something. When you were telling me about the OMEGA Days this afternoon you mentioned something called a new synthesis. I was wondering what it is."

"Ah, now there's a big topic," Betska said with a chuckle. "I bet you weren't expecting that!"

"Well, it's rather complicated...." Dezzy began. "Where do you want to begin, with Socrates or Marx?"

"It doesn't have to be that complicated," Leo said. "It's like a clear stream of water flowing through a forest."

"That's my Leo," Betska said proudly. "You'll be our Tolstoy yet!"

"No pressure there," Lucas said. "You can always join me in the woodwork shop if you want trade your books for a band-saw."

Leo, I later learned, had been home-schooled by his parents with a community of home-schoolers, and was studying political science.

"The easiest way to understand the new synthesis is through its three levels— syntropy, synthesis and solutions, and the way they change political philosophy," he said.

Now I was listening. Three years studying politics at university had taught me enough to be bored with the old philosophers from Cicero to Sartre, weary of the modern political division between left and right, and so frustrated with the post-modernists I wanted to tear my hair out. So talk of a new synthesis had me paying attention. And there it was again, that word *syntropy.*

"*Syntropy* is Satyanendra's new scientific principle that unites consciousness, energy and matter. It provides the thrust that has driven all existence to self-organize ever since the Universe began," Leo declared.

Woah! I wanted to stop him right there and have him explain that slowly, but that was a luxury I'd need to wait some time to enjoy.

"*Synthesis* takes the best aspects of pre-syntropic political philosophy and unites them into a single coherent whole," he continued. "And *solutions* represents the package of applications that are the logical outcome of the synthesis, creating positive, life-enhancing, evolutionary change."[158]

My mind was reeling. And he couldn't have been much older than I was.

"The new synthesis takes the best aspects of liberalism, expressing our innate desire for freedom going back to the enlightenment philosophers—Locke, John Stuart Mill and Voltaire—while leaving out the worst aspects, such as the neo-liberal belief that we should extend freedom to non-personal entities such as banks and corporations and demolish the regulations that control the market. We all know what that led to.[159]

"Next," he continued, "it takes the best aspects of the green movement, such as the belief that we need to live in harmony with nature, but it walks away from the judgmentalism and the belief some greens have that humans are a plague on the planet and we should return to a pre-industrial or pre-agricultural utopia."

"What's so bad about a pre-industrial utopia?" Lucas asked. "We were so much closer to nature then, and we had a far smaller ecological footprint."

"Can I come with you next time you go into the forest, Uncle Lucas?" Jake asked. "It was wicked the last time. Lucas showed me how to use his bow and arrow."

"We'll talk about it later, Jake," Dezzy said. "Leo's talking."

"I agree, Lucas," Leo said. "But you try giving eight billion people each a slice of rural bliss. They'd soon destroy whatever bliss they had. The new synthesis also takes the best aspects of capitalism, such as Adam Smith's recognition of the importance of the free market, provided it is properly regulated, but it adds the recognition that markets need to show the true price of all external costs, including nature's services, while discarding the worst aspects, such as the hyper-capitalism that people in the financial sector used to gamble and enrich themselves.

"Next, it takes the best aspects of socialism, such as the belief that every human has the right to a secure home, a good education, meaningful work and good healthcare, but it discards the worst aspects, including class warfare, militant unionism and too much state ownership and control. It transcends the left-right division that has dominated politics for the last two hundred years. The duality that matters today is different. It's Gaia versus Zeus; community, kindness and harmony with nature versus domination, control and the manipulation of nature."

"I can see that your philosophy studies have been paying off," Dezzy said.

"If that's so, it's not because of all the books we had to plough through," Leo replied. "Most of my professors turned their noses up at Satyanendra and didn't know enough science to understand the significance of the new integration. The new synthesis also embraces the best aspects of anarchism, such as the belief that humans thrive best when they have the freedom to self-organize in small groups and communities, while ignoring the worst aspects, such as the belief that the state is the enemy and the only way to get change is by violence and street warfare."

"I can see why you call it a synthesis," I said, hoping to get a pause in his explanation to digest some of these ideas.

"But wait! I'm not done yet!" Leo exclaimed, dashing my hopes. "It also takes the best aspects of science, such as the importance of observation, reason and experiment, and fuses it with the best aspects of spirituality, including humility in the face of the Universe and knowing that we have hardly begun to penetrate the secrets of consciousness and existence. But it walks away from the worst aspects, where science becomes the corporate manipulation of knowledge and dogma and dualism prevent the acceptance of new ideas, and the worst aspects of spirituality, where an absence of learning and a surplus of fantasy allow weird ideas to proliferate and people's brains to become mystical mush."

"What did I tell you?" Betska said. "He's our very own Spinoza and Pico della Mirandola rolled into one. Your mother would be so proud of you."

"Hey, enough with all the babushka stuff!" Leo shot back with a grin. "At this rate you'll be telling us you're descended from Trotsky's secret love-child."

Betska, I had to remind myself, was Leo's grandmother, and had grown up in Russia.

"So maybe I am!" Betska replied. "There was a rumor that my grandmother had a secret love affair. If Trotsky's my grandfather, that would make you his great-great-grandson. From Leon to Leo. How does it sound?"

"Betska's right about one thing," Aliya interjected. "When you consider how syntropy is shaking up science and opening new avenues of thought it's very similar to Spinoza's way of thinking."

"And who is this Spinoza?" Lucas asked. "Or am I the only one who's getting a bit lost? And who is this Pico character, too?"

"Spinoza was one of the world's greatest philosophers," Betska answered in her soft melodic voice. "He was a Jewish genius who lived in Holland in the 17th century. He was a pantheist who rejected the dualism of the monotheistic religions, which separated nature from God and God from humanity. To Spinoza, God was in everything and everything was in God. The entire universe was a celebration of divine unity. He was a bit like Einstein, who thought the same way.

"His views brought new life to an ancient way of thinking that goes back to our Neolithic ancestors, which is still held by many aboriginal people. I haven't understood syntropy theory very well, but I know enough to understand that it unites the realms of spirit and matter. And that's music to my Russian ears."

"It's music to my Muslim ears too," echoed Aliya, "though I suspect that Spinoza would probably be censored and jailed by the mullahs if he were alive today. I'd never heard of Spinoza until I read Dan Brown's book, *Einstein's Lover*, and saw the movie. It's still one of my favorite movies because of the way it links spirituality, science and love. I love the scene where Carlos is sitting by the ocean and in his imagination the waters part, separating science and spirituality and then circling back together, creating that incredible dance of the waters."

"I *loved* that movie," Betska said. "But I never did get the title. Who *was* Einstein's lover?"

"As I understood it, she was God, Science and Nature, rolled into one," Aliya replied. "That was the mystery that enabled Spinoza to fathom out the laws of the Universe and had the priests in the Vatican and the rabbis in Lisbon competing to control the world."[160]

"But Pico got there first," Leo said.

"Can someone remind me who this Pico is?" Lucas asked, with a degree of impatience.

"Giovanni Pico della Mirandola, flower of the Renaissance, lover of life, wisdom and humanity," Leo replied with a flourish.

"And lover of women too, I believe?" Dezzy commented. "Jake—you can go now if you want to. Can you get ready for bed?"

"But I want to stay!" he said. "I want to listen!"

"Well, okay. But soon, okay?"

"Well, yes. And why not?" Leo continued. "Women are part of God's creation—should we not love them too? Pico believed that there's a hidden unity behind all knowledge, whether it comes from nature, the ancients, Plato or Moses, and that if we studied long and hard enough we would be able to reconcile and unify all knowledge, enabling us to become masters of our own fate. He was an early seeker after the Theory of Everything, but he sought it in Plato and the Kabbalah, since modern science didn't exist in those days."

"How does this connect to syntropy, and the new synthesis?" Betska asked.

"The new synthesis became possible because syntropy allows us to reconsider important ideas that were previously dismissed as being vague or spiritual," Leo replied. "The new syntropic paradigm views the dimension of consciousness as an essential integrating field which permeates all reality, including matter and energy."

"That's fine, but what does it have to do with politics?" Betska persisted.

"If you look at the history of political thought," Leo replied, "the people who developed big political ideas all wanted to use science to lend support to their ideas. In the seventeenth and eighteenth centuries, the liberal thinkers of the Enlightenment drew confidence from the progress that scientific rationalism was making in an era of rapid exploration and discovery. In the 19th century, Marx and Engels and their followers grounded their theories of socialism and communism in evolutionary materialism, and the belief that revolution was an inevitable stage in a process of dialectical emergence. In the 20th century, many greens grounded their environmentalism in scientific materialism and the law of entropy. Before syntropy arrived on the scene the dominant belief among scientists was that the Universe was a random affair with no purpose or direction apart from biological self-replication. One of my professors used to argue that the drift from democracy to plutocracy that we saw before the OMEGA Days was related to the loss of direction in science. If the Universe is purposeless then life and politics are purposeless too, so why not act selfishly and pursue purely personal goals?

"In each era," he continued, "the political assumptions followed the scientific assumptions. Science's assumptions change, however. That's its strength. Scientists constantly gather new information and build new theories as they work to get closer to the truth."

There was silence in the room as we pondered Leo's words.

"So how does syntropy theory change things?" Aliya asked.

"Syntropy includes the recognition that consciousness is an omnipresent dimension in the Universe," he replied, "not just a neurological expression of a random, material world. That points to the need to explore the true nature of consciousness, and it provides a new dynamic for growth and change, which includes political change. Syntropy proposes that the experience of existence brings with it the impulse for cooperative self-organization at every level. All units of existence self-organize cooperatively to realize their higher potential, and it's as true for humans as it is for microbes. So when a movement like the OMEGA Days arrives with its powerful call for self-organization to achieve social justice and environmental harmony it's a direct expression of syntropy at work. So science and political theory have a new partnership."

"You're so good with the big ideas, Leo," Lucas said. "But may I offer another way of looking at it?" He had been quietly listening to the conversation.

"For sure, Lucas. I know when to shut up."

"Not at all," Lucas said. "You're doing great. There's something Jim used to say when we were in the thick of the OMEGA Days. He said that whenever a philosophy isn't grounded in a faith in humanity that includes our personal lives, its followers tend to become self-important and dogmatic, putting principles ahead of love. That feeds egoism and infighting, because when you remove love you remove the connection that unites all existence and you allow division to enter in its place. He said that whatever political philosophy you embrace, never forget the heart."

"That's so true," Leo replied. "Syntropy has given everything a new frame. We can unite around our love of life and bring it into whatever campaign we are working on, knowing that we do so on solid scientific ground."

"What does this mean in the real world?" I asked, still trying to understand what syntropy was. It was obviously a major new idea.

"For sure," Lucas replied. "At one point during the OMEGA Days the campaign for more bike lanes required that the parking be removed on a particular road. The plan was to go door-to-door, seeking support. But some people were impatient. They wanted a big rally with cyclists demonstrating their right to use the road—probably naked, for all I know. It was all very grand, but when we tuned into how the residents might respond we saw that rather than being grand it was grandiose, and likely to get their backs up. So we had to talk them out of it. By making our campaign slow and respectful, with individual conversations on the doorstep, we won the support needed to make it happen. Afterwards, we gave every household a bunch of flowers as our way of saying thank you."

"That's beautiful," I said.

"And highly effective," Lucas added. "I was one of the people who went around with the flowers. It goes back to Jim, and his insistence on the Middle Way, avoiding the rocks of extremism and ego."

"Anyone for more pie?" Dezzy asked.

"This is wonderful," Betska said. "I feel so wealthy to have such good friends."

"Yeah, TVH," Aliya said, and they all laughed.

"Excuse me, but what's TVH?" I asked.

"Total Vancouver Happiness," Aliya replied. "It's something Vancouver measures every three years. A score of a hundred would tell us that everyone was deliriously happy—eating like kings, loving their neighborhoods, enjoying their work and having passionate love affairs that would remain forever secret."

"Ahem," Lucas interjected. "I take it you're not expressing your personal wishes?"

"That might depend on how good you are at changing the diapers," Aliya replied, smiling at Lucas.

"You've cheered up!" Dezzy said.

"That's a good definition of happiness," Betska said. "But where is my prince to bring me breakfast in bed each morning and make passionate love until noon?"

"Eugh. That's gross, grandma," Leo responded. "You're almost a hundred!"

"I'm just happy to be in my workshop and to come home to my beautiful princess," Lucas said.

"And I'm happiest when I'm out on horseback with Lucas riding the trails at dusk," Aliya replied.

"What about you, Dezzy?" Betska asked. "What makes you the happiest?"

"Oh, don't ask," she said. "I've got too many bad memories. They haunt me like sad love songs."

The room fell silent. Then Betska spoke.

"Sadness is not the absence of happiness, Dezzy. Sadness is the memory of past happiness that clings to the soul. It is something to celebrate and then gently let go of, however great the loss."

"I thought I had," Dezzy said. "But they cling to me like limpets. My father, my daughter, Derek, Jenny, Thaba. They make me sing my happiness in a very minor key."

"It's a gift," said Betska. "To have your heart broken is a gift from the Universe, even when it happens tragically. If you embrace it, it will transmute you and turn you into one of God's angels."

"Ever since my father was killed so brutally I've seen pain and sadness I never saw before," Dezzy replied. "It doesn't bring me happiness, though."

"Our world is full of suffering," Betska said, "and grief only needs one friend. It's bitter medicine, but I do believe that it's not until our hearts have been truly broken that we can comprehend the depth of suffering that surrounds us, and know true compassion. The animals suffer too, and so often at our hands. Think of all the wild animals that are still being held captive in the world's zoos. Whenever you reach out to a stranger or to an animal you open a door to heaven. We should never be ashamed of loving someone, even if we grieve when love ends or is torn away. It's only by embracing grief in its fullest that we are most truly healed. And when you finally pick yourself up, you may find yourself in a much better place, and be grateful to whatever it was that gave you the grief."

Aliya then did something I'd never seen before. She walked over to Dezzy and did a slow dance with her hands above her head, while the rest of us watched. Then she reached forward, placed her cheek beside Dezzy's and held up her hands for us to take, forming a raised circle. After a minute, Dezzy started sobbing—and I found myself crying too.

Later that night, when everyone had gone, I found myself imagining small acts of healing like this happening all over the city in quiet, unchronicled ways. Had such a thing become normal? I couldn't recall a dinner party ending this way in my time, but it could have been going on for ages without my knowing. After all, the tiny slice of time I inhabited was just a bubble of nothingness compared to the ocean of consciousness the rest of humanity experienced.

After people had gone and we had cleared the dishes, Dezzy came up to me and said, "It is such a pleasure having you here, Patrick."

"I must admit, I find it a bit overwhelming at times. Your friends are very lively."

"Oh, Leo. He's just completed his degree and he can't stop talking about it. And Betska—well, she's just Betska. I hope I'm as lively as she is when I get to her age."

"And Aliya—does she often do a thing like that with her hands?"

"Yes. There's a lot more to that girl than meets the eye. I'm so happy Lucas has found himself such a good woman—*and* that she's pregnant. I expect there'll be a wedding soon. I can't see her mother allowing her to have a baby without being married. Lucas has quite settled down since they got together. But what about you? You said you were single, but is there someone perhaps you are waiting for?"

"No—well, not really."

"What do you mean, not really? There either is or there isn't."

So then I unburdened my heart to Dezzy about the girl in Jerusalem. I felt stupid, but Dezzy took me seriously. This was the first time I had told anyone about the way I loved her—not even Daria or my parents.

"I've no advice I can offer you," she said. "Life is so full of possibilities, most of which go unexplored. Maybe she's dead. Maybe she's the girl for you, if she's still alive. Or maybe you'll go to Jerusalem and she'll be happily married with six kids and hardly remember you. You've got plenty of time. Look at me. I was head-over-heels in love with Thaba, but I still had to leave him. I've given up trying to find reason in affairs of the heart."

The evening over, Dezzy invited me to take a bath and to help myself to a glass of the brandy I'd find there. Relaxing in the water, I saw a red disk on the wall. When I pressed it the light dimmed to darkness, revealing a full night sky across the ceiling accompanied by a chorus of peaceful music. The occasional meteorite passed by, and galaxies melted into distant nebulae.

My mind wandered back to that day in Jerusalem. I could see her gazing at me so intently. Was I drawn to her because she had been so badly hurt? Or was there something deeper that could never be erased?

The music ended, so I touched the button again. The ceiling turned into a kaleidoscope of changing fractals set to Pink Floyd's *The Dark Side of the Moon*. I wanted to turn the volume up, but I was concerned about waking Dezzy. After a while I pressed it a third time and it changed to a sequence that was distinctly R-rated, color-brushed in reds and maroons, but it gave me confused thoughts about the girl in Jerusalem, which felt both right and wrong. Why was life so complicated?

Afterwards, I investigated the rest of the bathroom. I was twenty years in the future, remember, and curious to see how things were done. The bath had a tap underneath that diverted the summer greywater onto the garden, and the faucets to the sink came on by touch and emptied into a sleek, dual-flush toilet-tank. On another occasion I had a chance to try her shower, called a Nebia, which had the most amazing flow of water, while apparently using far less water than every other shower.[161]

As I stood by the sink, a painting appeared on the wall showing four women sunbathing under a tree in autumnal red and brown colors. *The Four Bathers*, it said, by the Belgian impressionist Theo-Van-Ryseelberghe. Were bathrooms everywhere being transformed into temples and art galleries, I wondered? The possibilities were endless.[162]

9

The Health Revolution

I SLEPT IN on the Friday morning, and when I awoke Dezzy had gone, leaving me a note on the kitchen table:

> *Gone to work. Make yourself at home. I've arranged for you to meet*
> *Li Wei-Ping at the Green Economy Institute at 1 p.m. You'll like him.*
> *Aliya has invited you over for coffee, and she'll ride into town with you*
> *when you're ready. See you for supper, then we're all going to the Song*
> *of the Universe.*

Today I would go downtown and seek answers to some of my many questions, such as how much had Vancouver been able to reduce its use of fossil fuels? And how did its economy operate? Enough with all the philosophy for a while. I wanted some practical answers.

After breakfast I went to meet Aliya in the laneway house she shared with Lucas. Aliya al-Kuzbari—that's her full name.

"Good morning, Patrick!" she greeted me. "Fresh coffee? It's level four organic, shade-grown, fair trade, shipped here from Nicaragua on the SolarSailor."[163]

The what? Best leave till later.

She was browsing the morning news on her screen. "I like to start the day by scanning my personalized news," she said. "I get all the news I want in one place, including the daily news from Syria, internal news from the hospital, and the Positive News Network so that I can start the day feeling uplifted."[164]

"It also tells me about any new videos or radio programs that match my interests. There's a program on Radio Moscow today on the use of herbal medicine in Siberia, for instance, so I just click, and it's saved as a translated podcast."

I had often wondered why radio had been so slow to crawl into the digital age, and why radios didn't even have the means to pre-record a program, the way you can with television.

Aliya said she'd be happy to ride into town with me along the Eighty-Eight Elements Trail, which had been created to showcase the city's transportation initiatives. "I can go with you as far as the Future Café, and after that it's clearly marked. I gather Dezzy has set you up with a meeting?"

"Yes, and that's perfect. What's the Future Café?"

"You'll see when we get there. I've just got to do my weekend shopping, then we can chat for a while."

"You're going out? To do your shopping?"

Aliya looked at me quizzically. "No, I do it online. It'll be delivered tomorrow by bike. Take a look."

Using her screen she pulled up Sustainable Produce Urban Delivery, 'loyally serving Vancouver's residents for thirty-five years.' She clicked "weekly order" and all her regular items showed up. Each time she chose an item a widget gave her a score, flashing amber when she added a tub of blueberry vanilla ice cream.[165]

"Darn it," she said. She switched to frozen yogurt and the amber went away. "As a nurse, I love that widget, but as a human who likes ice-cream, I hate it!"

"What does your score represent?" I asked. It had gone from zero to eighty-five as she clicked on the different items.

"It's my weekly FoodScore. Everything I buy has been encoded for its health properties. Look, I'll show you." Aliya clicked on her score and it showed her accumulating tally for a wide range of essential nutrients from Omega 3 fats, iodine and zinc to calcium, iron, folate, niacin, magnesium, thiamine riboflavin and various vitamins. "It tells me if I'm running a deficit in any particular nutrient and gives me suggestions on how to fill the gap. It also tells me if I'm getting too much of something, since it's easy to overload on vitamin D or A, which can be as harmful as not getting enough. A lot of health problems are related to the absence of nutrients so this is a valuable service, especially now that I'm pregnant."

"Yes—congratulations!"

"Thank you. I wasn't expecting it. Lucas is very happy."

"What about the food you grow yourself or buy locally—does it score for that?"

"There's a space where I can add anything I'm harvesting from the garden or buying at the Saturday market. So I'll click here for peas, lettuce, spinach, carrots and chard, which we're eating from the garden."

"Does it tell you when to brush your teeth?"

"No," she laughed, "but it does tell me about calories, fair trade, food-miles, colorants and preservatives. It says that 64% of the produce I've ordered has been grown within 50 kilometres. That's good, because they'll have more phytonutrients."[166]

"Why is that?"

"Well, take salvestrols, one of many phytonutrients. Plants produce them when they're attacked by mold or fungus, which only happens when they're ripe and when they've been grown organically and not sprayed with fungicides. The salvestrols trigger a process that fights any incipient cancer cells. The further food is shipped the more likely it is to have been picked before it's ripe, before it's had a chance to be attacked by fungi and to generate salvestrols in self-defense."[167]

"So the salvestrols protect you against cancer?"

"Yes. Before, when the farmers used so many pesticides and fungicides, the salvestrols had almost entirely disappeared from our diet, so our bodies had one less defense against cancer."

"That would explain a lot," I said. "Does your FoodScore app work in the stores as well?"

"For sure. The information for each product is in its barcode. All you need is a phone and a digital shopping list."

"How much of it's organic?"

"All of it. All food grown in Canada is organic these days. It's not all five-star, but three or four-star is still pretty good."

I was really surprised. Back in my time, only two percent of Canada's farms were organic. "What made all the farmers go organic?" I asked. "And what's the five-star system?"

"Five-star means that as well as using no chemicals, the farmers build their soil, treat their farm animals with kindness, pay fair wages and take care of their wildlife. They get a star for each if they meet the requirements."[168]

"That's impressive."

"As to why the farmers went organic, it all happened quite quickly. Soon after the OMEGA Days the government commissioned a study to investigate the full cost of conventional farming. They looked at everything from farming's climate impact to the loss of habitat and species, herbicide-resistant super-weeds, nitrogen pollution from fertilizers getting into the water, soil erosion, the impact on bees, the abuse of antibiotics, and negative health impacts caused by the use of pesticides and fertilizers and the loss of essential nutrients from the soil. That includes cancer and dementia, which have been linked to the use of pesticides and nitrogen fertilizers. Maybe autism, too.[169] Pre-natal exposure to pesticides was also contributing to ADHD, and to a fall in children's IQ. A crop like celery was being sprayed with as many as sixty-seven different pesticides. Can you believe it? When the government saw the full social, environmental and healthcare impact they brought in a tax on pesticides and fertilizers, to recover the costs."[170]

"Just like that? Didn't the farmers protest?"

"Oh, for sure. There was a lot of complaining. But there were many benefits to going organic. It eliminated the cost of fertilizers and pesticides, and when the farmers realized that they could save money and get better yields it became a no-brainer. Conventional yields had been falling anyway due to pesticide resistance and the spread of herbicide-resistant weeds,[171] especially with the genetically modified crops, so the farmers didn't need much persuading. The government gave the money back to the farmers to subsidize their transition to organic, and the rest is history."[172]

"That's a really significant change."

"Yes," Aliya replied. "We have a family friend who farms in the Okanagan. Since she made the change they've been busy planting trees and hedgerows along their contour lines, and learning about permaculture. Instead of fighting the insects they're working to attract the ones that are beneficial. Do you know how many species of bacteria and microorganism there are in a single handful of soil? More than there are people on the planet! Previously, before the farmers went organic, so many of them were being killed by chemicals. Globally, I've read that the soil performs economic services worth *$20 trillion dollars a year*, yet it never sends us an invoice."[173]

"That's a great way to look at it. How did the big food corporations react?"

"They were furious. When Vancouver required all food served in city facilities to be organic, Vivendo sued for lost sales. There was huge pressure to back down, but the city hired a lawyer and put out an appeal to cover her costs. Vivendo lost, since the courts ruled in the city's favor on public health grounds.

"Vivendo also sued the farmers, demanding that they honor their contracts to buy genetically modified seeds, and accusing some of using GM seeds illegally. It was never proven, and when the farmers counter-sued, Vivendo had to pay a two billion dollar settlement to the Canadian Organic Growers Association.[174]

"Canada now has a Seeds Law that prohibits the use of genetically modified seed and specifically protects seed companies that grow heritage, open-pollinated seeds. Vivendo is still active overseas, but it's only a matter of time before they're gone. They kept telling us the world would starve without their seeds, but yields on organic and agro-ecological farms are going up, not down, and organic farmers benefit from increased carbon storage and soil moisture, which results in better yields in drought years."[175]

"When you trained to be a nurse did you do a course in nutrition?"

"Yes. I was in the first cohort to be trained in the new medicine. Our entire training emphasized the benefits of exercise, diet and attitude."

"What triggered the change?"

"You name it. Canada's healthcare system was costing more each year, fuelled by a fourteen percent annual increase in chronic diseases like obesity, diabetes, heart disease, cancer, respiratory disorders and dementia.[176] Together they were eating up most of the healthcare budget, and the drugs used to treat them were costing more each year. The whole healthcare system was going into crisis.[177]

"The OMEGA Days had a very liberating effect, opening up new ideas and possibilities. The health care community went through a period of intense review. There was a lot of tension when doctors were reassigned to new positions of responsibility in the Community Health Clinics and their regular work was taken over by nurse practitioners. There was talk of a strike, but the doctors were divided, with some supporting the move and others opposing it. In the end they developed the approach we have today, based on nutrition, prevention and community care.

"It's diet that's the big one," Aliya continued. "Right at the start of our training they showed us a video about an American doctor, Dr. Terry Wahls, who'd been a marathoner, ski marathoner and Tae Kwon Do champion. She had multiple sclerosis, and she was getting weaker and weaker in spite of all the chemotherapy and expensive drugs. In those days it cost up to $30,000 a year to treat a patient with MS, plus $34,000 a year in indirect costs. In the States, it drove many people into bankruptcy.[178]

"Dr. Wahls had declined so much that she was confined to a zero gravity wheelchair. But rather than give up, she started researching what caused her multiple sclerosis and she learned which vitamins and supplements were important for the health of her brain. She compiled a list, and when she started taking the supplements the speed of her decline slowed. Then she asked herself which foods contained those nutrients and she redesigned her diet to include the nutrients her brain needed. Within a year she went from being in a wheelchair to being able to complete an eighteen-mile bike ride. She had cured herself, using just food.[179]

"That was a huge eye-opener. She developed a diet to maximize her brain health, with nine cups of specific organic vegetables every day, combined with the rigorous elimination of foods she was sensitive to, plus neuro-stimulation and exercise. It was this insight, that she could cure herself of multiple sclerosis using mainly food alone, which laid the foundation for our training. There's a host of chronic diseases that can be prevented, cured or alleviated by a change of diet. Do you know how much it used to cost to treat diabesity with gastric bypass surgery? Thirty thousand dollars. And with diet, including peer support? Three thousand."[180]

"That's a huge difference!"

"Most people are still not getting their daily dose of Omega 3 fatty acids, iodine, zinc, calcium and magnesium. That's why I like the FoodScore app. Every new nurse, doctor and health-care practitioner is now being trained in functional medicine, based on nutrition, prevention and community care. We're changing from a disease-oriented to a health-oriented approach. Instead of having a family practitioner, you're attached to a Community Health Center where your primary care provider is a nurse practitioner, and the focus is on the prevention and management of chronic disease."[181]

"My parents used to work in the refugee camps in East Africa," I said. "There was a lot of sickness, especially among the children, so they often discussed medical matters. It got me interested. So if I lived in Vancouver and I went to a Community Health Center, what would I expect to find?"

"That's a very timely question, since I'm making the shift to community nursing this fall. If you were new, and I was your primary care provider, first I'd interview you to understand your life situation, your health, and any chronic diseases you might have. If you were sick, I'd take you under my wing, just as a family doctor would, and I'd do whatever it took to get you up and well again. As soon as you were better we'd discuss your diet and read your weekly FoodScore, just as I did five minutes ago, and we'd talk about the importance of getting enough essential nutrients, and how the right food can protect you against various chronic diseases.[182]

"Then we'd move onto exercise. I'd give you an aerobic fitness test and put you on a treadmill to measure your oxygen uptake. We'd work together to complete your Total Health Inventory, or THI, which includes everything from weight, body fat, blood pressure and cholesterol to diet, fitness, alcohol, drugs and sleep patterns. I'd do some basic cancer tests using the SmartSmell detection sticks, and I'd give you a questionnaire to assess your stress, brain health and mental health and your overall attitude to your health. Finally, I'd use the community toxics map to check for workplace health and safety, and we'd work together to estimate your cumulative X-ray exposure, since that's a long-term cancer risk. Your THI would be fully transparent—the lower your score, the better. When I went for my annual check-up last month the nurse recommended that I start using a brain-app to protect myself against future eye disease, since there's a history of it in my family."

"What would happen when I'd answered all those questions?"

"If you had an ailment or a disease I wasn't trained to handle I'd refer you to one of our physicians, or to a herbalist, acupuncturist or massage practitioner. We have all been trained in basic counseling, but if need be I could refer you to a professional counselor. For most ailments, I'd be able to help. Two weeks later you'd come back and we'd create your Personal Health Plan, primarily covering diet and exercise. I'd set you up with an app to give you daily or weekly feedback, and if you needed major changes I'd recommend that you use it to track your health and fitness, your insulin and other health indicators, and maybe suggest that you join one of the clinic's peer support networks."[183]

"How do they work?"

"They're based around shared interests, so that people have something in common, whether it's religion or a love of dogs. They meet regularly, and people help each other adopt new dietary habits, learn new cooking skills, grow more food, get more exercise, or break a habit such as sugar addiction. Joining a support group is twice as effective as trying to change a habit on your own. We're social animals, and we need peer support. That's how we have evolved over millions of years."[184]

"I can identify with that," I replied. "People in the villages in Sudan help each other all the time with everything from childcare to building each other's homes. But what if I just didn't care, and I wasn't concerned about my health?"

"Here in Canada we have a socialized health service, so every Canadian pays an annual healthcare premium. Under the new system if you have a low THI you get a discount, just as you do with car insurance. If you make good progress on your THI you can also win a prize. So there's an incentive. If you don't participate, you pay the full premium."

"That seems fair."

"It also makes allowance for environmental risk factors, though their influence is declining now that we've stopped burning fossil fuels, and the chemical industry is embracing green chemistry.[185] There's been a steady decline in heart attacks and lung cancer since we stopped burning diesel.[186] Do you want a cup of tea or a home-made juice, by the way?"[187]

I told her I was fine with my glass of water and urged her to continue. "You're a nurse in the general hospital—is that right?" Aliya nodded. "So what kind of difference has the new approach made there?"

"Not that much, surprisingly. We deal mostly with acute illnesses and injuries, which are past the point of prevention. Our biggest concerns are tuberculosis, malaria, and drug resistant superbugs, like the child who died on me yesterday for the lack of a workable antibiotic. It was wretched. She was only seven. It makes me sick that we've destroyed such an incredible aid, and all because the doctors handed out antibiotics so liberally and so many farmers dosed their animals to increase their yields. If they had known how much grief they were going to cause I'm sure they'd have stopped doing it."[188]

"How do you cope without antibiotics?"

"With great difficulty. We do bacteriotherapy, using live fecal transplants and synthetic stool.[189] It's like the worms we use against Crohn's disease[190] and the maggot enzyme healing gels we use to treat wounds.[191] They sound like they're out of the Middle Ages, but they're actually standard microbiology.[192]

"The crazy thing is that it was antibiotics that made it easier for some diseases to get a hold in the first place by wiping out the friendly bacteria. Most people have no idea how important bacteria are. Instead of feeling 'yuk' we should be grateful, since they're working to keep us healthy. There are literally trillions of them—there are three times more bacterial cells in our bodies than there are human cells."[193]

"Trillions?"

"Yes—around thirty-seven trillion, by the latest estimate.[194] Antibiotics have saved millions of lives, but their misuse may cost us just as many. Most people try to avoid coming into the hospital these days unless it's an emergency. People are self-medicating off the Internet, or going to a pharmacy where they print the drugs they want. The cleaners at the hospital get paid as much as I do, because their work is so important to keep the superbugs at bay."[195]

"Doesn't it make you feel a bit depressed, some days?"

"For sure, but there's a lot of great stuff happening to balance it. Now that we understand how important the gut microbiome is, for instance, our prenatal team is providing a microbiome optimization service for pregnant mothers who have compromised immune systems, using fecal transplants from children in rural Africa whose intestinal flora are genetically closer to the way our bodies used to be in the Stone Age."[196]

My jaw was literally dropping at her mention of fecal transplants, but I hid my surprise and pressed on.

"You mentioned drug-resistant tuberculosis. That sounds pretty bad."

"It is. A person with drug-resistant TB can infect ten to fifteen people a year if we don't catch them. That's potentially as many as 100,000 people over five years. Several hospitals have been converted into TB isolation hospitals, and there's no end in sight."[197]

"What about substitutes for antibiotics? Is there anything in the pipeline?"

"There's work going on to cultivate new viruses that can attack the super-bugs,[198] and biologists are exploring the thousands of species of flies, since their maggots thrive in infected material, which means they've had millions of years to evolve anti-bacterial enzymes. But you know what the real tragedy is?"[199]

I looked at her with a blank expression.

"It's the tropical rainforests. There are hundreds of common drugs that originated in the rainforests, like quinine and novocaine, cortisone and ampika, but we're still losing the forests to slash-and-burn farming, cattle ranching and illegal logging. It's like burning an ancient library full of texts that have never been translated. Tropical rainforest plants have evolved over millions of years, so they have remarkable properties, but they're still being destroyed. As for the Amazon, if there's another drought like the recent one the entire forest could burn to the ground by the end of the century, to be replaced by savannah. I can't begin to tell you how much that hurts. And to think that I'm going to bring a child into such a world."[200]

"What's being done to protect the forests?"

"Not enough," Aliya replied. "We did an initiative at the hospital last year called Nurses for Nature. We asked Vancouver's artists to create paintings of plants and trees from the rainforest that have led to essential drugs and we displayed them on the hospital walls, making the point that most rainforest plants have never been analyzed for their medicinal value. We asked people, 'Could this plant save your loved one?' and we encouraged them to support the Rainforest Action Network's campaign to stop illegal logging and slash-and-burn farming. We got a great response."[201]

"What a wonderful initiative! What about childhood brain disorders, such as autism? Are you seeing much progress there?"[202]

"Yes. We're seeing a decline in all childhood brain spectrum disorders, including autism, and also in gastro-intestinal and bowel disorders like Crohn's disease and colitis. The frustrating part is that we don't know *why* they're declining. It could be that pregnant mothers are eating healthier food and getting a wider spectrum of essential nutrients, including folic acid. It could be that doctors are prescribing far fewer antibiotics, which were destroying the bacterial microbiome in the gut, which has been linked to these diseases.[203]

"It could also be a spin-off from Canada's Toxics Use Reduction Act, which has banned the most toxic chemicals and imposes a fee on the others, reducing their use.[204] We do a cord blood analysis on one in every hundred babies and we've seen a steady fall in the number of chemicals in their blood. A newborn used to have as many as two hundred toxic contaminants in its cord blood, a hundred and eighty of which were known to cause cancer. Now it's down to forty, and it's falling every year. That's very reassuring, especially in my condition.[205]

"There are also various neurotoxins that were implicated with autism, including air pollution from diesel fumes, cadmium, and mercury from coal-fired power stations,[206] maybe also the pesticide glyphosate, or Roundup, that was used so extensively in conjunction with GM crops.[207] With 100% organic farming, no more coal-fired power, no more diesel and no more genetically modified crops, all of those risk factors have disappeared.

"Autism could also have been caused by a combination," she continued, "including zinc or selenium deficiency in the mother caused by soil depletion from conventional farming.[208] It's such a complex subject. It might even have been high fructose corn syrup, which reduced a mother's ability to absorb zinc.[209] I'm glad we've finally gotten that out of our diet."

"How did that happen?"

"When the evidence came out linking high fructose corn syrup to obesity, governments both here and south of the border sued the food industry to recover their healthcare costs and the industry went into a tailspin. The diet apps are having a impact, and the bans on high-fructose corn syrup and Bisphenol A put the final nails in the coffin. Today, if you want a soft drink it carries a warning that it will make you fat and flabby. Nobody wants that.[210] The government also hiked the tax on salt, sugar and saturated fats, which are the main culprits behind diabetes and cardiovascular disease, and they tightened up on the sale of alcohol, getting it out of the supermarkets, while imposing very visual warning labels.[211] There had been an epidemic of binge drinking fuelled by cheap prices in the supermarkets and it was causing an increase in fetal alcohol syndrome and all the other disasters that come from excessive drinking, including sexually transmitted diseases, some of which have become drug resistant. And I can tell you from my experience as a nurse, that's not a pretty sight."[212]

"I can imagine—but I'd rather not." Then changing the subject, I asked, "How much did the tax on salt and sugar gather?"

"At its peak, enough to cover a fifth of Canada's health care costs."[213]

"But that's enormous! Did it work?"

"Yes. There has been a big fall in sugar consumption. All sugar-products are now labeled on the front saying how many teaspoons of sugar they contain, and there was a lot of publicity around the lawsuits against Big Sugar. Before the decline, the average North American was consuming twenty-two teaspoons a day. No wonder we had so much diabetes and obesity and so many behavioral problems in small children."[214]

"Are you seeing progress with attention deficit disorder too?"

"Yes, that's also declining. It could be due to the restrictions on fire retardants and other household chemicals that were linked to ADHD. An ordinary couch used to be so laden with toxic chemicals, thanks to industry manipulation of the standards, and children would play on them with their noses right next to the fabric.[215] It might also be the ban on advertising in children's television. The programs were deliberately made with actions that were faster than real life to keep children watching, and the child's growing brain assumed that the whole world operated at that speed."[216]

Aliya paused. "I used to be ADHD myself," she said. "My parents weaned me off Ritalin by immersing me in nature. That was in Syria, before we came to Canada. The drug companies really went to town, persuading the doctors and parents that their kids needed to be on drugs."[217]

"You used to live in Syria?"

Aliya paused again. "Yes. Before the civil war. My father got us into a Turkish refugee camp just before the fighting got bad. I was nine. I'd had a very happy childhood until then. But my father was a doctor, and he returned to care for the wounded. It was his duty, he said. He was killed a few months later in an aerial assault on the suburb of Damascus where he worked."

Aliya was silent. I said how sorry I was, and then I told her about my own parents' death.

"I'm so sorry. You understand, then. We spent three years in the refugee camp before we were finally accepted to come to Canada. My father's a huge reason why I became a nurse."

"What was it like when you arrived in Canada?"

"I was just twelve, so it was all very new. We lived in an apartment block in Toronto. It was fine for a while but then the crash happened and the people we lived among were really struggling. The unemployment was already high among new immigrants and then the carbon rationing started, and the few people who owned a car had to sell them because they couldn't afford the gas. I was very happy when I was accepted into the School of Nursing at UBC, and was able to move out here.

"What was I saying? Oh yes, the importance of nature, and ADHD. The experience a child gets from TV only stimulates a fraction of the brain compared to climbing trees and playing outdoors. I'm doing a course on play in nature later this summer, for when I become a Community Nurse Practitioner."[218]

"Are you looking forward to it?"

"I am, but I do like the hospital work too. I'm fascinated by some of the new techniques we're using. Last week we had a patient with bladder cancer, and we had to remove part of his bladder. The biolab had grown a new one using his own cells, which we transplanted into his body. They're doing the same for damaged teeth, and retinal damage of the eye. They're printing 3-D blood vessels, livers and kidneys, and they're spinning proteins into flexible biofabrics that encourage the damaged cells to repair themselves and to self-organize into a new organ that can be transplanted into the patient's body. Isn't that incredible?"[219]

"Do you still need organ donors, then?"

"Yes, for sure. They've made it so that there's a donor check-box with every bus pass, driver's license and credit card application, and it's had a very positive effect on the number of people donating, for both organs and blood."[220]

"What comes next—brain transplants?"

"Since you ask—yes, kind of! We're seeing an increasing number of acoustic neuromas on the side of the brain where people held their cell phones.[221] It's difficult to operate, since the neuromas push up against the brain, so we try various tumor-shrinking techniques first. If we do have to operate, we use a technique that combines regenerative stem-cell tissue with quantum neurome latticework. It causes the brain to grow new neurons to replace the ones that are damaged during surgery.[222] The surgery itself is changing too. The surgeons use smart-knives and the nurses all wear smart-glasses. There's an overhead camera that communicates to the glasses and we can see the status of the tissue and blood vessels and record the operation in case there's a problem later."[223]

"What's quantum neurome latticework?"

"It's a spin-off from the human neurome project, where they're mapping sections of the brain. Some of the processes that govern neuron cell growth in the brain are uncannily similar to quantum processes, so someone suggested that it might be the key to atomic self-organization, and they're using it to grow new brain tissue. It's pretty wild. They say one day we may be able to use it for Alzheimer's and memory loss. Who knows what else might be possible?"

"Yes, it is incredible. What about cancer? Are you seeing much progress there?"

"Yes, lots. The incidence-rates for most cancers are falling, and we're hoping they'll continue to fall with the use of functional medicine, combined with toxics reduction and people eating healthier food. There's far more emphasis on cancer prevention these days, and we're making good progress with early detection.[224] We used to use dogs, and sometimes rats, which were trained to sniff out a cancer. Now there's a urine test that can be analyzed on a smartphone, an infrared biomarker test that gives almost instant diagnosis, and the SmartSmell detection sticks with their tiny DNA sensor molecules."[225]

I was puzzled.

"There are more than three thousand volatile organic compounds in your breath, and they change when you're sick. The SmartSmell stick can be tuned to read any smell, and it can reveal early cancer. We also lend them to parents to sniff out chemicals in the home that might pose a risk to a baby, and they're being used to screen people coming into the country who might be carrying a dangerous disease."

"That's quite something. How are we doing for time, by the way?"

"We should probably be going soon. The really exciting changes are happening at the community level, where the goal is to prevent most chronic diseases before they begin. One of the justifications Vancouver made for its big investment in bike paths was the healthcare savings. People who cycle more live longer, have fewer illnesses and take less time off work. They're also happier. The new medicine is really turning things around: we've had falling healthcare costs for the last five years in a row."[226]

"What about the drug companies? Are they supporting the changes or resisting them?"

"They fought them every step of the way, just like the tobacco industry did years ago. But they took a real battering when it came out that they'd been failing to publish studies that show negative results for new drugs. Ever since the OMEGA Days there's been a shift in favor of prevention. Under the new system, before a doctor can prescribe a drug, she or he must show you the listed side effects, and it has become normal for doctors and nurses to use Cochrane, which gives us clear, peer-reviewed evidence on the effectiveness or not of the various drugs and remedies, including holistic and preventative alternatives."

"What's Cochrane?" It was not something I had heard of.

"It's a vast body of evidence into the effectiveness of various remedies assembled by fifty thousand researchers and professionals around the world. It enables us to sidestep the influence of the drug companies, who have a vested interest in selling their products.[227] There's also a big campaign to get people exercising, eating more healthily and growing their own food. But speaking of exercise, we should get going ourselves. Do you want an electric bike or regular? And upright or recumbent?"

"Upright and electric, if that's alright."

Aliya picked up her screen and clicked a button. "Carl has a bike we can rent for two Deltas. Or there's a Bixibike station three blocks over if you prefer a public rental."

Deltas? Renting from a neighbor? Then I remembered Betska talking about their community currency, and I said that Carl's bike would be fine.

"I'll cover it on my Delta-card," Aliya said. "No problem. I like to use my Deltas for local services such as bikesharing and carsharing."

"You use Deltas to borrow each others' cars?"

"For sure. Someone did a survey and out of two hundred households in our neighborhood, only thirty own a private vehicle. The rest of us share twenty vehicles between us. Compared to how it used to be, it's a huge reduction."[228]

"And that's not all," she continued, fastening her ankle-straps. "The decline in fossil fuels has caused a big reduction in our ecological footprint, and almost all of our energy comes from renewables."[229]

"That's great!" I said. "Do I need a helmet?"

"No. It's very safe, thanks to all the separated bike lanes. I stopped wearing mine a few years ago when the requirement was relaxed. I can see why they were needed in the past, but that doesn't apply any more. Are you ready? Then let's go!"[230]

 10

Climate Compassion

WALKING OVER TO Carl's house to get the bike, I asked Aliya how she felt about being pregnant.

"I'm all mixed up," she replied. "I'm full of awe at the new life that's growing inside me, but I feel sad that my father didn't live to see his future grandchild."

"I'm sure he'd be so proud of you. How does Lucas feel about it?"

"He's nuts. He's already decided it's a little girl. He's going to love her to bits—or him, if it's a boy. I think he's a bit more intimidated about it being a boy. It probably reminds him of his dad, and makes him worry if he'll be a good enough father."

"He wouldn't be human if he didn't have those kinds of worries. I'm sure glad my parents created me. It means I can actually *do* something to make the world a better place. Maybe your baby will think that way when she grows up. Or he."

Five minutes later we were heading for the Eighty-Eight Elements Trail, which displayed Vancouver's sustainable transportation initiatives.[231] My bicycle was comfortable, and the separated bike-lanes made for easy relaxed riding. The handlebars had lights built into them and an electronic tablet with various functions including GPS, a map that showed where I was, a travel planner, and information on how much juice was left in the bike's battery. It had a top speed of fifty kilometres an hour and GPS to track it in case it was stolen. When we came to a hill I switched on the electric drive and the bike sailed up as if the road were flat, defying the law of gravity.[232]

Most of the streets were green and leafy. Not all had been reclaimed the way Dezzy's had, but most had a charm that showed someone was caring for them. Many boulevards had been planted with food and flowering shrubs, and many telephone poles had runner beans climbing up them.[233]

The trail went all around Vancouver, but we were on just one section. One of the displays explained Vancouver's transportation history going back a hundred years to the days before the motorcar, when Vancouver had a thriving cycling culture.[234] In 1967, the Chinese community had come out in force to oppose a plan to bulldoze their homes to make way for a freeway, with the result that Vancouver was never blighted with the ribbons of concrete and noise that make pedestrian life so unpleasant in other cities.[235]

Another display had a map of the complete bike network, crowd-sourced to show the best routes.[236] It told me that 40% of personal trips in Vancouver were happening by bike, 25% by public transport, 20% by foot and only 15% by motor vehicle.[237] Of all the innovations needed to make a great cycling city, it said, none was more important than safe, separated bike lanes. They were ten times cheaper to build than a new road, they created more jobs,[238] and they justified the cost by the health-care savings, since cyclists increased their fitness and overall health.[239]

For a short while we rode along the Oak Street Ferry Trail from Tsawwassen, thirty-five kilometres to the south where the ferries depart for Victoria. The bike lane was separated from the traffic by a strip of shrubbery in planter boxes and green lights along the trail told us that if we kept to an easy 15 kph we could sail through the lights without stopping.[240] The pedestrian crosswalks were clearly painted, marked by trees and shrubs, and there was a sign twenty metres before a crosswalk that flashed if a car was approaching too fast.

It was great to be able to ride on routes specifically designed for bicycles, with advanced waiting and painted green paths across the busy intersections.[241] As well as regular bikes there were recumbent bikes, bikes made from bamboo, bikes with super-cool designs that I'd never seen before, tricycles being pedaled by parents with young children, and cargo bikes carrying everything from plumbers' gear to girlfriends. On the road, as well as buses and electric cars there were some cute brightly colored three-wheelers called VeloMetros, human-powered electric tricycles that Aliya told me were manufactured locally in Vancouver.[242] I had never ridden among so many cyclists, so I didn't know I was supposed to signal when slowing down. As a result, I caused mayhem when I braked suddenly without signaling. The cyclist behind almost ran into me and another had to swerve to avoid the chaos. No damage was done, but I received some pretty ripe language.[243]

"Do you miss not having a car?" I asked Aliya as we got off the trail and were able to cycle side-by-side along a quiet residential street.

"Not at all. I can get most places I need by bike. The transit is great, and when we go away for a weekend we rent a car-share vehicle.[244] Owning a car feels so old-fashioned."[245]

"What about in winter, when it's cold and raining?"

"I've got a good rain-cape, and if the weather's bad I can always take the bus." Then she suggested we do a detour to show me a couple of things—and along the way I saw the first of what would be a series of posters, labeled simply *Change The World*. It had a photograph of the famous British leader Sir Winston Churchill, accompanied by some words he had spoken:

> *I am an optimist.*
> *It does not seem too much use being anything else.*

I liked it! We came to a piece of land that had twelve tiny houses around a village green with a barbecue pit. It was one of the micro-villages Lucas had told me about, but what was even more astonishing was the sign, which said that each house had been printed on a 3-D printer in less than a day for less than $10,000.

They each had a solar panel, a solar heat pump, a composting toilet, rainwater capture, a community greywater treatment system and wheels that made them easy to move.[246]

"They *printed* them?" I said with amazement.

"You can print almost anything you want these days," Aliya replied. "We could have printed our home if we had wanted to, but Lucas is very traditional. He said he would refuse to live in it unless it was built from wood." The sign said they had been printed using a carbon-negative liquid biomaterial that hardened into cement, surrounded with thick foam insulation. Each was basically a box with a roof, and their sides were covered with artwork.

"Aren't they cute? They are only temporary," Aliya explained. "They have twelve month's notice if the owner gets approval to develop the site. They're managed by the Community Land Trust, so they get help if they need to relocate. Come on—I want to show you something else."

We cycled another block and came to a building with six tiny apartments made from old shipping containers stacked three-up with stairs up the side. A sign said there were a thousand such homes around the city, and they cost $50,000 each. A business called The Container Co-op did all the conversions, wiring and insulation in advance of installation.[247]

"Do you want to see something else?" Aliya asked.

We cycled a few more blocks to a church and on the land behind it there was an urban farm, but not like any I'd seen before—it was a circular building three stories high filled with fish and salad greens being grown aquaponically in tanks.[248] Around the building a class of children was learning how to make a compost heap, while other children were playing inside a willow-tree dome, like an igloo covered in willow-leaves.[249] Another group was gathered around a structure made from old pallets, piece of wood, bricks and tiles labeled 'The Insect Hotel', which provided a home for beneficial insects and pollinators.[250] It was good to see the kids in the open air, out of the classroom.[251]

Returning to our bikes, Aliya took me on a route that passed a set of swings that made music as people swung on them, a bridge where the underside had been painted in bright colors, a musical sculpture that chimed each time the temperature rose or fell by a degree, and a bench with a solar roof that enabled people to recharge their devices. Along the street the merchants seemed to be competing to have the best bike rack, with some designed to look like fishes, trees and butterflies.[252] Later, I saw a parkade with space for six thousand bikes. You wheeled your bike up to the gate and a device scanned its barcode before carrying it away.[253]

The bus shelters were very creative—one was like a garden shed, with hanging baskets of strawberries and peas climbing up the side.[254]

"Aren't they great?" Aliya called out as we passed one that had ornate classical yellow pillars and a roof covered in flowers. "The neighborhood associations build them, using volunteer labor."

As we crossed West 16th Avenue, a floral arch covered in roses welcomed us to the Mount Pleasant neighborhood. "It's one of the ways we build a sense of neighborhood," Aliya said. "I'm the street rep on our Village Assembly."

"What's that like?" I asked.

"We meet once a month on a Monday night, starting with a potluck meal. We catch up on what's happening and then we break into groups—I'm in the health group. We are encouraged to practice deep listening, to go beyond the inevitable reading of reports. That was where I was introduced to the idea of four-level conversations."

"What's a four-level conversation?" It wasn't something I had heard of.

"It's the idea that a conversation can happen on one of four levels. It comes from a healthcare background, though this was the first time I'd heard about it. Level One is a simple exchange of information, when a doctor (for instance) discusses what's wrong with you and how he or she can fix things. In Level Two you go a bit deeper, and discuss what might be causing an ailment or a social problem, and how you might need to change your behavior. In a Level Three conversation we explore the attitudes and assumptions that govern our behavior. It goes deeper into the soul, is the way I see it. At Level Four you have a true meeting between two beings, laying aside your roles, just two beings, sharing heart to heart. I often wish that I could have that kind of conversation with the patients I deal with in the hospital, but there's never the time: there's always the pressure to get something fixed and move on. The great thing about the Village Assembly meetings is that we are deliberately encouraged to go deep, to seek a true meeting of hearts. And then when the evening is over we finish with circle dancing, so I always leave feeling happy and uplifted."

I laughed. I'd never been to a meeting that ended with dancing. But I *had* experienced deep listening: there were many times in the refugee camps when I had seen my parents engaging with people at a deep, heart-to-heart level, and they had always encouraged me to do the same.[255]

We had been cycling north up Yukon Street towards downtown Vancouver when we came to the intersection with Broadway. Back in my time, Broadway had seven lanes of traffic, making for a massive expanse of blacktop. It now had just two lanes of traffic, with bike lanes and bus lanes on either side, and a ribbon of trees and shrubs down the middle. There was a stop for the underground SkyTrain Extension to the university, and clear green lanes for bicycles. The monster has been tamed, I thought, as we sailed across.

It wasn't just the road that had been transformed. The buildings along Broadway looked completely different, too. Every flat roof carried a small rooftop dwelling, linked by sky paths and ribbons of greenery where the occupants were growing food, flowers and small trees. The whole street was delightful. What used to be an unpleasant strip of asphalt, speed and noise had become an urban paradise.[256]

A block further on a crowd had gathered in a small park. I was curious, so we dismounted, locked our bikes to a rack and eased our way in.

Five women were sitting silently on a pale blue blanket on the grass, dressed in white. 'We are Fasting for the Future of Our Planet,' their sign said. No-one was speaking. Behind them, a portable electronic screen told me they were part of Climate Compassion, a global action that was happening in three thousand cities around the world. They wanted ten million signatures by Sunday night on their petition to the United Nations to reduce the global carbon budget, make more funds available for disaster recovery, and accelerate carbon sequestration from the atmosphere.

The screen was showing footage from the climate disasters that had happened so far that year—the violent thunderstorms that had struck in May from Indiana to Virginia with hailstones the size of baseballs; the heat wave that was continuing across the US Mid-West, causing wildfires that were destroying homes and ranches across Colorado and in the hills above Los Angeles and driving more farmers into bankruptcy; the fires that had assaulted south-east Australia in January, burning several thousand homes in Canberra in temperatures above 50°C;[257] the three— yes, three—tropical cyclones that had slammed through the Philippines; the flooding that had inundated the Prairies in April and brought Calgary to a standstill; Hurricane Bertha that had just hit Florida; the forest fires that were burning out of control in Siberia fuelled by the summer heat-wave and the methane escaping from the permafrost; the forest fires that were burning right here in British Columbia, Alberta and Saskatchewan; and the villages that had been swept away in Costa Rica following the start of a tumultuous rainy season.

The display told the cost: the thousands who had died, the millions who had lost their homes, the millions of hectares of farmland now unfit for crops, and the billions of dollars the disasters had cost so far. I was appalled to see it all there, so real and so immediate.[258]

A second set of numbers listed the hundreds of thousands who were fasting for Climate Compassion, the thousands of places where fasts were happening and the six million signatures that had been collected so far. The Vancouver fasters hoped to raise $100,000 to help the survivors of a Costa Rica mudslide to rebuild their village on safer ground, designed for self-sufficiency and zero-carbon living.

"It's all so tragic," Aliya whispered, as we stood among the crowd.

I looked at her and nodded. I had come all this way to find a better future, but the disasters and the suffering were continuing. We could change the layout of Vancouver's streets, but we couldn't change the laws of physics, which controlled Earth's atmosphere and climate.

I knew there could be no short cut to climate tranquility. We had been releasing carbon from ancient fossil fuels for three hundred years, and a quarter of what was released in any one year could stay in the atmosphere for up to a thousand years, trapping heat, evaporating water and fuelling the disasters whose victims these people were trying to help.[259]

Climate Compassion. The words said so much. So many people were suffering because for years we had burnt fossil fuels and destroyed Earth's forests as if there was no tomorrow, ignoring the needs of the planet and its people.

"I would have joined the fast myself," Aliya whispered, "but I need my energy for my work. I've signed the petition and given $100 to the Costa Rica recovery effort, and I've asked my friends to help. It makes my work as a nurse seem so insignificant. I wish there was more I could do."

As Aliya was talking the women stood up and began singing a haunting melody. People joined in, and musicians appeared playing clarinets and saxophones. It was so moving. They gathered around the women, and when the music ended they urged us to take out our devices, go to www.climatecompassion.world and ask our friends to sign the petition.

The crowd hushed as a man appeared on the screen. He was in Costa Rica, speaking live to Climate Compassion groups around the world. He was clearly distressed, but he said a few simple words from his heart and thanked us for the efforts we were making. He showed us photos of his village before it was washed away, ending the lives of many of his family and his fellow villagers. "It hurts," he said, "but knowing you are there makes us feel better."

Then a woman came on from Colorado, her face filthy, the land billowing with smoke behind her. "Our farm is gone," she said. "I don't know what we're going to do. We're safe, but we've lost everything. Please urge your friends to sign the petition and to give whatever they can to help the people in Costa Rica. We'll get by, but they've got nothing."

"I can't take this anymore," Aliya said. "It's too close to home. My aunt and uncle lost everything in Syria when their village was bombed. Their daughter, my favorite cousin, was killed."

We pushed our way out of the crowd, but when the singing started again we paused to listen. Aliya started crying and had to sit down on the ground. I kneeled to comfort her, and a woman reached forward and put a hand on her shoulder. I put my arm around Aliya while she lost herself in grief.

"Wow—that was unexpected," she said when she had recovered. "This is all so real for me. I'm sorry."

"Don't ever be sorry for grieving," I said. "A world without grief would be a world without love."

When she had wiped her eyes we moved away from the crowd and reclaimed our bicycles. "Let's walk," she said. "It's only a couple of blocks to the Future Café and I don't think I've got it in me to cycle. I love what they're doing. It's easy to feel hopeless. They are like an inoculant, defending us against despair. We're making good progress, but the disasters keep happening. They're like a malevolent genie that has escaped from its bottle."

Climate change. Thirteen innocent letters that spell such a terrible future for so many.

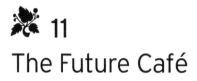 11

The Future Café

AFTER WALKING TWO blocks we pulled up at The Future Café. Above the door there was an inspiring quote in ornate handwriting:

> *The distance between the possible and the impossible is shrinking.*
> *It's just a matter of our imagination.*
> *Let's imagine a world where there's no poverty.*
> *Let's imagine a world where there's no unemployment.*
> – Muhammad Yunus

The café seemed normal enough, so I waited to see what might transpire. "What makes this The Future Café?" I asked.

"Wait and see," Aliya said. "Let's order a drink and see what there is to eat."

The menu was laid out electronically on a glass tabletop, and a waitress came over to see if we had any special requests. "Do you have any of your lovely hibiscus tea?" Aliya asked. "And I'll take a jasmine cake."

I ordered a black coffee and a rum strudel. The table display told me that while my coffee was organic and fair trade, the strudel was going to give me two hundred calories, zero nutrients and a day's dose of sugar.

Aliya laughed at my surprise. "You've not seen that before? That's so cute. It's almost as if you've arrived from another planet. Look, touch here." I touched the word 'Origin' and the screen showed the farm in Ecuador where the coffee had been grown, with a video of the farmer's family that I could watch if I put on earphones.[260]

"The café has a very colorful history," she said. "It was founded by hippies in the 1970s, but after fifty years it was sold to a man who set up a head office in Barbados where he took his chef each year with the excuse that they were developing new recipes. He used that to claim that the café operated out of the Barbados, where he had to pay almost no tax. When he was prosecuted there was a community boycott and he was forced into bankruptcy. Vancouver's Social Planning Council put together a community crowd-funding campaign on Wayblaze to take over the lease, and now they operate the café as a social enterprise.[261] There were over a thousand Canadian companies with their head offices in Barbados at the time. I doubt there's one today."[262]

"That's impressive," I said. "Another small step towards a prosperous green cooperative economy."

"You know about Derek?"

"Yes. Dezzy told me everything."

"Lucas keeps going on about it. But look! My friend Sophie is nearby. I'll send her a poke to tell her I'm here. Maybe she'll come and join us."

"How do you know?"

"It tells me right here on my watch. She's on my BB list. She's in a store just down the street."

"Your BB list?"

"Yes, Best Buddies."

"Isn't that, like, an invasion of privacy?"

"Don't be silly. Best Buddies are always mutual. I only put my closest friends on my BB list. It tells me when we're in close range so that we can meet up if we want to. I love it."

A minute later a tall, dark-haired woman came in wearing a long black coat revealing a beautiful silky dress in shimmering green and blue, as if the fabric itself was changing color.

"Aliya! I haven't seen you for *ages!* It must be at least a week!"

"My dreams have been calling for you!" Aliya replied. "My heart has been broken without you. Even the moon has been crying, ma Cherie. *Inta ilbalsam li-ijrouhi.*" ('You are the remedy for my wounds' in Arabic.)

Aliya held Sophie in a long, intimate embrace. She needs that, I thought. Then she whispered something into Sophie's ear and Sophie shrieked.

"That's amazing! What a miracle! Come here, you beautiful desert fox. You—a baby! Let me give you another hug!" I couldn't help smiling to see how happy they were.

"Sophie, this is Patrick. He's visiting from Sudan on a quest to learn how we've become so green and fancy here in Vancouver."

"'Tis all about the love, my dear," Sophie said, as she shook my hand. "I bet they never taught you that in school. Without love there's no hope, no future. Well, I suppose there is, but it wouldn't be much fun."

While Sophie and Aliya caught up, I pondered the table screen. I could use it to give feedback on the food, music, ambience, service and comfort; to pay by phone; and to donate to their Gift of the Week, supporting a school in Ecuador close to the farm my coffee came from.

"So what do you think of our city?" Sophie asked me.

"I think it's amazing from what I've seen so far."

"We've been cycling down the Elements Trail," Aliya explained.

"Did you take him to see the sculptures on Hastings?"

"No. That's downtown. Maybe you'll see them this afternoon," Aliya said, turning to me.

"Well, look, I must be getting on," Sophie said. "But wait, I must show you my latest work! Here, put these on."

Sophie reached into her bag and pulled out two pairs of glasses that were worn with a band around the head. The band contained sensors and when Sophie waved her hand a strange thing happened.[263] In the center of my vision there appeared a land full of creatures, faces, ferns and forests, as if the 15th century painter Hieronymus Bosch had produced a new *Garden of Earthly Delights*. As I watched, the painting came alive and the different components moved around, rearranging themselves. I sat up straight.[264]

"It's not complete yet, but it's coming along," Sophie said. "It picks up your mood through the aureatic connection with your skin."

I felt confused, especially when the creatures re-arranged themselves into a jagged, dissonant pattern and a fawn-like creature looked at me directly with hostility in its eyes.

"Remember love," Sophie said. "They find their best harmony when you flood them with love."

I made a conscious effort to center myself in the warmth of love and sure enough, the creatures rearranged themselves—I'm not sure how—and the picture took on a glow of harmony.

"That's amazing!" Aliya said. "You're a genius! It's so different from your last work."

"Thanks! That last one was a bit industrial, but it's selling well. They like it in the boardrooms because I programmed the connection to pick up on positive and negative predispositions. I'm told that companies are using it to score their managers' attitudes—they must hate that. It wasn't what I intended it for. But hey, I must be getting on. I love you, Aliya. Nice meeting you, Patrick. Good luck with your travels!"

"What was that?" I asked, when Sophie had left the café. "The glasses, I mean. How did she do it?"

"I've no idea. She calls herself an artistic shape-shifter. She's also a musical composer and a practicing magician. The high reality glasses are common, but I've never seen them used the way Sophie does. They usually do regular stuff like optical zoom, reading barcodes and checking a store's special offers before you go in.[265] But look, there's something I want you to see. I'll just pay; they take Deltas here."

Aliya pressed the pay button, paid the bill in the time it took me to get up from my chair, and walked to the back of the café where a series of framed panels displayed images of the future.

"Aren't they great?" Aliya said. "If you have an idea about the future you can present it to the café. If they like it, they'll find someone to illustrate it. The City Futurist helped set it up. Look, this is my favorite sequence."

The City Futurist? As a *job*? That's pretty wild, I thought.

In the sequence Aliya liked there were no cars, buses or trucks. The bikes and pedestrians were still there, but in place of the roadway it was all green, land-scaped with flowers, shrubs and boxed vegetable beds. Two aerial rails supported transparent oval pods carrying people. Another image showed a residential street where the road had been reduced to a single path that ambled its way down the street, ending in a place where you could catch one of the pod-like vehicles. It was like Dezzy's street, but taken to a whole new level.

"What do you think?" she asked. "It takes me back to the way villages were before the motorcar. I remember villages like this in Syria when I was a child, but without the travel pods. But look at this! It's a new sequence."

The images showed the back gardens of four houses without any separating fences. In the center there was a glass dome surrounded by a moat with fish in it. The second image, inside the dome, showed circular seating surrounding a crystal sphere. In the third image the sphere was filled with three-dimensional images of a rainforest—that turned into a desert—and then into some kind of soulscape with dreamy creatures floating around inside it. In the final image there was a raised platform inside the sphere and a woman lying under a mauve sheet....

My arms prickled and my hair stood on end. I lost all the breath in my body and couldn't say a word.

"Are you alright?" Aliya asked.

"Yes," I replied without thinking. In retrospect, my answer should have been a clear 'no.' I was experiencing a laser shaft of energy pouring through my head. My entire body was vibrating and I felt an urge to step into the image, to become the person lying there. It was definitely a scary moment. I felt so attracted to it, yet I knew that if I took that step there would be no stepping back.

"What are you doing?" Aliya called out with urgency. "You can't do that!"

When I stopped, I saw that I had begun to walk up to the image with my hands stretched out.

"They're not for touching. Sticky fingers, and all that."

I pulled myself back, which seemed to take an extraordinary amount of effort.

"Are you okay? You seem a bit dazed," Aliya said. "That's a very powerful sequence."

She had no idea. At least, I hoped she didn't. But what if... I forced myself to stop my train of thought.

"What do you think is happening here?" I asked Aliya, trying to be calm, as if we were discussing a cup of tea.

"It's a soul chamber," she said. "They're used for journeys to other realities. The people are guarding her as she travels. It's called hypnosyntropic imaging, but most of us call it soul-riding."

Looking more closely at the person in the image I could see the faint outline of her breasts rising and falling under the white covering. That tiny piece of movement affected me profoundly.

"It makes me want to go on a soul-journey," I said. What a ridiculous thing to say, given that I was already doing something just like that.

"Me too," Aliya said. "It takes me back to Syria. I think I left part of my soul there, and if I want to find it I'm going to have to return."

"Is that possible?" I asked, stepping away from the display. Aliya paused before answering. When she replied, her speech was calm and deliberate.

"The difficult times have ended now that ISIS has been defeated and there's a government of national unity working with Russia, the US, Turkey, Iran, Saudi Arabia and the United Nations to rebuild the country, after so many years of cruelty, horror and devastation. If it hadn't been for the women in Russia, Europe and America getting together and demanding that their governments work to depose Assad and set up a government of national unity, I dread to think what Syria might be like today. Those thousand Russian women who fasted outside the Kremlin, urging Putin to act: they were really brave.[266]

"I'm very attracted to The New Islam that is taking root, especially among the younger women," she continued. "It's such a break from the past. My parents were secular, but when I was in the refugee camp in Turkey I found myself being attracted to The Muslim Brotherhood, and I took to wearing the *hijab*. For a while I even wore the *niqab*, which covered my whole face.

"But when we came to Canada and I had a chance to study I was inspired by the Somali-Dutch-American scholar Ayaan Hirsi Ali, who wrote so persuasively on the need for Islam to undergo its own reformation, letting go of its determination never to be criticized, letting go of the belief that every word of the Qur'an is the infallible word of Allah, and abandoning its obsession with death, martyrdom and *jihad*, and controlling people's lives, especially women's lives.[267] I want to be able to pray in the land I come from. I want to be able to bow my head to Mecca and bury myself in the sublimity of Allah, may peace be upon Him. And I'd love to be able to help in some way to rebuild Syria, now that we've finally got peace. But I don't want to be treated as an inferior being or a sexual object just because I'm a woman. The very first Muslim, Muhammad's wife Khadija, was a woman—and a successful businesswoman, too. I don't want to live under polygamy or sharia law and have the Imams tell me the Qur'an gives my husband the right to beat me and force me to submit. I want the freedom to be myself, and to explore new ideas without being told what I should think by men with beards. I want the freedom to dance and play soccer without wearing a veil. I want to pray with those who experience the call to prayer as a call for unity, freedom and tolerance among people of all creeds and races, not just those who follow the Muslim path. I want to follow the soul intentions of the Prophet, may peace be upon Him. I want to know the peace and sisterhood of all beings, to merge my soul intentions with His."

"That's very beautiful," I said. It was more than beautiful. It was amazing.

"The Song of the Universe that we're going to tonight captures part of what I feel," continued Aliya, "but it's not the same as being in Damascus at dusk when the muezzin call the faithful to prayer at the Great Mosque, with the smell of spices in the warm evening air."

"Does Lucas want to visit Syria?"

"No. He thinks it's a desert. He's too much of a west coast boy. He prefers the forests, the mountains and the rain. And besides, it would take so long and it would be very expensive."

"And you?"

"I don't know. But if I don't go I think I'll always be pining. I could take the high-speed train across Canada, the Solar Sailor across the Atlantic, the train to Istanbul and then on to Beirut and Damascus. It would take three weeks, but it would be a great trip. If I went in summer I could take the trans-Arctic route from Vancouver to Arkhangelsk and down the Volga to the Black Sea. That would be pretty wild. But speaking of travel, I've got to get to work. You okay? The rest of the trail is very clear. It'll take you right into the downtown for your meeting."[268]

I assured her that I was okay and waved goodbye as she rode off to the hospital a few blocks away. For the first time since I'd arrived I was truly alone in the middle of the future. I felt the buzz of freedom and the delight of not knowing what might come next. I went to the men's room and laughed my head off when I saw a waterless urinal featuring a tiny hockey net with a puck hanging from the crossbar to encourage men to aim straight. Above the urinals, in huge italics, there was a quote by Teilhard de Chardin, the French scientist and priest from the last century:

The whole future of the Earth, as of religion,
seems to me to depend on the awakening
of our faith in the future.

Wow! This place really lifted you to another dimension. Before ordering another coffee I went back to the images of the future. One was like a science fiction future from my childhood, with skyscrapers linked by skywalks. Another was very dark. The green streetscape was still there, but the trees were haggard. The shrubs were wild and neglected, and looking down the road, the sea was lapping across the street, covering everything beyond. The people were lonely, not looking at each other. The shops were boarded up and covered in graffiti.

This is bad, I thought. I remembered the women fasting. Was *this* the coming reality? A future filled with unrelenting climate tragedy, for decades, perhaps centuries to come?

I was pondering these thoughts when an elderly man came up to me. He had wild white hair and a pure white beard.

"Mind if I join you?" he asked in a slow ponderous voice.

"No, be my guest. Can I get you a coffee?"

"Can you make it a hot chocolate? It's a cold day, even for June. They say the world has warmed, but I must admit, I can't feel it."

I ordered his drink and sat quietly, waiting for him to talk.

"All this greenery. I'm not complaining. Best thing could have happened to

Vancouver. But how much are we fooling ourselves? We can't control the climate. There's trouble coming. I can feel it in my bones."

"You mean in the future?" I asked.

"Yes, the future. Where else? We're like a beautiful island in a sea of trouble. But how long before the ocean takes us, just like it'll take everything? It's coming, surely as the ice-sheets are melting. Not even the Mayor of Vancouver can hold back the sea."[269]

"How soon do you think it will happen?" I asked.

"Who knows?" he said. "It's Wilf, by the way. The name's Wilf."

"Patrick. Pleased to meet you." His hand was large and rough, as if weathered by a lifetime of outdoor work.

"The scientists and ecologists, they say we're on the edge of the great descent. Last call for civilization, last call for Nature. When things break down I wouldn't hold much hope for the wilderness. Billions of people will still need to eat and feed their children."

"But there are so many signs of hope," I responded. "Aren't similar changes happening elsewhere in the world?"

"For sure, laddie, there's a lot of good things happening. But like I say, you can't argue with Mother Nature. Even if every blessed soul on Earth started living like Lady Greenpeace, try telling that to the glaciers. Last time I looked, Greenland wasn't reading no books. What are you going to do about that? What are you going to do about all them fires that are burning down the forests? What are you going to do about them? You can't control nature once she's got her angry on."[270]

It was true. I knew that the last time the world had been three degrees warmer the sea level had been twenty-five metres higher. And I knew that back in my time we were on track for a rise in temperature by the end of the century of almost four degrees Celsius, depending on how soon we stopped burning fossil fuels, eating beef and cutting down rainforests. It was one of the reasons I had come on this journey.[271]

"Anyway, don't let me be spoiling your day. You've got better things to do than listen to the ramblings of a grumpy old man. I just wish for your sake that we'd started living more sensibly back when you was born, and not left it so late. Way overdue, those OMEGA Days were. Way overdue. Maybe there's some way they can put the spirits back in Pandora's box, and get all that carbon back into the trees and soil, where it belongs. I hear they're talking about it, but they're also talking crazy schemes like building gigantic mirrors in space to shield us from the Sun, filling the sky with more chemtrails, and other scary ideas. Is that what we've come to? Lording it high and mighty over the whole of creation? Isn't that what got us into this trouble in the first place?"

"I agree. But I was hoping that if everyone started living in more harmony with nature, as it seems people here are doing, using far fewer resources, we could turn the corner. I like to believe there's still hope."

"Oh, another romantic. But a kind-hearted one, I must say. Who am I, a tired old fart, to deny you your dreams? If it hadn't been for the dreams of your generation we'd be far worse off today. There wouldn't be even a glimmer of hope."

"So you think there's still a glimmer?"

"Well, miracles can always happen. But they'd best start happening pretty soon if we're going to pull ourselves out of this one. But I must be getting on. You've been very kind. If I see you again, I'll buy you a hot chocolate."

After Wilf left I pondered his words. I still had so much to learn, and I was well into Day Two of my visit. For the first time, I felt confused and uncertain. Maybe it *was* hopeless. What about the rest of the world? If Vancouver acted alone, it would make no more difference than dog's piss in an abandoned alley. What use was an island of paradise if the ocean that surrounded it was becoming a whirlpool, bringing disintegration and collapse?

12
The Heart of Poverty

BACK ON THE trail, I passed a sign that told me Fourth Avenue was closed to cars every Sunday, and open only to cyclists, rollerbladers, runners and strollers.[272] I rode north over the Cambie Street Bridge, crossing the waters of False Creek. Below me on the water, there were—*what??* There were half a dozen people flying around on some strange kind of flying device, dipping and diving in a wonderful dance. Those, I later learned, were members of Vancouver's synchronized hoverboard club, enjoying a morning practice session.[273] Humans, flying like that! And then to see the banners of colored silk fluttering from the streetlights and the central median ablaze with rhododendrons and flowers—well, they set my soul ablaze. A banner at the end of the bridge proclaimed *'The Land that Ugly Forgot'* and welcomed me to the downtown.

I cycled to Wei-Ping's office on Water Street in Gastown and found a space to park Carl's bike in a bike rack designed like a red dragon. I had a while before my meeting, so I walked to the Waterfront station and turned up Seymour, enjoying the wide sidewalks, ample bike lanes and colorful food carts. Several buildings were covered with ferns and flowering plants that tumbled down their walls, as if a rainforest had taken up residence in the city.[274] At a crosswalk, instead of saying WALK it said DANCE and there was music that made it impossible not to—not just me but others too, laughing and smiling at each other.[275]

At the end of the street a cluster of people were gathered around a large electronic screen that showed the Eiffel Tower in the background. *'Paris: 10:20 p.m.,'* it said. People in Paris, where it was late night, were talking to people in Vancouver, using their phones to cross the language barrier. A group of Chinese girls giggled as they compared haircuts with a group of French girls. Close by, there was another *Change The World* poster, this time with a photograph of Helen Keller, the American woman who was born both deaf and blind, who went on to become a prolific author and a famous social change activist.[276] And printed on the photo were these words of hers:

> *Optimism is the faith that leads to achievement.*
> *Nothing can be done without hope and confidence.*

They had an effect, these posters. Next to the screen a vendor tossed and flipped her crepes before smothering them with chocolate and selling them to happy buyers. A man reached into his backpack and pulled out a package that unfolded into a bowl. I learned later that Vancouver had banned all Styrofoam, as well as plastic bags and plastic water bottles.[277]

And then, bam, right in the middle of the intersection at Hastings and Seymour there was this huge sculpture of a life-sized woolly mammoth, with a baby mammoth by its side. A sign told me that the mammoths had lived in North America for a more than a million years, surviving several ice-ages when much of the land had been covered with two kilometres of ice. They were standing on the back of a massive turtle, surrounded by bronze salmon in a circular river.

And then another surprise—the sign also said that scientists had successfully cloned a mammoth using DNA from a baby mammoth that had been exposed by the melting permafrost, and mammoths were once again roaming Russia's tundra. That took some pondering.[278]

Looking east, the entire width of Hastings Street was both car-free and bike-free. A green walkway had been laid into the middle of the road, and the street was shaded by trees. I stepped onto the path, joining several others, including a woman who was waving a feather while burning sweetgrass, casting blessings as she walked. The shops were busy, with outdoor seating and displays of flowers and shrubs. The path down the middle felt powerful.

At the next intersection there was a massive sculpture of a canoe in a vivid red and black design, carrying twenty people wearing woven cedar cloaks and hats, their paddles upright. The sign explained that humans had arrived here 15,000 years ago, paddling along the coast of Beringia (modern day Siberia and Alaska) when the sea level was much lower, enjoying the plentiful seafood along the way. '*In the beginning there was nothing but water and ice and a narrow strip of shoreline,*' the sign said, attributing the words to a Bella Bella First Nations oral tradition recorded by the anthropologist Franz Boas in 1898.[279]

The sculpture at the next intersection showed a family gathering clams. The sign told of the enormous variety of seafood the First Nations people obtained from the sea, and how they had built clam gardens to harvest the riches, piling up long walls of stones and boulders at the low tide mark of a bay to create a safe place for the clams, enabling them to harvest the clams with ease.[280]

Continuing east the trail came to four huge totem poles with trees growing among them. Three poles represented the people who had lived on these lands: the Musqueam, people of the River Grass; the Tsleil-Waututh, people of the Inlet; and the Squamish, people of the Sacred Water. The fourth pole represented the other First Nations who had been here, trading and sharing stories. This was long before there was civilization in Mesopotamia, long before the pyramids, long before Abraham, Moses and the philosophers of Greece.[281]

The sculpture at the next intersection was more intimate, and it told a tragic story. It showed a First Nations father carrying his dead child, his eyes looking up to the sky, asking '*Why?*' The sign explained that soon after the white people arrived on these shores they brought smallpox, which cut through the aboriginal peoples, killing as many as half, both here and all the way up the coast and into the interior. For the survivors, the pox brought horrible facial scarring, the loss of friends and loved ones, and a fear that the whole world had turned against them. It was the beginning of dark times, as it had already been for indigenous people right across the Americas, from Nunavut to Patagonia.[282]

The next sculpture told another tragic story—a First Nations mother desperately holding onto her daughter as an impersonal hand pulled her away. This was the period when their children were taken forcibly from their homes and sent to residential schools, usually far from home, where many suffered years of loneliness, hunger, cruelty, sexual abuse and sometimes even death. Their languages, cultures, spiritual rituals and sometimes even their families were driven from their hearts, and many were defeated, leading to years of poverty and despair, softened only by alcohol.[283]

The sculpture at the next intersection showed twelve people seated in a Healing Circle. They were framed by a square in the colors of the four directions: white in the north for healing of the mind, yellow in the east for healing of the spirit, red in the south for healing of the heart, and black in the west for healing of the body. In the center a powerful sculpture connected above to the heavens, below to the Earth and within to the Great Spirit.[284]

As I watched, the people I was walking with paused and placed their hands on the shoulders of the sculpted figures in the Healing Circle. As I joined them, I felt a wave of compassion fill my heart. There were ten of us, each behind one of the seated figures, then two more people stepped forward to take the empty places. A woman put her arms around the shoulders of the people on either side, and the rest of us followed. This sculpture was powerful. When we separated there were hugs between complete strangers.

Now I had eleven new friends. A man I had embraced was a social worker who had worked with First Nations families in the East Kootenays, beneath the Rocky Mountains, and he told me how moving he found the sculpture trail to be. He took off his glasses and offered them to me, explaining that they told the whole story as you walked along the trail. They were G-glasses, he said. When I tried them I saw that various spots on the sculpture had been tagged and when you focused your eyes on a tag it expanded to tell you more in pictures and words. The information felt overwhelming, however, so I handed them back to him.[285]

Walking together we arrived at the next sculpture, a number of cubes piled on each other with First Nations people working on and around them—building, nursing, teaching, fishing, farming, studying, parenting, filming, carving, each engaged in a fulfilling activity. The healing had happened, and new life had begun.

Finally we came to the intersection at Hastings and Main, in the heart of the Downtown Eastside, and the large old building known as the Carnegie Centre with its grand white pillars. Instead of a sculpture, the pathway led into a circular amphitheater in the middle of the intersection that had been excavated out of the ground and built up with seating facing inwards. Four carved wooden arches reached upwards, holding a roof that contained lights for evening shows. The perimeter sloped upwards from the road, with a break where the pathway entered. A saxophonist was playing and people were sitting around eating lunch, talking to each other. I had just completed a fifteen-thousand-year journey.

Back in my time, this area was Vancouver's poorest, most downtrodden neighborhood, attracting migrants from China and Vietnam and First Nations people who had been displaced from their land and culture. It was a community mired in poverty, mental illness, drug addiction and grief, constant unresolved grief.[286] Someone once said it was the land that love had forgotten, the place you fell to when there was nowhere else to fall.[287]

And yet it had also been a community where people found friendship and comfort. There had always been a determination to heal and make things better. And now, the old eight-story rooming hotels for people on the lowest incomes that used to be such dismal places had been painted in bright colors, and were obviously being cared for. Some had been demolished and replaced with new buildings, while others were being deconstructed and rebuilt, with large cranes towering above them.

Among the people sitting on benches or playing cribbage some looked much older than their years. Some used walkers or wheelchairs and some were missing a limb, but there was an atmosphere that said, 'This place is still ours. This is our home and community.' The neighborhood felt strong, while still being home to Vancouver's poorest. I saw a woman sitting on a bench surrounded by buckets of fresh cut flowers, so I went over and sat next to her.

"Hello, my name's Patrick."

"Hello. Where have you come from?"

"I'm travelling, visiting. I've come to learn about the changes that have been happening here in Vancouver."

She looked at me, pondering my words. My guess was that she was about fifty. Her hair was completely grey.

"Can I offer you a coffee?" I asked.

"That would be nice." I went over to a food booth and came back with two coffees and two cakes.

"That's nice," she said. "Thank you. Thank you."

"Do you mind if I talk to you?"

"You already are...." She chuckled and turned to face me.

Over the next hour I learned that her name was Emily and she had been living in and around the Downtown Eastside for most of her life, with the occasional trip to jail and to the interior of British Columbia. From a young age she had been a sex worker, a drug addict and a drunk. Until recently, nothing much had helped except booze and drugs, she said. The social workers had tried, but there were times when she had spat in their faces.

I told her about my childhood in east Africa, and how nobody wanted to talk when they suspected something immoral or bad was happening in a family.

"Oh, I know that," she said. "Shove it under the carpet. Hide it in the attic. Anything rather than talk about it. It's the silence that kills. My uncle, he used to abuse me, just as he was abused at his residential school."

"So what changed? You seem peaceful now."

Things began to change around the time of the OMEGA Days, she told me. The streets were full of joyful fury and there was a feeling of change in the air. They all joined in, the people of the street. They were more than willing to support those who understood their poverty, and the miseries they lived with.

About a year later the rooming hotel where she'd found a place to live had been taken over by the Carnegie Center, a community non-profit. There were meetings, and more meetings. Then a crew arrived and started to clean up the bug-infested hole. They cleaned the entire building, inside and out, and when they were done it felt completely new. She started going to the Neighborhood House, where she could eat good food and learn how to cook instead of diving for food in dumpsters and surviving off junk food, which tasted great but left you feeling like the crap the food really was.[288] And the Food Bank had changed into a Community Food Centre, where they taught you how to cook a good nutritious meal.[289]

A year or two later, she said, they tore up the road and installed all the sculptures and the amphitheater. When it was done there was a big ceremony, and people planted vegetables in the new boxes. It was good that their community had finally been recognized, she said, and that this was her rightful home.

But when they said they wanted to tear down her rooming hotel, the Hazelton, and replace it with a new one, she was afraid. She didn't trust their promise that they'd re-house her in a motel not far away with the people who were her neighbors. But they were true to their word, and after two years she came back to a brand new building, free of the damp, mold and rats. As well as her own space with its kitchen and fridge there was a shared space on the ground floor where they could cook and eat together and watch movies, and a library with computers and books. There were monthly meetings, and they even had their own currency, the Hazels, that they used to trade within the building.

But the thing that made an equally big difference, she said, were the changes to welfare. She now received the monthly Citizen's Income, no questions asked. Everyone who's Canadian gets it, she said. That had made life a lot easier for people in the Downtown Eastside. For those who also needed welfare, the office was run by people from the Carnegie Centre. "It's so much better than it used to be. Trying to get welfare used to be like a game of snakes and ladders, only it was all snakes and no ladders."

As we were speaking a man came up and gave her ten dollars for a bunch of flowers. She gave him his change and he bent down and gave her a kiss on the cheek.

"Nice customer, Franky. Comes every week. Where was I? Oh yes—the welfare office. It's got a fancy new name now: the Bud Osborne Centre for a Creative Future. He was one of our local poets.[290] I was even given a man to help me: Peter, God bless his soul. He stuck with me as I started to undo my troubles. Lovely man, he was. He had his own demons, but he found me my room in the Hazelton, and then asked if I'd like to help with the garden they were creating.

"I liked Peter, so I agreed to help. Then he asked if I'd like to join a prayer group. I'd been raised in Jesus, but I had a lot of anger against any kind of religion. I only agreed to join because I trusted Peter, but I was pleased to find that it was, like, different. We didn't use any books or formal stuff. We just talked about our lives, and he taught us how to pray. Why am I telling you all this? Here's you, just bought me a lovely coffee. You don't need to be hearing all this."

"Quite the opposite, Emily," I replied. "I understand about the church thing. There's a lot about religion that I don't like either, but we all need a connection to something greater, to the Great Creator, or whatever's behind all this. We're not made to live our lives on our own. It doesn't make sense to live like that."

Emily patted me on the knee and went on talking.

"I was a bit unsure to begin with, all the religious stuff, but Peter helped me to see Jesus as a real person, not some person the priests shoved down your throat, along with their...."

Emily spat, cleared her throat, and continued.

"Sorry. I shouldn't have said that. Peter showed me that Jesus could be a real friend, and I found I could really love him. Today, whenever life gets a bit crazy, He's always there for me. I can't explain it better than that."

Encouraged by having a home and a source of support, Emily had been persuaded to go to a rehab center to deal with her dependency on drugs and alcohol. Peter had promised that he'd be there for her when she got back, so she had gone away to a center in the mountains run by a First Nations community where she had received the support she needed to open up her box of fears, and bring out the hurt and anger she had kept locked away for so long.[291]

I was surprised at how willing she was to tell her story. The times she'd been abused, by her uncle, by a priest, and by one of her foster parents; her teenage binge-drinking; the beatings she'd taken; her three daughters, all taken away by child welfare; and the awful, awful emptiness she'd felt when they were gone. Her attempts at suicide; and the home she had found here in the Downtown Eastside among people who didn't judge her. But it had also become her prison, and she never knew when she might wake up in an alley, a stranger's car, or perhaps not wake up at all.

When she returned to the Downtown Eastside after her time at the rehab center, Peter was gone. He hadn't abandoned her; he had died. But she felt abandoned. When she went to his grave, she said, it was as if her heart had been ripped out all over again. She would have gone back to her dealer if the people from the Neighborhood House hadn't been there for her. They seemed to know what she was going through.

"They looked past how I looked, and they saw the real me that was still inside me," she said, looking at me with tears in her eyes. "How I cried, when that happened. Like a baby. They gave me a new woman to take Peter's place, an angel called Lorinda, one of Judy's Angels."

"Judy's Angels?"

"That's what they call them. It's like having a friend on the inside, someone who understands you. Lorinda would appear every few days and help with the things I needed. I don't know who that Judy is, but she must have been someone very special for the angels to have been named after her."[292]

Lorinda encouraged Emily to join Alcoholics Anonymous, where she found new friends and the support she needed to stay sober and drug-free. She also enrolled in a weekly circle where people from the Downtown Eastside discussed the neighborhood, and things they wanted to change.

"Lorinda even found me a job!" she chuckled. "But first, she helped get my teeth fixed." Emily grinned and showed me a set of good white teeth. "You should have seen them before," she said. "They were horrible, and they hurt when I ate. But who could afford a dentist? But Lorinda, she found this place where student dentists work, and where a regular dentist comes twice a month to give free treatment out of the goodness of his heart. Made a big difference, that did.[293]

"Then they gave me a job in a community garden. That was when things really began to change. I felt important for the first time in my life. We grew all sorts of food—salad greens, beets, carrots, onions, peppers, strawberries, even lemons and figs. Michael Ableman and Seann Dory—they were the ones who started it. Wonderful men. We grew food vertically too, up fences and frames. Never thought it would work, but as long as you water them the food grows just fine. And you know what we were growing on? Waste land that had sat empty for years. We grew everything in boxes, so if someone came along and said we had to go, go we could, boxes and all. Waste land.[294]

"In those gardens, I had work I was proud of. We were doing good for other people too, growing healthy food. I did that for a year, and we were selling flowers from the garden, and then I got this idea to start my own business. So I went to the people at the Carnegie Centre and they got me into a group with four other women, all starting our own businesses. We helped each other, and I was able to get a loan to buy a fridge to store my flowers and keep them fresh. So, here I am! Some of my customers pay me in Eastside Hours, which I like, because I can use them here in the neighborhood."[295]

Emily looked proudly at her buckets of flowers. "I'm finally learning to be content with my life. I'm tired of being angry all the time, blaming the rich, blaming the government, blaming the people who tried to help me...." Her voice trailed off. "And you know what? I'm even going to college! The community college—they've opened a branch where people like me who never finished school can learn things like philosophy, history and business skills.

"They've done a good thing here. I don't know who *they* were, but they've given me back my life. Every morning when I wake up I say a prayer to Peter and Jesus and Lorinda and I thank them for what they've done. I used to think I didn't deserve it, but someone must have thought me worth saving. Maybe God does still love me, after all."

I took Emily's hand and squeezed it. My heart was caught up in gratitude, but I could feel how close her pain was. I thanked her and gave her a kiss on the cheek, the way her flower-customer had done. She had given me more than I had given her, and I gathered the courage to tell her.

"You're a very sweet young man," she said, patting me on the arm. "Maybe there's hope for the world after all."

 13

Cape Danger

I HAD FIFTEEN minutes before my meeting, so I cycled further east along Hastings Street to see how things had changed. Back in my time it was a desolate, bleak part of the city, just a dull road with low buildings, few successful storefronts and no sense of care or affection.

The transformation could not have been more complete. The road had been reduced to two lanes of traffic, with bike-lanes and trees on either side. There was new mixed-use development, and among the new buildings some were 100% rental housing, with shops and community organizations at the street level. It was a pleasure to ride along.

Then outside a store I spotted another *Change the World* poster, this time with a photo of Vandana Shiva, the Indian woman who had done so much to make the world more conscious of the need for safe, healthy, organic food, accompanied by these words: [296]

> *We are either going to have a future*
> *where women lead the way to make peace with the Earth,*
> *or we are not going to have a human future at all.*

They were inspiring, these posters. Heading back to Gastown I parked my bike and climbed the stairs to the Vancouver Green Economy Institute on the second floor of an old building on Water Street, with a view across to the mountains of the North Shore. Li Wei-Ping was a jovial man in his mid-forties with dark-rimmed glasses and a mop of shiny black hair. He welcomed me in and offered me a cup of green tea.

"I'm fasting in solidarity with Climate Compassion, but I'm allowed liquids. How about you?"

Me, who had enjoyed a good breakfast, goodies at The Future Café and a cake with Emily? Guilty as charged. "In truth, I didn't know the fast was happening until a few hours ago. We passed the big gathering on Yukon St."

"Have you signed the petition? We need ten million signatures globally by Sunday night." Wei-Ping picked up his screen, found the page and handed it to me. The layout was simple so I entered my name, putting Sudan as my home country.

"We're really applying the pressure this weekend. It's so critical that we get a breakthrough. But before we get started, how's Dezzy doing?"

"She's great! She's been really helpful."

"Did she tell you about her brother Derek? It was before I arrived in Vancouver, but everyone was talking about it. It's the only political assassination Vancouver has ever known."

"Yes. She has a shrine to him in her home. Isn't it a bit suspicious that they've not arrested anyone after all these years?"

"That's what we all think. My uncle's a good friend of Liu Cheng, who was Chief Constable of the Vancouver Police Department at the time, so we've talked about it a lot over the years."

"And...?"

"From what he says the police were taken totally by surprise. They'd been droning the OMEGA protests and identifying the ringleaders, so Derek was well known to them, but they'd expected violence *by* the protestors, not *against* them, least of all from a precision-guided sniper. But that's no excuse. How can his killer leave no trace at all apart from the spent bullet casings? It's never sat easy with me. No wonder there are so many conspiracy theories."[297]

"Do you have a theory yourself?"

"I don't think it was anyone ordinary like an American gun nut or a pro-lifer. They're not professional enough for that, and besides, they like to boast. I'm more inclined to think CIA, or the Israeli Secret Service. I wonder if Derek had information someone really wanted to remain hidden. He spent time in Israel but he never wrote about it in his blogs. He was in touch with all sorts of people, and he might have learned something dark.

"You should talk to Laura MacGregor," he continued. "She's doing an internship with us. She's offered to take you over to visit the Triko-Op later this afternoon. She's the daughter of Donald MacGregor, a retired Scottish police investigator who's a good friend of Liu Cheng. They worked on Derek's assassination together. Liu Cheng asked Donald to help since he'd worked with MI5 in Britain and he had experience with this kind of thing. He was in Vancouver just last week, visiting Laura. Said it was a family visit, but you never know. But you didn't come here to talk about this. Dezzy says you want to learn how Vancouver became one of the world's greenest cities—is that right?"

"Yes. I also want to learn what's been happening to address the climate crisis, if you've got the time."

"Ah, yes. It's very confusing, global warming. It's the middle of June, and here I am still wearing a sweater."

"Why is that?"

"It's because it's so warm in the Arctic. It creates a mass of high air pressure that is disrupting the polar jet stream.[298] Last year it brought record heat and drought and the worst year for forest fires on record. There were weeks when the sky was creepy yellow from the smoke from all the forest fires and the Sun was a strange yellow orb, as if we were in a different solar system. Our family cabin burnt down, and hundreds of people in rural areas lost their homes to the fires. This year it's colder than normal. It's all out of whack. So while it's over 25 degrees Celsius inside the Arctic Circle it's a relatively chilly 15 degrees here. So I tell people, if you don't like the weather go up to Resolute and join a Trans-Arctic kayak expedition. Last summer the Arctic was ice-free for a full twelve weeks. At this rate it will be ice-free in *mid-winter* by 2050. The few polar bears that remain will have to spend the whole year at the feeding stations, and hope that their offspring evolve back into brown bears."[299]

"What does it mean for the world that the Arctic is melting so rapidly?"

"Well, it's looking really ominous for northern Europe, because with all the ice that's melting on Greenland, the added fresh water is diluting the salinity of the water in the North Atlantic, the turnover of which drives the Gulf Stream, which keeps northern Europe warm. So they've been having some chronically cold winters over there, caused, paradoxically, by global warming.[300]

"On the positive side, Arctic fish stocks are booming, since the sunlight on the ice-free water causes the phytoplankton to bloom. Apart from that, I wish I could be hopeful, but to do so I'd need to be blatantly dishonest, like the reporters on *Tiger News*. Last October they ran a story on how the Arctic ice was increasing. In October! Maybe their journalists are so badly educated they don't know how the seasons work.

"Let me put it this way," he continued. "There's an iceberg ten times the size of Manhattan in the waters off Nova Scotia right now, heading for New England. Some climate scientists think we passed the tipping point years ago, around the time of the OMEGA Days. That would be wretched if it's true, considering how much we've achieved since then. There's seven metres of sea-level rise locked up in Greenland, and if we have passed the tipping point we could see a three metre rise by the end of the century, with much more in the years to come.[301]

"The data from Antarctica is far more troubling. We passed the tipping point there twenty years ago, so there's a guaranteed five metres of sea-level rise coming over the next few thousand years, and as much as twenty metres if we don't get our shit together—and it could come a lot earlier."[302]

"Five metres?" I replied with shock. "Twenty metres? How will people possibly be able to cope with a sea-level rise that high?"

"They won't. They'll have to abandon all the low-lying areas of the world. Unfortunately, that includes many of the world's major cities—London, New York, Venice, Jakarta, Mumbai, Miami, Kolkata, Shanghai, Tianjin...."[303]

"I thought London had a tidal barrier."

"It does, but it's designed to stop the sea getting in on a temporary basis. If they left it closed the river Thames would have no way to get out. Here in Vancouver that much sea-level rise would put the whole of Richmond, Delta and Tsawwassen under water, along with the airport, all the docks along the Fraser and the highway to Seattle."

"Doesn't it make you depressed, that this much sea-level rise is inevitable?"

"Yes, it does. But the faster we stop using fossil fuels and cutting down the rainforests the easier it will be for future generations to cope.

"There's also all the methane that's leaking out of the melting permafrost," he continued. "And to think that some fossil fuel crazies were seriously proposing to mine the frozen methane hydrates. It's good that the Arctic has finally been declared a Global Marine Sanctuary, but we're not moving fast enough. We've shown that it's possible to live without fossil fuels here in Vancouver; now we need the rest of the world to do the same. You've seen how we live. Do you think we're suffering?"[304]

"Not at all. The city seems to be flourishing."

"We're making good progress globally, but nations should have started reducing their carbon emissions decades ago, and now we're dealing with the consequences. If we can accelerate the global effort we may yet make a relatively safe landing. If not, the temperature could rise by as much as six degrees Celsius, which is far, far outside every known boundary for human and ecosystem wellbeing. There's no evidence that we could adapt to that, no matter how many insulated bubbles people think they can live in.[305] We need to complete the global course correction we've started. Twenty years ago we were heading for the cliff, taking much of nature with us. Now that we're halfway through the course correction people can see a different future ahead, and they like what they see. More and more, they want it, but if we falter we'll still go over the cliff."

"Do you think we'll be able to make it?"

"To be honest, I don't know. But it's not about being optimistic or pessimistic. If you're a hockey player, there's only one thing that matters, which is are you determined or defeated? You don't sit around in overtime wondering if you're going to win."

"What about other cities? Are there others that have been as successful as Vancouver?"

Wei-Ping took a sip of tea and then said, "Let me show you. We've just acquired one of the new I-Balls. It took me a while to get the hang of it, since it works in three dimensions."

He cleared some papers, made a gesture with his hand and a black sphere descended from the ceiling.[306] He tapped his screen and a vivid image appeared of the Earth as seen from space.[307]

"Greenest cities," he said, and a bunch of green lights lit up around the planet.

"Starting in Europe, we've got Copenhagen in Denmark; Amsterdam and Groningen in Holland; and Oslo, Stockholm, Målmo, Vaxjo and Helsinki in Scandinavia. In Germany there's Berlin, Hamburg, Frankfurt, Munster and Freiburg; in Switzerland there's Zurich. In Britain there's London, Newcastle, Bristol, Brighton, and Kirklees in Yorkshire; in France there's Paris, Nantes in the west and Besançon in the east. Europe's also got Vienna, and in Spain there's

Barcelona and the little-known town of Vitoria-Gasteiz in the Basque country, close to the Pyrenees."

Wei-Ping waved his hand and the globe turned so that we were looking at Asia. "Here we've got Singapore, Seoul in South Korea, and in China there's Dongtan, Baoding and Tianjin Eco-City southeast of Beijing.[308]

"In Australia there's Brisbane and Melbourne, and in New Zealand there's Christchurch, now that they've recovered from the earthquake. Let's swing round to the Americas. In Brazil there's Belo Horizonte, Brasília, Curitiba, Rio and São Paulo; in Colombia there's Medellin and Bogota; and in Argentina there's Buenos Aires.[309]

"In the US there's Portland, San Francisco, New York, San Diego, Austin and Seattle. Here in Canada there's also Toronto and Victoria. Vancouver's a founding member of the Global Alliance of Green Cities, so we have good relations with most of these places. Do you want to see the other side of the story, the regions at the greatest risk?"

"For sure."

"Okay. We'll do heat-waves first." At Wei-Ping's touch, the planet lit up with yellow and red patches. "All of these places are experiencing a month or more when the temperature is above 40 degrees Celsius. Look what happens when I advance it to 2050, then 2080."

The result was alarming. The number of red and yellow areas doubled. Australia became a massive red zone. The whole Indian subcontinent was colored red and orange.[310]

"These are the places with extreme drought." He pulled up an image that showed large areas glowing red and yellow—including the entire North American west coast from Panama to northern British Columbia, most of south America's west coast, great swathes of western and southern Africa, the whole Mediterranean basin, the entire Middle East from Turkey to Afghanistan, most of southern China and most of Australia. It was really alarming.[311]

"That shows British Columbia, as well as California," I said. "How has it been affecting you here?"

"It's really rough on the farmers, especially on Vancouver Island, and it's a huge problem for the salmon. In some years some of the creeks and rivers have been so low and warm that the salmon fry can't survive, in spite of the volunteer efforts to save them. It's been far worse in California. Half of their agriculture has been forced to shut down and the other half is on a stringent new water conservation regime, with a new system of water pricing and tight controls over groundwater pumping. The Sierra Nevada snowpack has disappeared entirely, and it's possible that within ten years there will be no food exports coming out of California at all.[312] Groundwater is being drawn down all over the world in response to drought, and usually with no replenishment.[313]

"These are the projections that got the Chinese leadership so alarmed," Wei-Ping continued. "The Chinese know from their history that drought and famine caused the downfall of the Tang, Yuan and Ming dynasties. The leadership is very conscious of this, and they're really worried about the melting glaciers in the Himalayas that feed China's major rivers. They don't want the Communist dynasty to fall as well."[314]

Wei-Ping went on to show images for cities and regions that were at risk from sea-level rise, hurricanes and cyclones. New York was spending billions to protect itself from the rising ocean, he told me.

"It's really troubling. As I was saying, if the world fails to achieve the full carbon phase-out and the successful drawdown of the excess carbon from the atmosphere every coastal city on the planet will eventually face a sea-level rise as high as twenty metres. We know it's possible from the geological record—but who can handle a sea-level rise that high? It's just not doable."[315]

It was a shocking reminder of the future we faced if we didn't get the climate crisis under control.

"What makes Vancouver such a green city?" I asked, changing the subject.

"We're scored on a set of social and environmental indicators that include energy, ecology, food, poverty, homelessness, equality, democratic engagement and happiness.[316] There was quite an argument about making it so wide. Some people said it was getting too political."[317]

"How many points did you need to win?"

"We shared the top score of 72% with San Francisco and Copenhagen, so we still have a way to go. If you look at the world as a whole the data shows that we're only 40% of where we need to be if we're going to make it around Cape Danger and into the calmer waters of a peaceful green future. The population is still rising, though it looks as if it will stabilize at around ten billion and then start to decline. It's still a very troubling number, because every new baby puts an additional demand on the planet's strained resources."[318]

"What are the major successes that have enabled us to get this far?" I was feeling my way through the discussion, hoping to learn about the positive as well as the negative aspects of change.

"Well, for a start, most of the world's farmers are going organic. Thanks to the motivational power of the 2040 Imperative—the pledge that many nations have taken to get as close to 100% renewable energy as possible by 2040—global emissions are down to half of what they were twenty years ago, and we're steadily ramping up the eco-sequestration. It looks as if the CO_2 in the atmosphere will peak at 425 parts per million within a couple of years, and then start to decline. In the long run it means the temperature should start declining too, but there's a long time lag, and every indication that it's going to pass two degrees. If that happens, and if the methane and the other feedback mechanisms have the last laugh, it could still rise by four, five or even six degrees, which would cause the collapse of civilization and the loss of every species that can't evolve fast enough to adapt to the heat."[319]

This was alarming, but carbon dioxide peaking at 425 parts per million? That was very encouraging. Back in my time, if things continued the way they were, we would be approaching 450 parts per million by now, the level the climate scientists said we absolutely had to avoid. This was compared to just 280 parts per million in the years before the Industrial Age. So something pretty big had changed. I had no time to ask, however, since Wei-Ping was pressing on.[320]

"On the other hand, we've yet to get a handle on deforestation in Indonesia and Central Africa; we're still assaulting natural habitats around the world; the ocean[321] is still in peril from multiple causes; and we're a long way from solving the fresh water crisis, although water recycling and solar desalination are helping.[322] And we're still over-consuming, eating away at Earth's habitats. There are four billion people who are just one step out of poverty, and they all want to go shopping. And who can blame them?"

"It seems like a massive problem," I said.

"I've always been fascinated by the progress of civilization," Wei-Ping continued, shifting into a more philosophical mode. "It probably goes back to my Chinese ancestors, and our 8,000 years of civilization. Even the Mongols couldn't break us. We turned them into civilized beings who loved ceramics, silk and poetry. Ha! But when you look at the reasons why civilizations fail, I admit to having sleepless nights. I worry that we're just not changing fast enough, and that the shadow of collapse is still bearing down on us.[323]

"The great British historian Arnold Toynbee thought there were two major reasons why civilizations fail: excessive concentration of wealth in the hands of a few, and the inability of the elites to introduce significant changes—and that's exactly what was happening in the decades leading up to the OMEGA Days. But in spite of all the changes there are still too many people who have too much wealth, too much power and too much sense of their own importance. They're behaving like the aristocrats in France before the French Revolution, thinking they have the right to take as much as they want regardless of the social or ecological cost. I've a friend in Victoria who has developed what he calls The Guillotine Index, measuring people who are egregiously selfish and shortsighted. I still don't know if he's joking or not.

"There's another book I was reading recently called *Why Nations Fail*, by two American professors. They concluded that nations fail when their political and economic institutions fail to be inclusive, shutting people out of participation. I would add a third factor, that they must be ecologically inclusive, for by all the laws of physics it's impossible to have perpetual physical growth on a physically finite planet. Can you imagine what would happen if you or I kept growing forever? We'd be seven metres tall by the time we died. How crazy would that be?"[324]

"Yes, that would be pretty weird."

"Do you know the story about the man who lived in ancient Babylonia in the year 3,030 BC who had a cubic metre of all the goods he had accumulated? He wanted to know how big his pile would be if his ancestors continued to grow the pile at an annual compounded rate of 4.5%, so he asked a priest to calculate it. The priest told him that after five generations his pile would be 82 cubic metres in size. 'That's not enough!' he said. 'I want my memory to live forever!' So the priest went away and did some more calculations that showed that in 628 years his pile would be the size of the entire planet. 'But what about the stars?' the wealthy man asked. 'When will my pile reach them?' The priest bowed his head and admitted that he was unable to answer, fearing for his life. But today, with just a pocket calculator, we can work it out. After a thousand years his pile would be the size of 13 million Earths. After 1,600 years it would be the size of four galaxies. After 3,000 years, by around the beginning of the Roman Empire, it would be the size of 2.5 billion billion solar systems. That's the power of compound interest. And yet we still have leaders who believe we can grow the global economy forever at three or four per cent a year. It's a strange, stupid delusion. Here in the Institute we call GDP Gross Depletion of the Planet, instead of Gross Domestic Product. It tells a more honest story."[325]

Wei-Ping tapped his screen, waved his hand a few times, and his I-Ball showed a cartoon image of the Earth being eaten by an overweight man.[326] "That about says it, don't you think? It's good that so many countries are using the Genuine Progress Indicator (GPI) alongside GDP, since it includes real things like income distribution, education, leisure time, and social and ecological capital.

"Before GPI became widely understood it was impossible to have a discussion about economic growth without people assuming that you wanted to destroy the economy. That was the single biggest obstacle when it came to climate change. The politicians said, 'We'll tackle it as long as it doesn't disrupt the economy.' Now that most people understand that two degrees Celsius is such a dangerous line to cross, and that genuine progress matters far more than plain economic progress, the debate has become much more intelligent."[327]

"But if you don't have economic growth won't that mean more unemployment, more poverty and more people losing their homes?"

"That's what people used to think. But when The New Economy Team looked at the jobs that would be lost or gained in a planned transition to 100% renewable energy they discovered that during the transition there would be almost twice as many jobs created as there would be lost, and by the end of the transition there would be as many jobs as there were before."[328]

"What kind of jobs?"

"First there's all the jobs building solar, wind and geothermal installations to replace the use of coal and gas to generate power. Then there's the ongoing work to retrofit every building to make it more energy efficient and to replace oil and gas with heat pumps and district energy. There's all the work to electrify the railways, and install solar along the tracks. There's work creating all the new bike lanes and rapid transit lines, and there's been a boom in bicycle tourism now that there are so many safe bike lanes. It goes on and on. There are more jobs in farming too, because of the shift to organic farming.

"So about fifteen years ago, as the news about jobs began to circulate, there was a shift in attitude. Instead of fearing the transition, people began to push for it."

"That's great! But doesn't that mean the economic growth you were just attacking is continuing?"

"It's not as simple as that. What matters is that we end the growth of material consumption, which has such a heavy ecological footprint. It doesn't matter if there's growth in music and the arts, in intellectual work or ecological restoration—we *want* that. Our analysis shows that in a world that operates on 100% renewable energy, with 100% organic farming, 100% green chemistry and 100% materials recycling, with a predominantly vegetarian diet and a fully circular economy, the human ecological footprint falls way down to a sustainable level.[329] The issue we really need to address when it comes to jobs is automation. In a fair, rational world we'd have much more work-sharing, and more people would enjoy a four-day working week."

"But is there still time, given the danger of the climate threat?" I asked. "Or are we going to follow Easter Island on its road to ruin?"

"Easter Island's problem was its remoteness," Wei-Ping replied. "The first Polynesians to get there had to paddle or drift for more than three thousand kilometres on the open Pacific, and it's a further four thousand kilometres to the coast of South America. So for six hundred years it's possible that the Easter Islanders never met a single other human. For all they knew they were the only people alive on the planet, and when things started to go downhill they had no-one to turn to.

"There's an island on the other side of the Pacific called Tikopia that faced a similar crisis, but they handled it very differently. Instead of cutting down the last tree and reverting to tribal warfare and cannibalism they enforced a policy of zero population growth, banned all the pigs from the island, and developed forest permaculture. When the anthropologist Raymond Firth worked there in the 1930s he reported that they were very proud of their culture, and were holding together well."[330]

"So why the difference?"

"The Easter Islanders were so alone, they seem to have given up hope. Their syntropic response mechanism collapsed, so they turned to magic, building the Moai, the huge stone statues. The Tikopians, on the other hand, were only three weeks sailing from Vanuatu and the Solomon Islands, so they were an active part of the wider Polynesian culture. Their syntropic response mechanism kicked in and they did what they needed to survive."

"What's a syntropic response mechanism?"

"It's life's inherent positive response to threat, which propels the drive to increased cooperation and integration. It's as real for humans as it is for atoms and molecules. When you were at school did they teach you about the Club of Rome's computer models that looked at *The Limits to Growth,* back in the 1970s? The models predicted some pretty dire outcomes if we continued the way we were. The latest World7 computer model includes gradients for information, connectivity and responsiveness, which are the precursors of the SRM—the syntropic response mechanism. You can take countries with very similar data for food, population, energy, pollution and economic growth, but a variable SRM predicts highly divergent responses to an emerging crisis. Some countries go into collapse mode, while others rise to the challenge. Easter Island had a very low SRM, perhaps as low as zero. Tikopia had a very high SRM. The early environmental models that predicted ecological collapse only looked at the material factors. That was the way science was in those days: strictly material, to the exclusion of the spiritual and psychological dimensions of change."[331]

I found Wei-Ping fascinating, though I still wasn't clear what syntropy was. "So what do we have to do to make a safe passage around Cape Danger?" I asked.

"We've got to continue to raise our global game. The methane that's leaking out of the permafrost is a dangerous wild card. Molecule for molecule, in the short term, it's trapping eighty-four times more heat than carbon dioxide.[332] It's great that countries are reducing their use of fossil fuels, but most of the carbon dioxide that has been released over the past century is still up there. A fifth of it will continue to trap heat for hundreds of years to come unless we can speed the rate of sequestration, and suck the surplus carbon dioxide out of the atmosphere."[333]

Wei-Ping tapped a few keys and pulled up a graph that showed the atmospheric CO_2 rising steadily, then flattening off. "Like I said," he continued, "if the current progress continues it should stabilize at 425 parts per million and start declining by 2034. If the reductions in fossil fuel use continue and the eco-sequestration efforts perform as projected we hope that it could fall back to 350 parts per million by the end of the century, just as the organization 350.org has been campaigning for all these years. So things are much better than we thought they'd be twenty years ago."

"When I was growing up my parents were really worried about climate change," I said. "What caused such a big turnaround?"

"Well, I can start by telling you what happened here in BC," Wei-Ping replied. "It began before I arrived. The leader of the province's New Democrats had been reading some troubling reports about the unsustainability of the current model of economic growth and the likelihood of an economic collapse—similar to 2008, but worse, since this time there'd be no public willingness to bail out the banks. So he established a New Economy Team and invited some of the smartest people to join it, including Derek. He asked them to prepare the foundations for a new economy that would be sustainable and resilient, both financially and ecologically, that would get as close as possible to zero carbon by 2040. He was thinking ahead to the next election, which he was determined to win.

"They were deep into their work when the crash happened, making it a matter of urgency to win the election and get the New Economy Transition Plan into play.

"The Plan had a close fit with the OMEGA solutions, which were laid out a year later. Derek was involved in both. For climate change, the New Economy Team showed how we could get to 100% renewable energy for most heating, transportation and industrial needs by 2040. Electricity was never the problem, since most of our power here comes from hydro and wind. Solar contributes about 10%, tidal energy adds a bit and the new geothermal plant is adding base-load power."

"Dezzy was telling me how her solar pays for itself."

"The money that a solar homeowner saves over thirty years can be ten times more than the investment.[334] It was eliminating the fossil fuels from buildings and transportation that was the challenge, along with ending fossil fuel exports. When the New Democrats became the government in coalition with the Green Party, they ramped up the carbon tax and used the revenue to finance a huge drive to make building energy upgrades easy and encourage the switch to heat pumps and district energy. They also made a big investment in cycling and transit and increased the incentives to buy an electric vehicle, including free parking and free use of the ferries and HOV lanes. These are all part of the reason why Google chose to locate its new zero-waste, zero-carbon, circular economy data-centre here, cooled by seawater.[335] They also joined with the other provinces in persuading the federal government to ramp up the carbon fuel standard, requiring all new cars and light trucks to produce zero emissions of carbon dioxide per kilometre by 2030. Since the average vehicle's life-span is ten years, by 2040 most cars in Canada will be electric."

"That's really positive," I said, struggling to take it all in.

"The policy changes were important," he continued, "but the real breakthrough came with the huge waves of public engagement that followed the financial meltdown and the OMEGA Days: the home-groups and street-groups; the Green Teams in schools, condos and offices; the Green Living Co-operatives; the rallies and meet-ups. A group of non-profits toured the province with a Better Future Roadshow, taking it to the rural communities that were the heartland of conservatism and organizing community circles to help people self-organize. People finally woke up. They understood how dangerous the climate crisis was and they realized that if they pulled together, they could build a new economy that would be in closer harmony with nature, while still providing jobs and improving the quality of their lives and the communities they lived in. There was a huge cultural shift as the new vision took root in people's hearts.

"There are not many people who want to walk a solo path through life," he continued. "Do you know the African proverb? *If you want to travel fast, travel alone. If you want to travel far, travel together*. The community circles and the climate clubs enabled people to take to the road together. I'm sure Dezzy has told you about the things they're doing on her street. Now imagine that happening all across the province."

"So what made the difference?" I asked. "Why did people engage *then*, but not before?" It was a really critical question, given the reality back in my time, where none of this stuff was happening.

"That's a good question. The OMEGA Days were the biggest thing. Maybe we needed the financial crash to motivate people to get off their backsides and start doing something. Overall, though, I'd say there were three things. The first was the personal leadership by activists, business leaders and politicians before, during and after the OMEGA Days, without which nothing would have happened.

"The second was the public's active embrace of the vision of a better future, without fossil fuels. Have you seen that movie, *Hope's Sister*?"

"No. Is it good?"

"It had a big effect. It ignored Hollywood's obsession with cataclysmic dystopias and laid out a powerful vision of a better future. It's the story of two sisters who were separated when they were young and raised in different families. One grew up to be sad and defeated while the other was inspired by her parents to make a difference. It showed their dreams unfolding in parallel worlds, and as well as the dark side, it presented a compelling vision of a positive, sustainable world. I still love that movie. It's my generation's equivalent of *The Sound of Music* that my grandparents loved so much."[336]

"I'd love to see it."

Wei-Ping chuckled. "The third factor was the roadmap. People needed to know there was a practical path that could get us there. That's where the New Economy Team came in. When you combine the leadership, the vision, the roadmap and the community organizing, the transition really took off, not just on the climate front but on everything: the economy, healthcare, education, growing more food, homelessness—you name it."[337]

"What about climate activism? Did that play a role?"

"Absolutely. The movement's biggest successes happened when there was a clear target, such as the pension funds that were invested in fossil fuels, or the proposed oil and gas pipelines. Have you seen what they've done with the coal terminal at Roberts Bank, which used to ship coal to Japan and Korea?"

I shook my head.

"The Tsawassen First Nation has turned it into a Renewable Earth Center with a solar farm and an underwater ocean heat recovery project that's pumping heat to new developments in the Tsawassen area. It's all built on stilts, to cope with the future sea-level rise. There's a beautiful dome with an exhibit that tells the story of the coastal First Nations going right back to the last ice age, and a powerful sculpture where the land sticks out into the ocean. They've opened up gaps in the causeway to let the ocean currents return and they're revitalizing the eelgrass beds to provide food for the salmon and the orca whales. They get loads of visitors, thanks to the new SkyTrain extension to the ferry."

"Are you saying there are no longer any coal exports?"

"None whatsoever. Most of North America's coal industry shut down several years ago, following one of the more dramatic moves of the twenty-twenties."

"Closed down?"

"Yes. Four American billionaires worked with Norway's Sovereign Wealth Fund to buy up the entire industry and close it down, taking care of the affected communities. It cost them $50 billion over ten years, and it sent a shockwave through the world that this kind of thing was possible.[338] Many of North America's big charitable foundations also prioritized action on climate change, providing the funding that helped fuel the transition.

"Canada and America had no real problem adjusting to the end of coal. They just had to ramp up their wind, solar and geothermal power and increase their energy efficiency and power storage. BC's coal exports were a bit different since they were metallurgical coal, which was needed to make metallurgical steel. But the Koreans and Japanese were keen to stop burning coal, and they pioneered a high temperature process for making steel using hydrogen that is now being adopted around the world.[339]

"The billionaires were just one small part of the change," he continued. "There was also a big 'Don't Bank on Climate Change' campaign that persuaded investors to move more than two trillion dollars out of fossil fuels and into clean tech. I remember my eight-year-old daughter coming home from school one day all excited because she had been part of a classroom protest demanding that the teachers move their pension fund out of fossil fuels."[340]

"When she was eight?"

"Yes. That really impressed my wife, Jiao."

"So is British Columbia close to its zero-carbon goal? And what about the rest of the world?"

"We're making good progress—we're about 80% of the way there. As to the rest of the world, that's a much bigger question. Have you got the time?"

"I do if you do. Otherwise, Vancouver's just an island of grace in a sea of trouble." I remembered the old man at the Future Café, and his fears about the future.

 14

Climate Progress

"WHERE TO BEGIN?" Wei-Ping said. "What's happening is enormous. It's a complete overhaul of the way the world obtains its energy and looks after its forests and farms, and far from being a drain on the economy, as the climate deniers and many others feared, it's renewing economies and creating jobs."

"There are sixteen forces that are driving the change: four motivators, four agents of change, four practical methods and four wild cards—at least, that's the way I present it in my talks. Give me a second...." He said something to his device and projected a diagram onto the wall, which I've done my best to recreate.

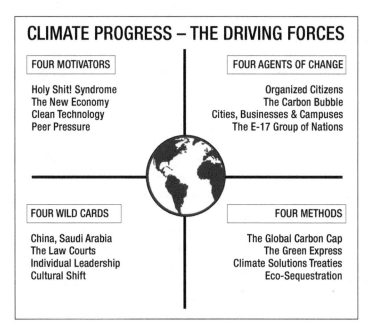

"Looking first at the motivators, the big one was what psychologists call 'Holy Shit! Syndrome.' We are good at emergencies. Whenever there is a fire, flood or tragic accident, almost without exception people abandon thoughts of comfort and do whatever it takes to help.

"When an emergency has a long fuse, however, as the climate crisis does, our emergency response mechanism becomes very muted. It runs more like this:

Deny, delay, procrastinate,
Dither, dither, blame,
Dither, delay, deny some more,
Delay some more, then—
HOLY SHIT!! What do we have to do?"

"That's so true!" I said with a laugh.

"Most humans hate fundamental change. We like our habits and routines, our Sunday walks and our evenings with the kids. There's an invisible notice pinned to most of our minds that says, 'Please do not disturb with unsettling information or ideas.' But when unsettling information does finally register, we scramble: we roll up our sleeves and get to work, since deep down we want to survive."

"So what finally fired the 'Holy Shit Syndrome,' as you put it?"

"It was the dawning realization that a 2°C temperature rise would be far from a safe climate landing, that it was going to be extremely dangerous and enormously expensive—and with the way we were burning fossil fuels there was no way we could halt the temperature rise at 2°C. That was when people began to get that awful sinking feeling in the belly as they realized that we were on track for a rise in temperature that would cause immeasurable suffering, followed by possible global collapse. What was the logic in protecting the economy from supposedly expensive climate solutions if the absence of those solutions was going to kill it?[341]

"The danger was easier to accept thanks to the second key motivator, which was the growing realization that the *faster* you phase out fossil fuels the *better* it is for your economy, and that the new economy is such a positive development. Not only does the transition off fossil fuels generate almost twice as many jobs as it loses; it also eliminates the healthcare costs caused by air pollution and the future wars that would have been fought over oil and gas.[342]

"The realization that a green, renewable energy future would be a *better* future made people *want* to change. It created the attractor that had been missing in the early years of the climate movement. It has been happening in the developing world too, where people have realized that a zero-carbon clean-tech world is a far better bet for tackling poverty than the continued subsidization of fossil fuels, with all their hidden costs. All in all, there has been a huge turn-around in the public's attitude.

"The third motivator is the inherent attractiveness of solar power, electric vehicles, green buildings and clean tech. Young people know instinctively that the new technology represents the future, while fossil fuels represent the past. So they are excited. They want to be part of it."

"What role does peer pressure play? You have it listed as the fourth motivator."

"Peer pressure exists all around us, even when it's unconscious. As the tide changed it became cool to install solar panels and to have a membership in a carshare co-op, and it became uncool to shop with businesses that made no effort to green their operations and to support banks that still invested in fossil fuels. It created a cultural shift in what's considered normal."

"What about the next box, the four Agents of Change?"

"The most important agent of change has been organized citizens, with their tens of thousands of protests, green teams and climate clubs in schools, campuses, churches and neighborhoods. It's they who have provided the vigor and the determination to speed the transition.

"The second agent of change was the carbon bubble. When the bubble burst, helping to trigger the crash, $5 trillion of carbon assets were written off, stranded underground, where I hope they'll remain. Shell lost $4 billion in its ill-fated fantasy to drill for oil in the Arctic. Petrobras, the big Brazilian operation, lost $300 billion. Most people—including many bankers—had no idea how many credit derivatives and junk bonds had been created to leverage the carbon delusion. Zombie investments—that's what so many oil and shale gas projects had become. Some of the big oil and shale gas companies were investing $4 in capex—that's capital expenditure—for every dollar earned. They were hit by a quadruple whammy, which helped trigger the crash: capex up, price of oil down. Price of renewables down, climate policy up.[343]

"The third agent of change has been the organized response by the leaders of cities, corporations and universities around the world. When they combined forces to launch a huge global campaign with advertising that showed why a new green economy would build a better future, and why they were divesting from their investments in fossil fuels, it carried a lot of sway. And they used their connections to pressure governments and persuade other cities, businesses and colleges to get on board.

"The fourth agent of change, finally, arose when the leaders of the most environmentally-minded nations put their heads together and formed the E-17 Group of Nations. Today it's become the E-70. The original seventeen were Denmark, Norway, Sweden, Finland, Iceland, Germany, Holland, France, South Korea, Japan, Ethiopia, the Philippines, Brazil, Bolivia, Ecuador, Costa Rica and Uruguay. They agreed to work together, and they took the lead in the global effort to build a greener world. Canada joined later, once we had a change of government. The US is still a holdout, but a number of states have participant observer status, including Washington, Oregon, California, New Mexico, Colorado, Nevada, Illinois, Vermont, Massachusetts, New York and Hawaii.

"The E-17's leadership was invaluable. They gave the world fresh hope regarding the ancient First Nations prophecy that told of a time when the planet would be poisoned by man's greed, and when people from all nations would unite to defend the Earth and its animals. One of their first actions was to order 17,000 electric buses, using the power of bulk purchasing to drive down the price. They made other countries wake up to the fact that they were accelerating in the race to build a new green economy. They broke the ice, which allowed other nations to follow."[344]

"How did the G-7 and the G-20 nations respond?"

"Three of the G-7 nations were part of the E-17, but no-one really remembers the G-7 or the G-20 any more.

"So next we have the four practical methods that have made the most difference. The most important by far has been the global agreement to place a cap on the future use of fossil fuels, based on the need to limit the cumulative loading of carbon in the atmosphere. The physics is simple: each additional tonne of carbon traps heat, so if we are to limit the overheating of the planet we must limit our carbon emissions. That was the real wake-up call, the sign that the world's leaders had finally agreed to stop pouring carbon into the atmosphere.

"Before the *Houston Accord on Climate and the New Economy*, where most of the world's nations agreed to adopt the global carbon cap, the coal, oil and gas barons had been happy to go along with vague commitments to reduce emissions. They had even been happy to pay a small carbon tax, knowing that they could pass it on and the effect would be small. But if a price on carbon was like a small fine, the carbon cap was the guillotine. The cap means that eighty percent of the Earth's known reserves of fossil fuels will remain underground, unless some future boffin can figure out a way to extract the energy as pure hydrogen, leaving the carbon underground."

"How did the big fossil fuel corporations respond?"

"They had begun to lose the fight when their investors realized that they would have to leave most of the reserves in the ground. When the crash happened, caused in part by the carbon bubble, there was a stampede to get out before share values totally collapsed. The market doesn't have a brain. It has a few intelligent antennae, but the rest is simply herd instinct. The big fossil fuel corporations had already seen their share values collapse as their values were downgraded by fund managers around the world, so they didn't have much fight left, and their investors were increasingly demanding that they plan for a steady transition, instead of trying to bullshit their way out of trouble."

"What was the cap that the nations agreed to?"

"It was originally negotiated at 600 gigatonnes of carbon dioxide, which was twice as much as the climate science said it needed to be. It was the best they could agree to at the time. Later, when politicians saw how well the public was responding, they reduced it to 300 gigatonnes—and it falls every year. It's still too high—that's one of the demands that the Fast for Climate Compassion is after.[345]

"It was the public's acceptance of the signs saying EXTREME DANGER AHEAD that drove the politicians, but most people would not have admitted the danger if they had not been attracted to the vision of life beyond fossil fuels. People like their comfort. So the moral of the story is don't threaten people's comfort unless you have something better to replace it with.

"The power of our vision must be so much stronger than the power of our fears," Wei-Ping continued. "What drove most of the denial was that the fear was stronger than the vision. Yes, the climate denial was being financed by big coal and oil and led by right wing libertarians, but it got its support from the simple fact that people didn't want to think about change until they knew there was a better place to go to. As long as people thought that accepting global warming meant giving something up—driving less, turning down the lights, having to ride a bicycle—the climate denial message was enticing, at least as an excuse for not acting. But as soon as the vision became stronger than the fear, people found it easier to accept the science and embrace the change."

"You make it sound quite cathartic."

"It was, especially for the committed deniers, who had been determined not to be dragged somewhere they disliked by a government they disliked. For those who failed to move their savings out of fossil fuels, it was quite traumatic. But the rise of the cleantech funds has made the transition easier. On some nameless day the deniers woke up to find that the wind had gone out of their sails. It was probably the same when the printing press replaced the quill and the railway replaced the horse. It's been scarcely ten years since the cap was imposed, but already the new solar reality seems as natural as the dawn.

"Do you understand—I mean do you *really* understand—that once we make the shift to solar energy, along with its derivatives wind and hydro, that the Sun will continue to send us the energy we need for not just a hundred but a *billion* years? And with every passing year, the solar technologies will improve and fall in price. They say that by 2050 solar PV could cost as little as two cents a kilowatt-hour.

"It's such an incredible transformation. Never in the history of humanity have we known true energy security. The firewood would always run out as the forests were cut down; the fossil fuels would be depleted. The Sun, by contrast, will never cease to give us what we need, not until she begins to turn into a Red Giant in something approaching two billion years, and then she'll be sending us too much energy, not too little. It's a total miracle when you compare it to the energy scarcities, the wars over energy and the freezing in the cold we've had to put up with in the past."

Wei-Ping stopped. I sat in silence, pondering the power of his words.

"It's pretty cool, eh?" he said with a laugh.

"It's more than cool. It's a complete transformation."

Wow. But what did I expect, travelling twenty years into the future? I was familiar with the global climate talks and how each year they ended in disappointment, but back in my time there was no talk of anything like a carbon cap, and no stirring vision of an amazing future.

"The cap was the most important component of the *Houston Accord*," Wei-Ping said. "And with the cap came the logical consequence, as each nation developed ways to ration its use of fossil fuels under the cap, and an agreement to seek full global decarbonization by 2050. That's too late—we need it sooner, by 2040.

"So every coal, oil or gas corporation that wants to extract or import fossil fuels now has to bid for the right to do so at a national auction," he continued, "where permits for a share in the declining carbon budget are sold to the highest bidder. That puts an additional price on carbon, supplementing the carbon tax."

"Was there any distinction made between the developed and the developing nations when the caps were assigned?" I asked.

"No. The proposal for sharing based on population failed to get support, since the budget for the developed nations would have been so small, making it impossible to decarbonize in time. China would have got 20% of the cap while America got only 4.5%. So it was never going to fly. Instead, there was an agreement to base the caps on actual emissions. Five years earlier the developing nations would never have agreed to that, but with the new understanding that the faster your economy gets off fossil fuels the better things will be, the resistance disappeared."

"What happens to the money governments earn from selling the permits?"

"In Canada, at first, 75% of the money from the auction and the carbon tax was returned to the public in reduced taxation. The rest went into a Climate Solutions Fund where it was invested in renewable energy, bike lanes, electric vehicles, public transit and so on. As the urgency grew so did the pressure for a more rapid response, and the proportion going into the fund increased. Two years ago they scrapped the tax cuts entirely and all of the money now goes into the fund."

"What was it like for you personally during the carbon rationing?"

"For us as a family, it was no problem at all—it was rather exciting. We invited our neighbors round and we made sure they knew about the Zero Carbon website, which gave easy information on how to reduce your footprint by the 10% a year that was called for. The real challenge was for businesses. The cap didn't come into effect for a year, which gave people time to prepare. The Canadian government looked at what British Columbia had been doing and it launched The Green Express, modeled on America's New Deal from the 1930s. The Green Express had twelve carriages, each containing the critical advances and policy changes needed to get to 100% renewable energy for transportation, heating and so on. They used the Canada Earth Bank to provide zero-interest loans for everything from electric vehicles to solar heat pumps, adding tax credits to sweeten the change. One of my neighbors ran a small trucking company. He was able to reduce his carbon footprint first by making a variety of fuel-saving initiatives, then by hiring the EV Conversion Co-op to convert his fleet to electric drive. He could never have done it without support from the bank and the public investment in fast-charging stations."

"What's the Canada Earth Bank?"

"It's a branch of the Bank of Canada that issues zero or low-interest loans to help the transition. Ever since the Bank of Canada reclaimed its ability to create money in the public interest it has been doing cool things like this. The private banks hate it, since it undermines their monopoly over the creation of money, but it's been a godsend when it comes to speeding the transition off fossil fuels."

"Does that mean the Bank of Canada can create the money needed to cover Canada's debts, too?"

"Yes, within limits. That's how Canada financed its participation in World War II, and how it built key parts of Canada's infrastructure like the Trans-Canada Highway and the St. Lawrence Seaway. It was only in 1974 that the Bank signed away its ability to create money, following pressure from the International Bank of Settlements, which represented the world's private bankers. So yes, the Bank is now able to finance major infrastructure projects itself, creating the money that's needed without need for taxation. The same applies in the provinces, through set-ups like the Earth Bank here in British Columbia."[346]

This was fascinating, but I didn't want to interrupt Wei-Ping's narrative on the climate solutions, so I let it pass for now.

"All this must have taken a huge amount of planning," I said.

"It did. I have a friend in the civil service, and she said her work had never been so chaotic—or such fun. But you're right: we've needed major planning to create the organized transition."

"What about other countries? Did they have their own versions of The Green Express?"

"Yes. The E-17 Group of Nations gathered the best practices and shared them with the rest of the world. The whole decade of the Terrible Twenties—or the Transformative Twenties, as I prefer to think of them—felt like a race against time. You asked how it was for me personally, and I answered that it was a lot of fun. But my parents in Shanghai had a very hard time. The neighborhood where they lived was flooded by a typhoon, and both my grandparents on my mother's side drowned. They couldn't get out fast enough. My parents were lucky to escape with their lives, but they lost everything else. They live with me now, here in Vancouver."

"I'm really sorry to hear that."

"Thank you. It was a really difficult year. As well as the extra-strength typhoons, the monsoon failed in India, causing crops to dry up, which led to food riots, violence between the Hindus and Muslims, and then famine. The price of grain doubled, so the relief agencies' budgets could only buy half of what they needed. We saw people starving on our screens, and the line-ups of women carrying children with bloated famine-stomachs. It was awful."[347]

"My parents worked with refugees and famine relief in East Africa," I replied. "I've seen what it's like. It tears your heart out when there's nothing you can do." Then pausing, I asked, "How have your parents adjusted to life in Vancouver?"

"They appreciate the care we're giving them, but they miss their home and their neighbors. It's something we're all going to have to get used to in the coming crisis years."

"The crisis years?"

"Yes—the next fifty to a hundred years. I should probably have said 'The Crisis Century.' Even if we are successful in every initiative that's underway we're going to have to live with the consequences of global warming for many decades to come—the continuing storms, floods, forest fires, heat waves, droughts and hurricanes. We should have started reducing our emissions forty years ago if we wanted to avoid all the turmoil and distress. But profits came first. But I'm getting diverted. Where was I?"

"You were explaining the four methods. You've covered the carbon cap, the rationing, and the Green Express. What are the Climate Solutions Treaties that you have listed next on your chart?"

"Okay, moving on. The Climate Solutions Treaties have been central to progress now that the cap is in place. Let me backtrack to the period before the 2020s. The fundamental problem with the United Nations climate talks was that they had always been framed in the negative. They used words such as 'share the burden' and 'mitigate.' I ask you: when did you last use the word 'mitigate' around the dinner table? 'Honey, can we mitigate this soup? It's a bit bland.' There was no vision of how people could thrive in a green economy using 100% renewable energy. You can never motivate by fear alone; you must provide vision and hope. That's what made the Climate Solutions Treaties so powerful. They switched people's mental framing from negative to positive, from despair to hope."

"How did they come about?"

"I was very involved with them when I was a student in Shanghai. It was the green city leaders who took the lead. Their work was already based around vision, since they were working to build a green future. They wanted to apply the same approach to the planet as a whole, so they asked the universities for help.

"I was doing my degree in climate change and international studies at the Shanghai Centre for Global Studies and we set up a monthly call with our colleagues at Germany's Wuppertal and Potsdam Institutes, Copenhagen's Climate Council and the University of East Anglia's Climatic Research Unit in Britain. It was an incredible experience, working with some of the best brains on the planet. It was during one of our monthly sessions that someone proposed shifting the whole framework of the UN climate negotiations from negative to positive, from problems to solutions.

"We discussed the idea of specific treaties to accelerate the uptake of geothermal energy, wind energy and electric vehicles—things like that. We organized a big conference here in Vancouver that brought together people from the green cities, green business and green campus movements.

"Working together," he continued, "we assembled proposals for fifteen climate solutions treaties, covering everything from cycling and solar to ending fossil fuel subsidies and ramping up global energy efficiency standards for appliances.[348] Standards and regulations may sound boring, but they are the single most effective climate solution, after the carbon cap. When you get global agreement on fuel efficiency standards for vehicles, it impacts automobiles all around the world. The *Fuel Efficiency Solutions Treaty* required all new cars and light trucks to not exceed an average 100 grams of carbon dioxide per kilometre by

2025, falling to zero by 2030. Europe had already shown that it was possible, so it was not a big stretch for the rest of the world to follow. As I pointed out before, since the average vehicle is on the road for ten years, by 2040 there will be very few cars or light trucks anywhere in the world that are not electric. There are some pockets of hydrogen vehicles in Germany and Japan, since that's the path Mercedes, BMW, Toyota and Honda chose, and there are some biogas cars and buses in Sweden, but otherwise they're all electric.[349]

"It's good that cleantech has made such fast progress—the solar panels, electric cars, LED lights, enhanced geothermal and so on. The progress of the 100-percent-renewables club has been amazing. They showed that countries can get all the energy they need from the sun, wind and water, without fossil fuels or nuclear.[350] And they demonstrated that the health care savings alone that come with the end of air pollution are greater than the cost of the shift to renewable energy?[351] The fossil fuel producers had never come close to paying their full cost to society; all along we'd been simply absorbing it as citizens and taxpayers.

"Anyway, that summer was the devil, with its hurricanes, cyclones and tornadoes, its floods and forest fires, and the first complete summer Arctic meltdown. The Intergovernmental Panel on Climate Change published its first Climate Emergency Report, and the growing alarm bells brought some great volunteers. We invited some of the big global charities to join us, like Médecins Sans Frontières, the World Wildlife Fund and the Red Cross, and activist groups like 350, Greenpeace and Avaaz.

"In October we put out a global appeal for a hundred countries to start work on ten Climate Solutions Treaties in December, and a further five treaties the following December. To back our appeal we announced a Global Fast for a Better Future during the big UN climate conference. The fast took off, with millions promising to participate. The conference almost never happened, due to a major terrorist attack just three weeks before it was due to start, killing many people. It was horrendous, but the organizers persisted, and the fast took off, with millions promising to participate. We had half the members of the German Bundestag, two thirds of the deputies in the Brazilian National Congress, more than three hundred deputies from the French General Assembly, and thousands of priests, mullahs and religious leaders around the world. We also had global celebrities, film stars and Olympic champions.[352]

"On the opening day of the conference a million people began the fast, including several official delegates, urging the world's leaders to negotiate the Solutions Treaties in addition to reducing emissions."[353]

Wei-Ping paused. "I can see that you're following this with interest."

"It's like a drama," I replied. "I'm waiting to see what happens next."

"The delegates were all over the place. We had set our goal at a hundred signatories to give it the force of moral persuasion. Some said they had no mandate to discuss this kind of thing, and our proposals were a distraction; others said they were fed up with thirty years of treaty-making that had achieved almost nothing and it was time to try something different.

"The conference was due to end on the Friday night, and by Wednesday night only forty nations had said they would participate. This grew to seventy-five on Thursday, and all day Friday we ramped up the pressure to reach the hundred mark.

"The hours ticked by, and around the world millions of people were doing whatever they could to persuade their delegates to support the new treaties. At 4:05 a.m. Beijing time on Saturday, Georgia signed up as the hundredth country and we all went crazy! By 4:30 a.m., another twenty-one countries had come around and we ended up with a hundred and twenty-one countries committed to work on the details, with the intention that they would sign the treaties at the next global climate conference in Beijing.

"It was amazing. A group of observers at the back of the room started singing the global climate anthem, with its powerful lyrics: *'Do it now, do it now, do it now, now, now!'* Do you know it? It's sung to the tune of the popular Italian song, *Bella Ciao*. Then the delegates joined in, so the entire room was singing. We did the same in Beijing, all three thousand of us. What a high that was!"[354]

"That must have been amazing. How old were you during all this?"

"Twenty-four. Yes, I know what you're thinking. How old are you?"

"Twenty-five."

"So, it goes to show what young people can do."

"Didn't you get hungry?" It seemed like a stupid question, but that's what I asked.

"Yes, but only for a day or so. After that the body adjusts, as long as you drink enough juice or water. It must be a genetic inheritance, since our ancestors had to survive long periods without food. But do you know what our biggest frustration was? It was that when the fast ended we couldn't go and pig out. We'd been told to break it gently, so we waited a week and then we had a big celebration when we really let our hair down. There were three thousand of us who had been fasting in Shanghai, and a thousand of us squeezed into the Art Scene Warehouse in the Suzhou Creek artists' quarter. They had held China's first exhibition of contemporary art on climate change, and they organized China's first exhibition of Visions of a Better Future. So when we celebrated we were surrounded by all that visionary artwork. It was an incredible high."[355]

"I wish I could have been there...."

"It was the most amazing three weeks of my life. It shaped my career, and put me on my current path. Me and hundreds of thousands of others across the planet. Something shifted that day. It felt like a huge planetary 'click,' as if the global attitude had shifted from negative to positive. Things that had previously seemed unthinkable began to be possible, like the climate emergency warning signs at gas stations, the carbon labels on everyday products, and the normality of the fact that that all new buildings are zero carbon and all new vehicles are electric."[356]

"How many Solutions Treaties are there today?"

"Fifty-eight, at the latest count. The most recent came into effect in May, to accelerate habitat protection for the top predator birds: the eagles, vultures and owls. They control the ecology of so many other species, all the way down to the bugs and beetles, so it's essential to protect them."

"So they're not just about climate change?"

"No. The model has been adopted by groups all around the world. Nations are finally working together instead of arguing."

"Where does China fit in? You have it listed as one of the wild cards."

"When I was a student in Shanghai we were burning so much coal and our economy was growing so fast that you only had to say 'China' and it would take the wind out of the sails of the most dedicated activist. I often visited Beijing, and the air pollution was awful. It made my eyes run and gave me a burning sensation at the back of my throat."[357]

"So what happened to change things?"

"The Chinese government had already been making an effort to encourage greater energy efficiency and install more wind and solar. They were very aware of the dangers. The region around Shanghai is close to sea level, and China has a painful history of recurring famines, so the increasing frequency of drought was a huge concern. As I said earlier, entire dynasties have fallen for that kind of reason.

"I'm not sure if there was a particular moment when things changed. It was hard to read the Central Committee or to know what went on behind the scenes, but the Chinese people are very aware of global warming, and they're very worried about it. Less than one percent of China's people have ever been climate deniers—that was a peculiarly Western disease pushed by Big Carbon to defend its interests.[358]

"It was while Xi Jinping was Head of State that the changes started to happen. First it was the carbon tax, which started small and grew steadily.[359] Then the government raised the price of coal-fired power, increased the feed-in tariff for renewable energy and set a very ambitious target to phase-out all coal-fired power plants that did not capture their carbon emissions. There has been a massive amount of solar development, with the big floating solar arrays and all the plants in Qinghai province and other parts of the northwest.[360] There has also been a huge amount of wind energy development, especially in Inner Mongolia and offshore.[361] And there have been big initiatives related to electric appliances and energy storage. When the *Electric Vehicles Solutions Treaty* was adopted China doubled its starting goal to 10% of new vehicles each year.[362]

"China was already making most of the world's solar panels, the cheapest LED bulbs and the cheapest electric cars. Then the government announced plans to install solar on every south-facing roof. It required every new building to meet the passive house design standard and they established effective energy rationing by hiking the price of electricity for households that use more than 5,000 kWh a year.

"When China made such significant changes, the world woke up. People realized that China was winning the race to be green. China's economy is currently on track to be zero carbon by 2045."

China, zero carbon by 2045? That was a *huge* change. "How are they getting their electricity if they're closing down all their coal-fired power plants?" I asked.

"From a combination of solar, wind, hydropower, geothermal, ocean energy and bioenergy, combined with efficiency and energy storage. The same mix that everyone is using. There are no big secrets. It's just a matter of scaling up. China's progress with efficiency alone has reduced their future estimated demand by 50%."[363]

"So by 2045 they'll be using no more fossil fuels?"

"Correct, except for a few exempted industries such as steel, aluminum and chemicals. They're electrifying all their transportation and they're using stored solar and heat pumps for heating."

"Is this happening in Tibet as well?" It was a bit of a cheeky question, since back in my time most of the Chinese people I knew were very defensive about China's occupation of Tibet. But Wei-Ping continued without so much as a blink.

"Yes. There's been a rapid uptake of renewable energy ever since they signed the China-Tibet Goodwill and Harmony Agreement, a few years ago."

"That's good," I lied. In reality, I was full of surprise. How did *that* come about? From everything I knew back in my time, China's rule in Tibet had been pretty oppressive.

"That must be very gratifying," I said.

"It was. From what I've learned from conversations with my contacts there was a split within the Politburo, the Standing Committee where a handful of China's top leaders make all the key decisions. There was mounting pressure to act faster to solve the chronic problems of air and water pollution, and there was a growing concern that the continued policing of Tibet and the repression of its people was a financial burden that was sucking resources out of China's state coffers. One faction in the Politburo wanted to continue as usual, but the other faction, who ended up winning the argument, wanted to release the same tigers of entrepreneurial zeal within Tibet that had enabled China to make such rapid progress. Apparently, there was a group of top-tier Chinese venture capitalists who had the ear of Xi Jinping, and they were the key influence that led to the signing of the Goodwill and Harmony Agreement. It literally came out of nowhere, taking the world by surprise. That's one of the virtues of a tight little dictatorship: you can at least make some decisions quickly!" he joked.

"What kinds of thing were in the Agreement?"

"It was an agreement by the Chinese to treat all Tibetans equally, with kindness, trust and respect. They agreed to grant them equal opportunities, to respect their human rights, including freedom of speech, religion and political expression, and to allow them to speak the Tibetan language freely and to study it in school alongside Chinese. They allowed the return of the Dalai Lama and the Tibetan refugees, and they agreed to respect Tibetan cultural and religious practices. In addition, they agreed to free all Tibetans who were in prison for cultural or political reasons, to work together to restore Tibet's forests and grasslands and protect her endangered species, and to allow outsiders to visit every corner of Tibet without restriction, just as we can in Canada. It was a very great thing that China's government did, and the Dalai Lama was very gracious in brushing away all the years of difficulty."

"Do you think Tibet will ever become an independent nation?"

"Ah, that's not for someone like me to have an opinion about. Maybe it'll be more like the European Union, with open borders and shared policies. But moving on, we should look at the wild cards. I've already covered China. Saudi Arabia has been another surprise. Twenty years ago they invested $100 billion in solar to meet a third of their rapidly growing power needs. Then they switched their desalination plants to solar and declared that they would turn Saudi Arabia into Solar Arabia. It's not just Saudi Arabia: every country in the Middle East is doing it—Egypt, Jordan, Iraq, Iran.[364] Israel/Palestine is now running on 100% renewable energy, thanks to the electrification of transportation and the advances in battery storage, and Qatar has been a leader too. What's especially encouraging is the rebuilding of Syria: it's all based on 100% renewable energy, with every building and city being designed to maximize natural cooling, with a walking and cycling culture.[365]

"To give you an idea of scale, the world uses 200,000 terawatt hours of energy a year, and in the Middle East a thousand square kilometres can generate a hundred terawatt hours of solar electricity a year. Saudi Arabia has two million square kilometres of land, so in theory they could generate twice as much electricity as the entire planet needs. So far they have installed solar on ten thousand square kilometres, all cleaned of dust by waterless robots,[366] and they're producing a thousand terawatt hours a year—four times more than they consume. The surplus they're converting into hydrogen using seawater, which they're selling to the shipping industry at their new terminal at Jeddah, and some to SpaceX to fuel their three-day tourist trips into space."[367]

"That's incredible—I had always assumed they were so loaded with oil they wouldn't make the effort."

"A thousand square kilometres of desert in the Middle East produces solar energy equivalent to 1.5 billion barrels of oil a year. Twenty years ago, Qatar had proven oil reserves of 25 billion barrels. Today they are generating that much energy every sixteen years, using nine percent of their land. Saudi Arabia can generate the same amount of energy from solar that it used to get from oil on slightly more than 1000th of its land."[368]

"I can see why people have been getting excited about the solar revolution."

"Yes. The second wild card has been the courts. It started many years ago when a Dutch court ordered Holland's government to reduce its carbon emissions much faster than it had planned to, drawing on strong legal precedents.

It came as a shock, and the Dutch citizens group Urgenda that brought the case celebrated wildly; they also had their costs repaid. The Dutch government accepted the ruling, after some protest, and in the years that followed there have been legal challenges in many other countries, some of which have succeeded, some not. But the principle had been established, and it has played an important role in accelerating action."[369]

"Why Holland?" I asked. "Was there a particular reason?"

"I gather it had to do with Holland's historical tradition of polder politics: the polders are the low-lying lands that the Dutch people drained and reclaimed from the sea. To maintain them, everyone had to act together, for if one farmer failed to build a dyke or pump out the sea everyone would get flooded. So in this sense, there was a deep historical understanding of the common good and the need to cooperate to tackle a shared emergency.

"But moving on, the final two wild cards have been the amazing examples of individual leadership we have seen, and the cultural phase-change. Take Sally Bingham, in California. She inspired the congregations of thousands of churches to go zero-carbon through a program called Cool Congregations.[370] Then there's James Hansen, the climate scientist. I can't begin to count how many times he has been arrested. He inspired a whole generation of children to have the courage to speak to their senators and congressmen and to pursue legal challenges to force the pace of change. Climate courage—that has been their call."[371]

"And the larger global shift…?"

"I have come to think of culture in terms of its central DNA. In the decades leading up to the 2020s the DNA in most Western countries said, 'Be an individual. Go out and make something of yourself. Get what you can out of life.' It sounds great, but the individualism fostered selfishness, greed, and a consumer culture. The new DNA sends a different message. It says, 'Care for each other and for the planet. Help other people, and they will help you.' It's a fundamental change, and it feels so much better. The old DNA feels tired and lonely—it's exhausting having to prove yourself as an individual all the time, and it sends the message that if you fail for whatever reason it's your own fault. Personally, I am really glad to see that change. But we should press on, since the other side of the climate equation is equally important."

"The other side?"

"Yes—the other side. The fourth of the four methods. Sucking the carbon out of the atmosphere. Eco-sequestration."

 15

Calling All Carbon

"ECO-SEQUESTRATION?" I HAD not heard the term before.

"Yes. It's nature's way of sucking carbon out of the atmosphere. It's essential if we're to have a chance of ending the climate crisis. Nature already absorbs 55% of the carbon, storing it in forests and plants—and in the ocean, but that's also making it more acidic, killing the oysters, shellfish and coral beds. That's a whole other problem—it's climate change's evil twin. However, it's the 45% that stays in the atmosphere that we're concerned with here.[372]

Wei-Ping pulled up an image in the I-Ball that showed New York with a pile of blue balls completely burying the Empire State Building. "These represent New York's annual CO_2 emissions." Then he pulled up an image of the Earth showing piles of blue balls covering most major countries. "This is our planet's output: just over five gigatonnes of carbon a year.[373]

"Twenty years ago we were releasing ten gigatonnes a year by burning fossil fuels, making cement and destroying rainforests. Today, we've got it down to five. Look—it's all laid out here." Wei-Ping showed me a table with the actual and the hoped-for carbon emissions to the end of the century.[374]

"Thanks to all the recent changes we are close to the turnaround point, and as the eco-sequestration efforts kick in the level of CO_2 in the atmosphere should start to decline. The other greenhouse gases are more complex, since their lifetimes and breakdown processes vary. But there's still a huge urgency, and the financial costs of the climate crisis will be enormous. That's why eco-sequestration is so important.

"There are eight or nine ways in which we can capture the carbon and bring it back to earth. Five involve the forests and farmlands, three involve the ocean and a possible ninth involves mineralization. If they all succeed we could in theory capture as much as six gigatonnes a year, but there's a biological limit to nature's ability to store carbon, so we can't assume that much. We need to get close to zero emissions for the planet as a whole by soon after 2040, and for that we'll need to get every laggard nation on board, including Russia. Once we get down to zero, if the sequestration continues to reduce the carbon, we could get back to 350 ppm by the end of the century. But it's a very big if. Maybe we need a fresh cup of tea to refresh our brains. Green tea with gingko and jasmine? Or do you prefer chai?"

"I'll take the green tea; that will be delicious."

While Wei-Ping was making the tea I looked around his office. At the end of the room there was a long, tri-paneled electronic wall-screen with a series of classical paintings of Chinese landscapes—misty waters, willow trees, red-crowned cranes and mountains. I once had a crush on a Chinese girl who wore a black silk blouse with cranes. She said they were a symbol of fidelity. There were also a variety of awards on the wall including the David Suzuki Award, the Katerva Green Cities Award and Vancouver's Best Chickens Award.[375]

"You've found our awards!" Wei-Ping chuckled as he returned with the tea. "Best Chickens Award. How many economists can boast a thing like that? Our daughter Lijuan used to make earrings from the rooster's feathers. She wasn't at all happy when we ate him because his crowing was waking the neighbors."

"And the beautiful paintings?"

"They're my choice. I get to enjoy them on Fridays. On Mondays, Laura usually picks something with a Celtic theme. They're a cover for our Telepresence suite for long-distance meetings.[376] On Tuesday I'm meeting with a group to discuss ways to help North Korea accelerate its solar development. So where were we? Let's start with farming and forests."

North Korea? Solar development? Maybe I'd find out later.

Using the I-Ball, Wei-Ping pulled up the image of the Earth and said "Eco-sequestration categories." Various colors showed up.

"First there's the temperate rainforests. They're shown here in pale green." Large areas of Canada and northern Europe showed up, and parts of Russia. "Selective, ecosystem-based methods of forestry that mimic the ancient forest store much more carbon than short, sixty-year cut cycles. If nations honor the new *Global Forests Treaty*, the temperate rainforests will gradually sequestrate an additional gigatonne of carbon a year, without any long-term loss of timber harvest. It's a challenge when they're in private ownership, since it means changing the investment horizon from short to long-term, but Sweden, Canada, Germany and Switzerland all now require long-term ecosystem-based forest management by law, which makes it easier."[377]

"What about the tropical rainforests?"

"They're shown in dark green. Twenty years ago the loss of forests in South America, Central Africa and South-East Asia was causing the release of one and a half gigatonnes of carbon a year. Today it's down to half a gigatonne, thanks to better policing by the South American governments and Global Forest Watch's network of spotters who use drones, satellites and computer analysis to pinpoint the illegal logging and burning. They can track every logging truck that leaves the forest, right down to the dock. They track the shipments, then organizations like Greenpeace organize boycotts of the illegal timber purchasers. We are steadily cutting off the market for illegal timber, but there's a long way to go before we end it completely."[378]

"We also need to plant more trees, by the billion. A tree can absorb around three kilograms of carbon a year and a hectare can absorb 1.6 tonnes, so a big global tree-planting effort could store a third of a gigatonne of carbon a year.[379] China has made great strides with its huge new forest belt,[380] and Africa is making progress with the Great Green Wall of trees that runs across the continent from Senegal to the Red Sea, but we're only a third of the way there, and we need to ramp up our efforts."[381]

I wanted to ask more, but Wei-Ping pressed on.

"Next there are the farmlands, in purple. Historically, their soils stored an immense amount of carbon, but we have lost a lot since we started farming. When farmers manage the land organically they rebuild the carbon in the soil by up to a tonne per hectare per year. The world has almost two billion hectares of farmland, so if every farmer went organic they could suck one and a half gigatonnes of carbon a year out of the atmosphere without any loss of yield, and with all sorts of benefits for wildlife and the soil."[382]

"Do you really think it possible that the whole world would go organic?"

"Why not? Canada's farmers are doing fine, and their costs are lower than they used to be. It just needs government leadership to make farmers pay for the hidden costs of conventional farming. There are even apps that tell farmers what kind of soil they have, and how they can start storing carbon optimally.[383]

"Next we have the grasslands." The I-Ball showed patches of yellow across large areas of North and South America, Africa, Central Asia, Russia, and northern and eastern Australia.[384]

"The grasslands?"

"Yes. The grasslands are fundamentally about two creatures: the mammoth and the wolf. Historically the Russian tundra was grazed by tens of thousands of woolly mammoths and other herbivores. They converted vast amounts of grass into dung, which got mixed with dust and became soil, locking the carbon away in the permafrost. Scientists have estimated that the permafrost contains more than half the world's soil carbon, but with the rapidly rising temperature it's leaking carbon and methane at an alarming rate.[385]

"Two Russian scientists, the brothers Sergey and Nikita Zimov, are trying to save the Siberian permafrost by creating a Pleistocene park, bringing back as many herbivores as they can—Yakutian horses, reindeer, elk, moose, musk oxen, bison shipped over from Canada, and fifteen woolly mammoths, thanks to the success scientists have had in cloning them from ancient mammoth DNA, which I'm sure you've read about."

Yes, I *did* know they had brought back the mammoth, thanks to the sculpture on Hastings Street.

"In the past, the mammoths and other herbivores trampled and scratched the snow to get at the grass underneath, exposing the permafrost to winter temperatures averaging minus 24 degrees and colder. That kept it frozen, along with its carbon. During today's winters, due to the absence of those ancient herbivores, the land is covered in a thick blanket of snow that insulates the permafrost at a comparatively warm minus 10 degrees Celsius, speeding its breakdown. Their work is hugely important, and thanks to support from the Russian government, they have been able to expand their park to 50,000 square kilometres. It needs to be several million square kilometres to make the difference we need, but it's a start."[386]

"That's incredible. What a vision!"

"Yes! The rest of the world's grasslands are primarily about the wolf. Take a look at this." Using the I-Ball he pulled up an image of a grey wolf looking directly at us against a background of snow. It was stunning.

"The wolf is the guardian of the grasslands carbon. Wherever there's a healthy population of wolves, the cattle and sheep cluster together to defend themselves. They pee and shit in the same area, impacting the soil in a way that causes the grasslands plants to grow root systems twenty feet deep, storing carbon all the way. When you kill the wolves, the grazers cease to worry about safety and become picky delicatessen eaters." He pulled up a photo of buffalo clustered tightly together with two or three wolves circling; then an image without the wolves, with buffalo roaming all over the place. "When they wander they no longer impact the soil. The animal nutrients get scattered instead of being concentrated and the grass seeds have no divots to hide in, so they get blown away on the wind. Within a few years, the ecosystem collapses. Reverse the process, and the soil stores carbon again."

"How do you do that?"

"Ideally, you bring back the wolf and pay farmers for their losses. In reality, the farmers prefer to fence the grazers and move the fences, replicating the way cattle grazed when the wolves were around. It's called rotational grazing. As well as storing more carbon the deeper roots store more moisture, so they produce fatter cows.[387]

"Personally, I'd love it if we could bring back the wolf," Wei-Ping continued. He looked out of the window towards the mountains, and then went off on a tangent. "Wolves are such powerful creatures. My father, Li Wan-Po, was sent to Inner Mongolia during the Cultural Revolution. He had so many stories about the wolf, and the way Mongolian nomadic culture was being destroyed by the Han Chinese farmers who hated the wolves and killed them any way they could, even with machine guns from the back of a truck.

"The wolf was the nomads' connection to Tenger, the sky-god, who controlled the shamanic world. If your body wasn't eaten by a wolf when you died, your spirit couldn't go to heaven. It was a huge loss when they began to kill the wolves in earnest. Ever since then my father has been a Tengriist, a follower of Tenger. He has a great reverence for nature. He had to keep it secret, for he could have been jailed if anyone informed on him. That's all changed now, and at the age of eighty he's able to have a shrine to Tenger in his home without raising any eyebrows. When he dies I'm going to have to find a way to take his body out to the grasslands and leave it for the wolves to consume, now that they are returning."[388]

"That's quite the story."

"Yes, I suppose it is! So getting back to the grasslands, if we can persuade the world's ranchers to adopt rotational grazing the grasslands will start storing carbon again, up to two gigatonnes a year. Between the forests, tree-planting, farms and grasslands, nature could absorb almost five gigatonnes a year.[389]

"There's also biochar, when you burn farm and forest wastes without oxygen to create charcoal and then bury it. It was done by ancient forest civilizations to increase soil productivity, and there's a move afoot to deploy it worldwide. It could potentially sequestrate a gigatonne a year, but it would be a massive logistical challenge because of the sheer volume of tree growth, harvest and burial needed without cutting into existing forests or farmland. So I'm only assuming a very small amount until we see if it can be scaled up."[390]

"Next we come to the oceans. First, there's seagrass." Wei-Ping pulled up an image of lush green underwater seagrass and a map that showed seagrass restoration projects in bright blue.

"Acre for acre, a seagrass bed stores three times more carbon than a forest; that's why we call it blue carbon. It's been under threat throughout the world, but thanks to the *Seagrass Restoration Treaty* there are a lot of restoration projects underway, including here in the Salish Sea, parts of which were recently declared a Marine Sanctuary. Globally, the results are too small to show, but we're hoping for a quarter of a gigatonne of carbon a year."[391]

"Then there's ocean iron fertilization to stimulate phytoplankton sequestration. We think it has the potential to absorb a further half gigatonne a year."[392]

"Isn't that supposed to be dangerous?" I had read about geo-engineering projects like this, and others such as pumping sulfate aerosols into the atmosphere to shield the sun's heat, or spraying aluminum dust by means of airplane chemtrails, and I knew that some people thought them extremely risky.[393]

"People used to think so," Wei-Ping replied, "but the projects are plotted with extreme accuracy in areas where the ocean currents pull phytoplankton down to the depths. So I'm comfortable with it.[394]

"There's a third ocean sequestration method that has some people excited, while others are skeptical," he continued. "The ocean has areas where the carbon-rich plankton are drawn down to the deep, where the carbon gets locked away. Twenty years ago, a neurophysiologist from the University of Washington had the idea of using floating ocean wind turbines to pump ocean biomass and dissolved organic carbon down into the ocean depths in the waters off the Greenland and Labrador Seas, and the first batch of turbines have been pumping away for the last few years. He—William Calvin—calculated that with enough turbines we could sink six hundred gigatonnes of dissolved carbon and plankton into the ocean depths over twenty years. If that's true, it would remove the entire carbon surplus that has accumulated since the start of the industrial age, reducing the atmospheric CO_2 to 280 parts per million, but the numbers are really hard to believe, which is why people are skeptical. Calvin compared it to plowing under a cover crop on a farm, which we do all the time. Personally, I don't see the risks, since ocean downwelling has been happening forever. It's going to be a while before we get reliable results, but I think it safe to assume at least a small number. If it can be scaled up, it could be very useful."[395]

"That brings us to the final possible method, mineral carbonation, but I'm not yet convinced. Nature has been mineralizing carbon dioxide for millions of years, converting it into limestone and dolomites. Olivine is a naturally occurring sedimentary rock that captures carbon as it weathers, and the world is full of sedimentary rocks, which store a vast amount of carbon. It's also known as magnesium silicate—here, let me show you." He pulled up a series of images in the I-Ball that showed a dull green rock, then a bright green polished crystal made from the rock, then a pile of ground-up rock, and finally a farmer scattering ground-up olivine on the land.

"Olivine captures carbon dioxide and turns it into carbonate by weathering, which takes place over millennia. But if you grind it up and scatter it, it speeds the rate of weathering enormously. Each crushed cubic metre absorbs a tonne of carbon dioxide, so if olivine mines around the world were to crush and distribute four cubic kilometres of olivine a year, the weathering could absorb a gigatonne of carbon, but you'd need more than five thousand industrial crushers and all the mining, grinding, trucking and spreading would require energy, and if it's not renewable energy it's not going to make sense."[396]

"You just grind the rock up, spread it on the land and it absorbs carbon dioxide?" It seemed unreal.

"Yes, but it only makes sense if the energy used in the crushing doesn't produce more carbon emissions—and it would be an enormous logistical challenge, which is why I'm skeptical.

"When we put all this together here's what it looks like." Wei-Ping pulled up a chart that summarized the various methods. "If everything works as expected, nature could capture almost six gigatonnes a year—but we're in a race against time before the tipping points push us over the cliff."[397]

"What about machines that capture CO_2 directly from the air?"

"There are some, but their contribution is tiny. And they too need energy to manufacture them, and to transport and dispose of the harvested carbon dioxide so that it doesn't return to the atmosphere. There are a few thousand Lackner machines at work, as they're called, but they'll need massive scaling up if they are to have an impact.[398] You've got to imagine that we're in a vehicle rolling downhill towards a cliff with one foot jammed on the accelerator and very poor brakes. The foot on the accelerator is the fossil fuels we're still burning; the brakes are the various methods of eco-sequestration. The faster we stop producing new carbon emissions and the better the brakes, the greater the chance that we'll be able to slow down enough to turn the corner and not go over the cliff."

"That's a very alarming metaphor. Do you ever look back and wish that we'd never started burning fossil fuels?"

"My ancestors in China have been burning coal for more than ten thousand years, and if we hadn't discovered how to use the incredible store of energy that's locked into fossil fuels, we would probably have killed the last sperm whale, cut down the last forest and gone back to the Stone Age. It was coal, in effect, that saved the last of Europe's forests, and it was natural gas that saved the whales. So we should have huge respect for fossil fuels. Without them, we would never have been able to develop the scientific and engineering know-how to develop modern medicine, and to build solar panels and electric vehicles. I think of fossil fuels as the launch pad for the Solar Age. They gave our civilization a blast of ancient energy, enabling us to invent our way out of the drudgery of manual labor and build a better future.

"The problem is not fossil fuels as such," he continued. "It's that we carried on burning them for too long. The big fossil fuel corporations were allowed to misuse their influence to confuse the public and buy the politicians, all so that they could go on profiting at everyone else's expense."

Wei-Ping sat back, stretching his hands behind his head.

"Well, life goes on. We do what we can. More tea?" As Wei-Ping topped up our cups, I pondered what he had been saying.

"How are these eco-sequestration initiatives being funded?" I asked.

"Aha. You're not as stupid as you look! I'm sorry—that's something my mother used to say. She meant it affectionately, so no offense, I hope."

"Not at all. I'll just bill you for five years of therapy."

"Ha! You're funny. But to your question: they are being financed by a slice of the global financial transactions tax. The money goes to the World Bank's Carbon Restoration Fund and is distributed according to the effectiveness of each method. It's appropriate, don't you think, that the big financial speculators, who have done so much harm to the Earth and her people, should pay to restore things back to balance?"

"Yes—very appropriate. What needs to happen next?"

"We can't afford to relax our guard. This summer the Arctic will be ice-free for twelve weeks. By 2045 it will be ice-free all year round, even in the dead of winter, and the jet-stream disruptions will make today's problems seem like a pleasant tea party. We are nowhere near out of the woods. Now if the E-70 Group of Nations became the E-170, we might start getting somewhere."

I paused, wondering where to take the conversation in the short amount of time remaining. "Can I ask you a question on a different topic?"

Wei-Ping raised his eyebrows, so I continued.

"What is your take on the second financial crash? What caused it, and what came out of it?"

"Well, you *are* a sucker for punishment!"

"Yes, but you're an economist. You must have studied it."

"Yes, I have. And I also give lectures on it. But this calls for a fresh cup of tea...."

 16

Ethical Entropy

"SO," WEI-PING SAID, as he returned with a fresh pot of green tea. "You were asking about the last financial crash. There were three fundamental causes, three consequences, and three lasting lessons. At least, that's the way I teach it. Will that be sufficient?"

"Definitely."

"Well. You'd best put your thinking cap on, since this stuff is even more complicated than resolving the climate crisis. So first, we have the three fundamental causes, which were ethical entropy, the absence of oversight, and the carbon bubble.

"The first, *ethical entropy*, came from the failure of balance between our individualistic, selfish natures and our cooperative, caring natures. By exaggerating free market economics as an ideal, mankind's selfish impulse had taken over and installed itself in the brain of the economy and the heart of the big banks—not that they had such a thing in those days. With communism and state socialism both seemingly dead (but don't tell the Danes and other Scandinavians), there was no further reason for the plutocrats to worry that an excess of greed might lead to political retribution. So the bankers lobbied their governments to remove the regulations that had held their selfish impulses in check—things like the main street banks not being allowed to operate as investment banks, and banks needing good capital buffers as insurance against failure.[399]

"By then there had also been a change in ownership. In the old days, the owners of the banks and lending houses knew that if they made a bad investment they'd lose their money, so for every ounce of risk they added two ounces of caution. Now, the banks were owned by distant shareholders and pension funds, so any sense of personal responsibility was gone. Instead, there was just a relentless drive to maximize profits, with senior management demanding that their workers produce ten percent more revenue every year, using someone else's money to play with, and firing every worker who didn't turn in enough profit. As one banker told me in an unguarded moment, it was like playing Russian Roulette, but with someone else's head.[400]

"I worked in one of the big banks in London for a year," Wei-Ping continued. "My father thought it would be a good career move, so he got me recommended for a junior intern position with one of the quants. I was good at math, and I managed to fake it through the interviews. But it wasn't my thing. I kept wanting to talk about the climate crisis, and the bankers looked at me as if I was weird."

"Excuse my ignorance," I asked, "but what's a quant?"

"A quantitative analyst. They were the ones who dreamt up the complicated new financial products based on complex algorithms designed to predict the behaviour of the markets. They typically had a PhD in math or physics. They made the banks a tonne of money, but they were also partly responsible for the crashes. I was shocked, coming from my friends and family in Shanghai, to discover how amoral, aggressive and deceitful the world of the bankers was, and how short-term their horizons were. Some of them were completely mercenary, and they had zero loyalty to the bank. Why should they, when they had zero job security? They were performance junkies, obsessed with the next kill. Climate change? Ethical or moral concerns? Who cared? That was in London. I gather it was not so bad in Vancouver and Toronto.

"So anyway, with the weakening of the regulations, the banks were free to invest in whatever they thought might bring a return using whatever new scheme seemed profitable, even if it was so obscure that only the quants could understand it. And remember, if you're a bank, you can in effect create money out of thin air with the click of a mouse. The only constraint was the bank's capital requirements, which had been either eliminated or reduced to almost zero in many cases.

"The result was an explosion of credit, which created asset bubbles and a huge pile of unresolved debt, including trillions invested in fossil fuels and in complex financial products that contained toxic assets—investments bundled up to hide their weaknesses.[401] And the banks had so many shadow banking arrangements to bypass their taxes and their leverage requirements that they found themselves unable to trust each other—and yet they were deeply interconnected, including through a host of complex offshore loan structures.[402] Andrew Haldane, Chief Economist at the Bank of England, said that the balances at the big banks were 'the blackest of black holes.' Even their own managements didn't understand the risks they were being exposed to, let alone the risks that other banks were exposed to. How could they? A CEO could be in charge of a hundred thousand employees, with multiple layers of hierarchy, and his bank could be making millions of trades a day. And they had learned nothing from the near-meltdown of 2008, when the world came to close to a complete financial implosion.[403]

"The surge of new money built a global time-bomb of debt, including in fossil fuels, in the emerging market economies like India, Brazil and China, and in personal and government debt. The average personal debt in Canada at the time was more than 160% of annual household income, far higher than it had been before the crash of 1929. In America it was as high as 370% – higher than the equivalent debt in Greece. That was just crazy, and at the time Americans weren't even allowed to declare bankruptcy for student debt.[404]

"In a well-governed society, the alarm bells should have rung and this kind of thing should have been regulated by the government, which brings me to the second fundamental cause, which was the absence of oversight. The financial regulators, ratings agencies and accountants who should have been waving red flags did not have a clue what was really going on, because of all the complexities and the shadow banking arrangements. The banks imposed an absolute code of silence on their workers, and there was structural corruption within the banks that compromised their accountants and their internal compliance branches, as well as the ratings agencies, which were paid by the banks. And the bankers knew that at the end of the day, if everything went belly-up, or 'tits up' as they say in London, the government would be there with the taxpayers' money to bail them out. So why worry? It wasn't as if they were risking their own money."

"But why wasn't the government able to step in and stop the abuses?"

"That's a very good question—why didn't they? The answer is that they were equally in the dark, and many politicians had been accepting campaign donations from the very same banks, and attending fancy dos where the bankers wined and dined them, creating a culture of regulatory capture, as we call it.

"Before the OMEGA Days it used to be considered normal for lawyers and accountants to advise their clients on the best ways to avoid paying taxes, and for corporate lobbyists to push for more deregulation. When people are focused so selfishly on short-term horizons, it becomes hard to achieve solutions that call for long-term thinking and shared sacrifice. It's good that some of the culprits are finally in jail, where they can contemplate the mess they got us into. The bankers knew full well what they were getting into. When a *Financial Times* journalist interviewed one, he said it felt like the last days of Pompeii, with everyone wondering when the volcano was going to erupt. As a friend put it, they were too blinded by their collective nastiness to do anything but march together over the cliff."[405]

"That's pretty grim. But wasn't there any global oversight, to keep the whole thing under control?"

"Only the oversight provided by the Bank for International Settlements, which was a private company owned by the world's central banks, not a public body answerable to any democratically elected forum. And most of the central banks were private banks too, with the notable exception of the Bank of Canada. So no, there was not any global governance to speak of. That's one of the big reforms being addressed now.

"But the failure of oversight goes a lot deeper," Wei-Ping continued. "I hope I'm not boring you?"

I assured him that he was not. Growing up in East Africa, and then living in Vancouver, I had no exposure to the way the world's banks worked, or how they were regulated. I didn't even know that banks could create money out of thin air until some of my friends told me about it and explained how it worked. And from what I gathered, none of Canada's schools offered any kind of financial education; the best students knew more about Greek history and Homer than they did about the world's finances, on which all prosperity depended.

"The problem goes right back to the crash of 1929," Wei-Ping continued, "which led to the Great Depression of the 1930s. That was a truly difficult time, and in many ways it was only the process of re-arming for World War II that pulled us out of it. So in 1944 seven hundred and thirty delegates from forty-two allied nations spent three weeks together at a hotel in Bretton Woods, New Hampshire, where they thrashed out new financial arrangements for the world as a whole, to prevent such a thing from ever happening again.[406] It was as a result of their work that the International Monetary Fund and the World Bank were formed. One of the critical needs was a means to balance global trade imbalances, and create a way to recycle financial surpluses that might accumulate in one country, while causing deficits in another. That could cause production to grind to a halt, since the buyers in the deficit countries would not be able to afford the goods being produced in a surplus country. For the world economy to continue to develop, money has to keep circulating, giving investors the confidence that their investments won't lead to bankruptcy. All economic activity is fundamentally premised on trust, the belief that the future will be a positive place, not a negative place where you will lose money, and in order to achieve that there needed to be a global surplus recycling mechanism that would prevent disastrous trade cycles such as the Great Depression.[407]

"One of the British delegates, the globally famous economist John Maynard Keynes, shocked the conference by proposing the establishment of an International Currency Union, with its own central bank, which would have the ability to grant an overdraft at zero interest to any country that was struggling financially, followed by further loans at a fixed interest rate. His purpose was simply to keep the money moving, and prevent the build-up of surpluses in one country at the expense of another. To achieve this, he proposed that any country with a persistent surplus should be charged interest on its surplus, the income from which would finance the zero-interest loans, while the added cost would force its currency to appreciate, acting as a brake on its exports. It was a very bold globally syntropic proposal, by which prosperous countries would accept a degree of short-term sacrifice in return for long-term global financial stability. Are you following so far?"

"Yes," I replied with a degree of uncertainty. "So what you're saying is that if one country accumulates all the money, and doesn't recycle it, the other countries won't have the means to buy their produce, so the whole system could fall into a depression."

"Exactly. The problem, however, is that the American delegates hated the idea. They were coming out of the war as the world's leading nation, with a very healthy manufacturing sector, and they wanted to maintain their power and influence. So Keynes's proposal was rejected, and the Americans proposed instead that the world's exchange rates be pegged to the dollar, with a fixed exchange rate, which is the system that was adopted."

"But how did that ensure that the surpluses would be recycled?"

"It didn't, which is the core of the problem. The world never developed a global mechanism—there was a failure of global syntropic nerve, and a reversion to national dominance. The Americans knew there was a recycling problem, however, and they addressed it by pouring their accumulating surpluses into rebuilding Germany's and Japan's economies, with the intention that they would develop balancing manufacturing strengths, with regional markets in Europe and Asia to absorb their goods. It was an incredibly bold move, which saved the world from a heap of post-war woes, but over time, things unraveled, and by 1971 the fixed exchange rate was abandoned.

"So jumping ahead to the 1990s and the current century, the world economy developed in the absence of any organized method to recycle the surpluses, and the vacuum was filled by two ethically challenged creations: the offshore tax havens, where the rich and the big corporations stored literally trillions of dollars of tax-free surpluses, and the banking centers of Wall Street and the City of London, which managed the surpluses, creating such good returns thanks to all their complicated financial products that even countries such as China chose to invest their surpluses in government bonds to finance America's massive government debt. That's what I mean when I say the absence of oversight went right back to 1929."

Wow. I had no idea that the world's financial troubles went so far back. But I also knew that Wei-Ping still had a way to go.

"So what was the third fundamental cause?" I asked, hoping not to get lost down a rabbit-hole of confusion.

"The third fundamental cause was fossil fuels, and our human propensity to live in the present and not think about the future. For the last two hundred years we have enjoyed the incredible power of fossil fuels. People never paused to think where their gasoline came from when they filled up, or where it went when they burnt it. People knew about the climate crisis, but the leaders of the big coal, oil and gas corporations and the investment specialists who negotiated their financing were so far inside the bubble that they were in full denial of the reality that if we were going to stop the climate crisis, most of the remaining fossil fuels would need to stay in the ground, and as a consequence, trillions invested in coal, oil and gas were going to vanish. Most of the oil and gas investment specialists were as ignorant of the sub-prime carbon crisis as the mortgage specialists had been about the sub-prime mortgage crisis before the previous crash in 2008. And when the stock market panicked, they panicked over the carbon bubble as well, tearing great holes in the balance sheets of banks that were heavily invested in coal, oil and gas."[408]

"So taken together, these three fundamental causes were like a triple weakness in the global economy?"

"Yes," Wei-Ping replied. "Their combined effect was to make the world's financial architecture seem like a three-legged creature trying to balance on three pieces of floating ice. It didn't take much to topple it."

"So what did topple it?"

"The immediate trigger was a collapse in the stock market, fuelled by fears that the emerging market economies in places like China and India were not going to be able to sustain their economic growth, and that the carbon crisis would bring down some the world's major banks—which it did.[409] Then amid the chaos, a rogue trader made a series of calamitous collateralized complex product investments which ended up costing her bank several billion dollars, causing a cascade of losses in financially interconnected banks and the complete collapse of two major banks. As salt in the wound, she tried to cover it up by causing an internal IT crash. That caused a high-frequency trading program at her bank to issue a cascade of computerized trade-orders that no-one could control, which caused the Dow to fall by 5,000 points in a single day. Talk about a house of cards.

"But here's the thing: in contrast to the previous crash in 2008, the public's antipathy to the bankers because of their perceived greed and ill-gained bonuses was so strong that it made a taxpayer bailout politically impossible, in spite of the consequences. And then the collapse in confidence pushed Greece over the edge, causing it to finally declare bankruptcy, triggering a collapse in the Eurozone bond market, which pulled down Cyprus, Portugal, Spain and Ireland."[410]

"So quite the mess."

"It was a total mess—and with the interest rates already being close to zero there was little that governments and central banks could do to stimulate their economies apart from pour in new money through what they euphemistically called public quantitative easing, which really meant printing money, pushing it out into the economy for all sorts of public works programs like retrofitting buildings and building new bike routes. It was a good thing that they did—I can't imagine how bad things would have been if they hadn't. It would have been like the Great Depression all over again, perhaps even worse.

"So that brings us to the three consequences. The first was the collapse in investor confidence, which caused credit lines to freeze, economic growth to grind to a halt, banks to collapse, companies to go bankrupt, and jobs to disappear. That brought record unemployment, and a dramatic increase in poverty, bankruptcies, soup kitchens, divorces, suicides, street protests, riots, anger and general distress.

"The second consequence, which led to the OMEGA Days here in Vancouver and around the world, was the explosion of local organizing and self-help at the local and the neighborhood level, as people found new ways to help each other during the hard times. The vast majority of today's community currencies, cooperative banks and business support networks had their origins in the crash, and the OMEGA Days that followed.

"But it was the third consequence that has been the most significant. The crash caused a transformational shift in public sentiment, causing most people to abandon their support for the neo-liberal policies that had caused the crash and to embrace instead a very different vision, based on solid ethics, long-term sustainability, and global financial governance. Syntropic sustainability is the way I like to think of it, in contrast to ethical entropy."

Syntropic sustainability. There was that word again. It sounded great, but I had no time to ask, since Wei-Ping was in full-stream.

"As a result of the shift in the public *zeitgeist*," he continued, "the world was finally able to move on global financial reform, and in Canada the government was able to lead on two very significant initiatives, the *Restoring the Balance Initiative* and *The New Economy Initiative*."[411]

"What were they about?"

"Globally, the crisis led to the Dhaka Agreement, which included the closure of the tax havens, the new global reserve currency, an International Currency Union similar to the one Keynes proposed back in 1944, along with very similar mechanisms to recycle the surpluses, the widespread adoption of public banking, and the replacement of free trade agreements that did not respect democracy, social justice and the environment with Fair Trade Agreements that did. Did you know that the free trade agreements negotiated before the OMEGA Days set up secret tribunals where corporations could sue governments, forcing them to allow the exploitation of their resources without any public accountability or transparency, that they outlawed things like public banking and the public control of common resources, and that they turned state-owned enterprises over to the private sector? It was outrageous, yet hardly anyone knew about them. Without Wikileaks, even fewer people would have known, since there was an attempt to keep the whole thing tightly secret. It took a lot of legal wrangling and defeating a massive pushback from the corporations to get the trade agreements scrapped and the new Fair Trade Agreements brought in to replace them. It was the same with the tax havens, which were such a blight on the global economy. The Dhaka Agreement would never have been possible without support from the majority of the world's nations, which tells you how far the pendulum has swung away from neo-liberalism.[412]

"The changes also triggered major changes for the world's banks, especially in Europe and America; less so in Canada, where our banks were actually quite well regulated. Globally, there has been a change to the entire mental and legal framework within which the banks operate, to their essential DNA. In America, they broke up the larger banks into smaller units that were no longer 'too big to fail'; they ended the cozy arrangements within the banks that had mixed risky investor banking with conventional commercial banking; and they made the main street banks more like regular utilities. They also banned the use of complex financial products that the regulators could not understand, and they required every employee at a bank or credit union to sign an Ethics Oath, following a practice begun in Holland several years earlier.[413] They placed a ten-year moratorium on issuing bonuses; and they prohibited banks from punishing employees for being financial whistle-blowers, ending the code of silence that had kept many bad deeds secret until it was too late. Some state governments also gave a tax break to any financial institution that converted its legal charter to become a Benefit Corporation. And across the world, countries laid the legal foundations to allow public banking and the public creation of credit. Here in Canada, the government reclaimed the original purpose of the Bank of Canada, enabling it to give interest free loans to governments for human capital and infrastructure expenditures."

"Wow—that's a lot of change. How did the bankers take it?"

"The top bankers kicked up an enormous fuss. They made all sorts of threats about how they would leave for better jobs elsewhere, but with the code of silence broken and the whistle-blower protection in place there was an organized initiative by bank employees who came out in support of the changes, some of whom shared how crazy their lives had been, and how their obsessive, profit and fear-driven work had been destroying their marriages and relationships. Many bank employees, it turned out, were as disgusted with the way the banks had been run as the public was, and they were more than happy to embrace the changes."[414]

"That's very refreshing to hear. What were the two initiatives that happened here in Canada?"

"Right. The first was the *Restoring the Balance Initiative*, which included the National Strategy to End Poverty and Inequality, with its $20 minimum wage, raising most taxes by a small amount, and increasing the rate for the top tax bracket to 65%. It included the Citizen's Income, financed by the tax increases, and the Citizen's Endowment that people now get when they turn twenty-one, financed by the new inheritance taxes. The legislation that accompanied the initiative closed a variety of personal and corporate tax loopholes, and it took a much tougher approach against money being hidden in offshore tax havens—this was before they were closed down entirely. It also eliminated student debt by paying for the cost of college education through a small percentage of a student's future earnings, instead of through student loans. Some people call it the Picketty Initiative, after the French economist Thomas Picketty, who showed why the gap between the rich and the poor would always increase under capitalism, since without active intervention, capital would always grow faster than the economy, and inherited wealth would always grow faster than earned wealth."[415]

"How successful has it been?"

"I'd say very successful: the wealth gap between the 1% and the 99% has been narrowing every year. It's not enough on its own to create a more egalitarian society, but it's a critical part of the change. The second big initiative, the *New Economy Initiative*, has been equally important. That laid out the plan to move Canada's economy towards 100% renewable energy, including all buildings, transportation and electricity; it established investment incentives for community-based economic development; it re-asserted the entrepreneurial role of government in supporting science, research and development; it paved the way for more employee share ownership and co-operatives; it established the tax on pesticides and fertilizers which encouraged Canada's farmers to go organic; and it fast-tracked the conclusion of First Nations treaties across Canada, enabling the First Nations to own their own land and control their own resources. It also began the legal challenges against the free trade agreements, with their investor state agreements and their secret tribunals that could overrule a country's democratically chosen laws, arguing that they infringed Article 21 of the *Universal Declaration of Human Rights*, which states that 'the will of the people shall be the basis of the authority of government,' which the world's nations signed almost eighty years ago."[416]

"It must have been exciting to see all these changes happening."

"It was—it still is! The bottom line was that the global economy had been operating from a deeply flawed operating manual, and it needed a complete overhaul, with new DNA. It's important to see it in context, however. The economic changes are just one dimension of a far deeper sea-change that's happening around the world."[417]

"Now I'm really curious...."

"Ha! I believe we are at the beginning of a critical new period of human history, comparable to the Renaissance in the late 15th century and the Enlightenment in the 18th century, which did so much to change the way people thought and felt. Some of us call what's happening Global Empathy. People are allowing themselves to feel more love—love for nature, love for each other, love for their communities, love for our amazing planet and love for their democracies.[418]

"The old attitudes are falling away—the ones that said it was good to exploit nature, and it was good to think primarily about yourself and getting ahead in the world. Instead there is a new syntropic willingness to think globally, and to self-organize for the good of the planet, not just selfishly for the good of your nation or your family. The economic changes are just one expression of this far deeper change. The various OMEGA changes and reforms are another. The desire by people to rediscover their connection with nature is part of it. It's very encouraging. But where was I?"

"You were talking about the consequences of the crash, and you got sidetracked onto the bigger picture."

"Right. I remember. I've done the causes and the consequences, so then there were the three lessons. The first was the need to turn away from the short-term ethical entropy that had been allowed to dominate our cultures, and to embrace instead a culture of long-term syntropic sustainability.

"The second was the need to end our obsession with unrestrained economic growth at the expense of nature, driven by the banks' need for capital to turn a profit whichever way it could. These days, most economic development goes hand-in-hand with pricing, policies, regulations and indicators of wealth that include progress in restoring nature, and genuine progress for humans.

"And the third was the need to extend the emerging culture of syntropic sustainability to the world as a whole, not just financially but also in terms of functional, uncorrupted democracies, deep ecological sustainability, and an organized commitment to lessen the global inequality that had created such a rift between the plutocrats and the middle classes, and the middle classes and the absolute poor.

"And that, I'm afraid, must be that. I've lined you up to spend some time with

Naomi Stern, who will tell you about the work we're doing in the local economy to put the lessons into practice. I think she's going to give you a tour. After that I've asked Laura if she'll take you to visit the Triko-Op, and tell you about our green business activities. That should keep you busy for the afternoon."

I was overwhelmed by Wei-Ping's generosity, and his willingness to help. I thanked him as best I could.

"No problem. But I need you to do something in return. Whenever you meet people over the next few days, can you ask them to sign the Climate Compassion petition? We need as many signatures as possible by Sunday night to put the pressure on for a reduced global carbon cap, more compensation for climate damage in the developing world, and faster sequestration. It's becoming very critical."

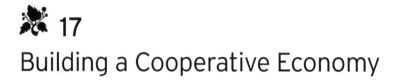 17

Building a Cooperative Economy

"PATRICK WU? PLEASED to meet you."

Naomi Stern was a good-looking woman in her thirties, wearing a black top and skirt with an orange waist-sash and a headband. As she shook my hand I noticed a large scar across her left cheek, which I later learned she had acquired while living in Israel. Naomi had history.

"Do you fancy getting out for a stretch and a cup of tea? I've been cooped up all day and I could do with a break."

She took me to a tiny restaurant owned by an Arab family from Jordan. "I feel at home here, and Fazil's an old friend." The tea was served on a traditional brass tray with a silver engraved pot, a brass sugar container and piping hot glass.

"I'm Jewish. On my father's side my family comes from Poland; they emigrated to New York in the early 1900s to escape the growing anti-Semitism. On my mother's side we come from Denmark. My grandmother escaped Hitler's clutches by the skin of her teeth by being boated across the water to Sweden one dramatic night in 1943, along with most of Denmark's Jews. I grew up in New York, where my grandfather was a labor movement organizer. I used to sit on his lap while he played *The Ballad of Joe Hill* on the piano, the house full of friends, talking and singing long into the night. Do you know the song? It's about the songwriter and labor activist Joe Hill who was framed for murder and executed by a firing squad in 1915."[419] She sang the chorus very quietly:

> *I dreamed I saw Joe Hill last night,*
> *Alive as you or me.*
> *Said I, "but Joe you're ten years dead,"*
> *"I never died," said he,*
> *"I never died," said he.*[420]

"That's powerful. How do you come to be here in Vancouver?" I asked.

"I was doing an economics degree in New York when the financial meltdown happened, followed by the OMEGA Days. OMEGA Wall Street was in full swing, and I became very involved. Do you know about Derek Brooks, from here in Vancouver? I'm a big fan of his book. He inspired me to take a year off to study the Emilia Romagna region of Italy, where I learned Italian and got a good understanding of their cooperative economy, which has shaped the green economy here in Vancouver. I'll get to that when we do the tour.

"So anyway, I decided to go to Israel/Palestine to learn about the situation first hand. I had heard so much from my family, but the conflict and the killings had gone on for so long and I found it hard to believe there was still no solution. I lived in a village called Neve Shalom/Wahat Al-Salam (Oasis of Peace), which was founded in 1970 by a Christian Dominican monk to promote interfaith dialogue, and as a place where Christian, Jewish and Muslim families could live together side-by-side, learning from each other.[421] It was one of the few places in Israel where there was a regular daily dialogue between Jews and Muslims, Israelis and Palestinians, both in person and by Skype. I lived there for three years, helping in the kitchens, on the land… whatever needed doing. My uncle in New York paid for my keep. He said nothing else was working to create peace, so it was a good investment. That's where I met my husband, Ben. He comes from an old Iraqi Jewish family. His great-grandfather moved to Israel in 1948 only to discover that Iraqi Jews were not welcome, so he left and brought his family to Toronto in 1967 after the Six Days War."

"So you must have seen a lot."

"It was while I was there that the Israeli women's movement began to come together, overcoming the differences and disagreements among women between the Ashkenazi Jews, the Sephardic Jews, the Arab Israelis and the Palestinians. I was there when the first All-Women's Peace March happened, converging on Jerusalem from all over the country with its powerful message *'We are learning to trust each other.'* Aviva Mir had recently become leader of Yesh Atid, The Future Party, and she was calling for Israeli and Palestinian women to come together to craft a better future.[422]

"'The men have been trying for almost a hundred years,' she said, 'and we are no closer to peace. Our future will be terrible if we don't let go of our old hostilities and antagonisms and come to a fair settlement. We all love our children in the same way; we all hurt in the same way; we all bleed in the same way. We all want security. We all want justice. We all want peace.' Aviva had a strong commitment to deep listening, and to mutual respect and non-violence. It was only by deep listening, she said, that we could hear the pain that was hidden behind people's words, as well as the hopes. We had to learn to listen from the heart, she said, not just the mind. We had to overcome our self-imposed tribal limits of empathy and allow our hearts to reach out into the hearts of others, including those who we might sometimes think of as our enemies, or our political opponents. That really impressed me. It's something I have carried into our work to build the new economy here in Vancouver.[423]

"So anyway, we were holding regular interfaith women's workshops at Neve Shalom, and also in Gaza, which was unheard of at the time. I had a friend who was a sculptor and she created a beautiful Cup of Peace, which we drank from as a symbol of our commitment to work together. When the Cup was stolen and smashed, she created another, and then another, and then we set up a workshop at Neve Shalom where we created thousands, so great was the demand.

"While I was there, Aviva and a hundred women leaders from Israel and Palestine met together for two weeks in Cyprus and I was able to snag a place as a cook, which let me hear some of their discussions. They dug deep into the thorniest political disagreements—about borders, the right of return for Palestinian refugees, political prisoners, the illegal settlements on the West Bank, the security barrier, the future of Jerusalem. They began to realize that instead of a one-state or a two-state solution they could find consensus around a United Federation of Israel-Palestine with two self-governing states, each with its own territory, united in one federation with an undivided Jerusalem as its capital, in which everyone would have freedom of movement, employment and commerce."[424]

"That's an exciting vision."

"Yes. Both the Israelis and the Palestinians have deep roots in the same land, and the same profound claims to territorial sovereignty, none of which can be negotiated away. By uniting in a single federation, Israelis and Palestinians can exercise self-government within their respective states, while sharing in the protection of fundamental human rights and enjoying protection against violence through a single federal military. Each state will have its own state police and determine its own laws and elect 50% of the members of the federal parliament, with the President being elected by the Members of Federal Parliament, alternating each year between a Palestinian and an Israeli.

"This realization that there could be a workable solution released an extraordinary energy, and the women started to share their hopes about the kind of future they wanted to build, putting their grievances and mistrust behind them. They developed new ideas for water justice; for shared water conservation and restoring the River Jordan, which has lost so much water it's hardly a trickle; new ideas for seawater greenhouses and solar energy; for permaculture, agriculture and tree planting; for housing and education; for building a cooperative economy; for bilingual education in elementary schools and equal access to healthcare, colleges and universities. The water thing is really important. It's not just in Israel/Palestine; it's all over the Middle East. There are whole cities in Egypt that are running out of water.[425]

"They wanted to share their music and dancing, and to hold joint prayer circles. The New Islam was generating interest as an alternative to Islamic fundamentalism, and there was a growing desire among Israelis for a more egalitarian Judaism. I overheard fascinating discussions around the possibility of paradise. The word comes from the ancient Greek and Persian words for a walled garden. Among the Jewish people, paradise is equated with the Garden of Eden, and the promise of a land flowing with milk and honey, both here on Earth and later in Heaven. We are instructed to be *shomrei adamah*, caretakers of the Earth and God's creation, and we have the all-important concept of *tikkun olam*, which means repairing or healing the world. There are many injunctions in the Torah to love and care for the stranger as if he or she was your neighbor.[426]

"And here's the thing: Muslims have an equally strong love for the promise of paradise described by Allah in the Koran: an eternal garden of joy with rivers of fresh milk and pure honey. The great Islamic gardens, from the Alhambra in Spain to the secret gardens of Sana'a in Yemen were all inspired by the desire to create a glimpse of the heavenly garden to come. And when we read the Koran, there are many concepts that speak of the importance of treating God's creation with respect. These include *tawhid*, the Oneness of God, who includes and transcends all creation; the concept of balance (*mizan*), which involves not upsetting the equilibrium of nature; the concept of mercy (*rahmah*), which extends to our treatment of all living beings; and the concept of trusteeship (*amanah*), through which humans have been appointed by Allah to be custodians of the Earth."

Naomi paused for a moment. "You must forgive me if I get excited, but it's so important when you consider our history, and all the violence there has been between two very similar peoples."

"Indeed," I replied, and she continued.

"The Muslim concept of *taharah*, of spiritual purity and physical cleanliness, speaks to the need for an environment free of pollution and an economy free of greed and corruption; the concept of *haq*, of truthfulness and the rights of others, extends to animals as well as humans; the concept of *ilm nafi* speaks to the value of knowledge and scientific research, which includes developing best practices for sustainable energy and farming the land. And the all-important concept of *amal salih* is about practicing good deeds, which includes repairing, mending and improving, all of which have a critical meaning for the Earth.[427]

"Jews and Muslims have been fixated on their history for far too long, whether in interpreting the words of God or fighting over past grievances. When we change our focus to the future, and look at the Earth in the light of *shomrei adamah* and *tikkun*, *amanah* and *amal salih,* a vision emerges in which we work together to create paradise, and we are all trustees of God's Earth.

"There was a group of women at the gathering who had put a lot of thought into this, and they presented their vision of Israelis and Palestinians working together to transform Israel, Gaza and the West Bank into a social, ecological and economic paradise, using the best practices of sustainable resource management and community-based economic development. They had developed a practical vision of sustainable communities where Jews, Muslims and atheists could live together in harmony. That's when I grasped the importance of future vision. It's the power of paradise."

I was listening intently, visualizing those villages, and the possibility of ending so many years of pain, conflict and grief.

"It was a very exciting time to be there, but the women faced deep hostility from the ultra-orthodox Jews, the fundamentalist Muslims, the rabbis and the mullahs, the Palestinian free-state activists, and often from their own husbands

and families. On the other hand there was growing support among many mainstream Israeli Jews, Palestinians and Israeli Palestinians. We had a sign above
the entrance of the Neve Shalom Centre in Tel Aviv with these words that have
been attributed to the Buddha:

> *In separateness lies the world's great misery,*
> *In compassion lies the world's true strength.*

"Ben and I became very active supporting the women's movement, including mobilizing Israeli men. We moved to Tel Aviv, where Ben got a job as an
engineer. It wasn't long before we got our first death threat, however, warning
us to stop our politics and leave the country immediately. We ignored it, which
was a big mistake. We were cycling in a suburb of Tel Aviv on a sunny afternoon
when a group of men rushed us, threw us off our bikes and kidnapped us. We
were blindfolded and taken to a basement somewhere in the city where we went
through three weeks of hell, separately, being beaten every day. I was told that
we had to produce five million dollars for the Jewish ultra-orthodox cause if we
wanted to live! Feh! I won't go into the details. Let's just say it was the worst
three weeks of my life. Then just as suddenly, we were freed. Our friends who
had been asked to raise the money used their sources to locate where we were
and they organized a raid. There were three young men guarding us, and two
were sleeping. There was a sudden rush of activity and we were shoved into a
car, our heads covered in black sheets. We were rushed out of the country, first
to Cyprus, then New York."

I felt sick to my stomach as my memory of the girl in Jerusalem returned. So
much violence: would it never end?

"Whew! Do you still have nightmares about that time?"

Naomi stared into her tea.

"They say you can get over it with the right treatment, but I'm not so sure.
And to answer your question, yes, I do still have nightmares."

I saw an image of landmines waiting to explode. I waited silently, but I desperately wanted Naomi to continue talking.

"And Israel? What's happening there now?"

"You don't know?"

Damn! I had fallen into the trap I had been trying so hard to avoid.

"I'm sorry," I said. "I should be better informed. I've been so immersed in the
difficulties we're having in Sudan. I'm not up on the latest progress."

"Yesh Atid won the election five years ago and Aviva became the Prime
Minister of Israel, governing in coalition with the Labour Party. She launched a
new round of peace talks, and both sides agreed to send primarily female negotiating teams to Washington where they reached agreement on an outline proposal for
a bi-national federal state with a dozen subsidiary peace and development treaties, committing the two groups to work together on practical matters of border
security, dismantling the walls, equitable sharing of the water supply, economic

development, housing, farming, tree planting, solar energy, and building a 100% renewable energy economy.

"Israel and the Palestinian Authority are in a five-year transition period. A Constitutional Assembly has been established to draft the constitution for the new federation, and the details are being negotiated regarding trade, borders, the West Bank settlements, annual quotas for the right of return, the new joint Israeli-Palestinian federal army, the rights of conscientious objectors, the release of prisoners, and dismantling the West Bank Barrier, the hated concrete wall. Whenever there's been a bomb attack or a kidnapping, groups of Israeli and Palestinian women have been going to the victims' families to offer support, to try to stem the instinct for revenge."

"Is it working?"

"Yes, by and large. Two factors have helped: the way the media has changed, and children's activism. The media is far more scattered than it was, so the big right-wing newspapers can't whip up anger the way they used to, fuelling the desire to avenge spilled blood whenever there's a bombing or a missile attack from Gaza. And people are telling their personal stories on YouTube, which is building empathy. And the children have emerged as a source of hope. They're incredibly well networked, sending messages signed jointly by Israeli and Palestinian children to pressure the politicians. There's still no agreement on many details, but most people know the new federation has to happen. It's either that, or another hundred years of hatred and bloodshed. All of the surrounding states support the plan.

"But look, you didn't come here to talk about this. Wei-Ping said you wanted to learn how we were building a new economy here in Vancouver."

"Yes, but I'm so grateful to learn about your experience. How did you come to be working with the Green Economy Institute, after living in New York and then Israel?"

"Ben is Canadian, and when we returned from Israel we needed a total change. OMEGA New York was struggling, but OMEGA Vancouver was going gangbusters so I wrote to Wei-Ping and offered my services. So here we are! I was able to work on a permit until I got my Canadian citizenship. Now I feel that I'm a triple citizen, an American Israeli Canadian."

"I feel that I'm a Chinese Sudanese Canadian."

"Maybe that's the way we're all heading. I'm so done with tribal loyalty, and the hatred and killing it brings. But then look at me, marrying Ben, such a good Jewish mensch. How tribal can that be? But let's get down to business. Where do you want to start?"

Where *did* I want to start? The OMEGA Days? The steps they had taken to achieve sustainability in the economy? The fundamentals?

"If you had to explain the *essence* of the changes that have happened, how would you describe them?"

"That's interesting. Before I got into this, economics seemed so complex and confusing. It was easy to criticize, and there were so many conflicting theories as to what we should do. That's what I liked about that book by Derek Brooks, his *Love Song to the Planet*. The stories he told about new ways of doing business and banking made such sense, and the conclusions he came to were so simple. It had been sitting there all along, but few had seen it. None of my economics professors, not even the great economists—the Keynes and Schumpeters, the Hayeks and Friedmans, the Fishers and Minskies had seen it. Not even Thomas Piketty, who explained so convincingly why the rich were going to get richer, becoming patrimonial capitalists blessed by inheritance because capital grew at 5% while the economy grew at 1% or 2%. None of them expressed it as clearly as Derek, bless his heart.[428]

"Fundamentally, Derek said, a society has five basic needs:
- The individual need to look after yourself and make something of your life;
- The human need to care for each other and to be cared for in a spirit of community;
- The transcendent need to pursue something greater, beyond the self;
- The practical need for good governance to keep our selfish impulses under control; and
- The ecological need to live in harmony with nature.

"Ever since Adam Smith realized that private enterprise flourished best in a free market we've had liberal economic thinking that emphasized personal initiative and transcendent yearning, to the neglect of community care, good governance and ecological harmony. This created rapid progress and enabled some people to get extremely rich, but it also caused immense suffering and misery among millions of working people, and it attacked Earth's ecology, though few people thought about that at the time apart from poets like William Blake.

"So along came its opposite, socialism, which emphasized community care and governance, but to the exclusion of personal initiative, transcendental yearning and ecological harmony. The first, capitalism, was a 'ME' economy—as in 'it's all about me'—that excluded 'WE'. The second, socialism, was a 'WE' economy that excluded 'ME', and they both excluded nature. The ME economy involved private land ownership, private business, private profit and private tax havens, along with some WE aspects such as taxation, regulation, public education and policing.

"In the ancient past, most tribes had a WE economy, with no concept of private land ownership and very little private gain that was not shared through feasting and gifts. As we spread out across the planet, we began to fight each other for the best land. The more aggressive tribes acquired new land by war and conquest, and as the warriors took possession they assumed private ownership, with the conquered becoming their slaves and peasants. If you have just invaded a lush valley, raped and enslaved its women and killed its men, the last thing you are going to do is call a community meeting to discuss ways of sharing. You give the best tracts of land to your most loyal warriors and their descendants cling onto it, becoming powerful barons and landowners, living in fancy estates and reading Jane Austen. From that time onwards, the assumption that property and business were a private affair became deeply embedded.

"It was in reaction to the misery and exploitation caused by the division between the land-owners and the landless, the factory-owners and the workers, that the impulse for socialism arose, starting with the French Revolution. When it was done well, a socialist state took care of its people, but centralized socialism deprived people of the freedom of individual enterprise, of turning dreams into reality. Neither the ME economy nor the WE economy could produce lasting success, since they both turned their back on other fundamental needs, and they both ignored the need to live in harmony with nature. Hence all our troubles, ecologically.

"We need an economy that can express all five basic needs, Derek said. We need a WE economy that includes ME, and that also cares for nature. And when you *do* meet all five basic needs, you get an economy that is more successful, more stable, more resilient and more sustainable. It's as if we had never quite gotten the formula right before."

"What do you call this new economy?"

"We call it, quite simply, a cooperative economy."

"This is fascinating. What are its main features?"

"Rather than tell you, I'm going to show you. Come on—we're going for a tour."

After Fazil waved her away when she attempted to pay for the tea, Naomi led me down the street to a store called Musiceum that sold musical instruments.

"See these three stickers?" Naomi said. "The circular one with the Oak Leaf and the five stars tells you this is a five-star certified green business. The eight-sided sticker tells you it's a member of the Business Network, and the sticker with 'B the Change' in a circle tells you that Musiceum is a Benefit Corporation."

"What's a Benefit Corporation?"

"Well, I just happen to think that it's the single most significant development in the world of business since the invention of the joint stock company in the 17th century. A B Corporation's legal structure requires a business to provide a tangible benefit to the community, nature or the world, as well as a financial benefit to its owners. It ends the requirement that company directors must maximize the return to investors regardless of the grief they cause, and that when a business gets into trouble it must be sold to the highest bidder, regardless of ethics. It ends the requirement that a company director must ignore all social and ecological costs and act in effect like a psychopath."[429]

"Do you know what community benefits this store provides?"

"Yes, I know the owner well. As well as a number of social, environmental and workplace commitments, Musiceum provides free music tuition for low-income children in the Downtown Eastside, not as an act of charity but as part of the core function of the business."

"That's impressive. How many Benefit Corporations are there in Vancouver?"

"Around 17,000, or about 70% of the total number of businesses."

"But that's huge!"

"When Vancouver joined the Global Alliance of Benefit Corporations, the city said that B Corporations would get priority treatment when it came to zoning or permit changes, and they'd get a tax break in recognition of the benefits they brought to the community. It's the larger companies that have been harder to persuade. There's a law being proposed that would require every business operating in Canada to become a Benefit Corporation within five years. That's the kind of change I like to see. But we must push on: this is just the first of twelve initiatives that are shaping the new local economy."

Twelve? Must there be so many? But I was all ears.

"The second sticker tells you that Musiceum is a member of the East Vancouver Business Network. When I was growing up, no-one paid much attention to local economics; we certainly never studied it during my economics degree. It was up to each business to swim or sink. They were ME businesses in a ME world. In theory, it was a perfect Adam Smith world, but in practice it didn't always work that way, especially when things went bad. That's when it became gold-plated socialism for the rich and tough, on-your-own capitalism for the poor. Neo-liberalism should have died in 2008 after the first financial meltdown, but its ghost continued to rule large parts of the world for another ten years, pulling the strings and making things worse.

"There were two places that did things very differently, however: Mondragon in northern Spain, and the Emilia-Romagna region of Italy, south of Venice, where I spent my research year before I went to Israel. What a wonderful place that is— it's all pasta and parmesan, vino and Verdi! They have a history of co-operatives that goes back to the 1850s, and following World War II they built a very robust economy under consistently left wing governments with about ten percent of their wealth coming from co-operatives. But here's what's important. Their privately owned businesses also work cooperatively to support each other. They operate in a free market, but they help each other, and by doing so they have created Italy's most successful region, economically, with the lowest unemployment. There are half a million businesses and co-ops in the region, far more than we have here proportionally, and they pay a levy on their sales to local inter-business organizations, in return for which they receive support with things like financing, training, research, development strategies and export efforts.[430]

"When the Emilia-Romagna region pulled through the 2008 meltdown and then the second crash in much better shape than other parts of Europe, people started to pay attention. They wanted to know what they were doing differently that enabled them to weather the storms when so many other economies were crumbling. After the year I spend there I came away persuaded of three things.

"First, local businesses need to work together and support each other, just as we do in a family, and to do that you need organized structures, financed by a small levy on a company's sales. There's a very deep understanding of the value of reciprocity in the Emilia-Romagna region. The businesses help each other and sub-contract to each other, knowing that they are building a network of strength and mutual obligation.

"Second, successful co-operatives are very beneficial to an economy, since their workers are committed to their businesses and they think entrepreneurially. The co-ops have their own support federations, financed by a 3% levy on profits. The money goes into a Co-operative Development Fund and is used to start new co-ops, to convert existing businesses into co-ops and to expand existing co-ops.

"The third thing I learned was that local and regional government has a critical role to play with legislation, taxation and development initiatives. You can't have big businesses without first having small businesses, and successful locally owned businesses bring many benefits to the economy that justify the intervention. They recycle more money back into the local economy, they are more reliable, they're more accountable, they generate a strong sense of local identity, their owners and staff engage in more civic activity, and they give more back to local charities.[431]

"By doing these things, Emilia-Romagna went from being one of the poorest regions of Italy after World War II to one of the richest. By combining the individual strengths of the ME economy with the community strengths of the WE economy, they ended up being more successful than either."

"It sounds quite obvious when you put it like that."

"Yes," Naomi replied. "So it always surprises people when I tell them that it was a communist local government that helped build the most successful business region in Italy. It's a similar story in Mondragon, in the Basque region of northern Spain, where they followed the same principles and got very similar results. The origins of their work lay in anti-fascism too, following the Spanish Civil War. The local Catholic priest, José Maria Arizmendiarrieta, was confronted with very high levels of poverty, hunger and hopelessness following the war, so he founded a technical college to provide training. He then asked himself a crucial question: "What does Jesuit teaching tell us about economic development?" The answers he got were 'solidarity' and 'helping each other,' which led him to the work of Robert Owen, and Britain's co-operative movement in the 19th century. Today, Mondragon has over 100,000 people who work in three hundred businesses, all owned by their workers. They've got their own bank, their own university and their own welfare system. That's what's possible when you work cooperatively."[432]

"So now you are applying these lessons to the economy here in Vancouver?"

"Precisely. Emilia-Romagna had a long tradition of working cooperatively, and they became very determined after World War II when Mussolini and the fascists tried to smash them. Vancouver's co-op movement didn't start until after the war, with Vancity credit union and some housing and food co-ops. The second crash demanded a very rapid response, however. The people who were involved in the OMEGA Days took the most important ideas and shaped them into the five solutions for Meaningful Work—the M in OMEGA."

"What about the E for a New Economy?"

"I won't have time to get to that. It's mostly to do with the larger economy."

I was disappointed, but it was only Friday so I still had time. "Okay. Don't let me interrupt you any more." Musiceum was on one of Vancouver's many car-free streets, and we sat down at a street table with a mosaic top.

"The first OMEGA M solution was titled *Let's Help Each Other*," Naomi explained. "It called for Benefit Corporations, Business Networks and Co-operative Support Networks, similar to those in Italy. The Chamber of Commerce was lukewarm to the idea, but a Vancouver group called LOCO was totally into it. They were part of a wider North American movement called the Business Alliance for Living Economies, or BALLE, which had been thinking this way for years."[433]

"What do the Business Networks do?"

"They provide their members with a free annual business check-up; they provide mentors and coaches for new start-ups; and they assist with innovation, training and financing, funded by a 0.4% levy on sales. Compared to non-members, the bankruptcy rate among the Network's members has fallen by 85%. Donations to local charities and community groups are up by 15%, 35% of the businesses are doing profit sharing, and 90% are reporting a higher rate of staff satisfaction. The average wage differential between the highest and the lowest paid workers has also fallen.

"The Networks also play an important role in Vancouver's youth enterprise movement. They helped us start a youth enterprise program in every school, and to launch the Youth Enterprise Centre where young people get help starting their first businesses. One of our millionaires, Wai Yeng Chong, got started this way when she launched her home delivery bicycle service, which she later expanded into a range of urban deliveries.

"But I need to press on. We need to walk a block for the next solution. I'll just slip in and say hi to Ruben."

I followed Naomi into Musiceum, a fascinating wonderland of guitars, lutes, oboes, horns, dulcimers and balalaikas. She chatted briefly with the owner, an older man with a bushy white beard, then gave him a kiss and dragged me out.

"Ruben emigrated from Russia fifteen years ago," she said as we headed down the street. "Within a year of being here he had started the Russian Folk Orchestra. It's amazing what you can do when your soul is big enough."

We walked into Chinatown, where the stores displayed open boxes and baskets of vegetables, herbs and spices—all without any packaging. We came to a building that announced itself as Vancity OUR Bank, with the same three stickers in the window. Inside, there was an area with comfy seating and an electronic diagram that charted the bank's services in two intertwined squares. Naomi chatted to a woman she obviously knew and then began to tell me about the changes that had taken place in banking.

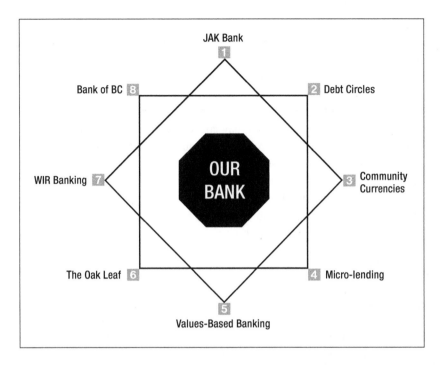

"The second of the five OMEGA M solutions is *Put Your Money Where Your Heart Is*. It's about money: the way we create it, circulate it, store it and lend it. Did you know that under the old financial system, between 1970 and 2010 there were 145 banking crises, 208 monetary crashes, 72 sovereign debt crises and 425 systemic crises? On average, there were ten countries in financial crisis every year. And they called that *success*?[434] Before the changes most banks were owned and controlled privately for the benefit of their shareholders, and new money was created by a bank whenever it issued a loan.

"To create new money, the banks just clicked a button and the money appeared on their balance sheets as a debt—on which you had to pay interest, of course. Most people had no idea it was that simple. They assumed there was some kind of government printing press, and a control on how much money was printed. Not so. If the banks had been required to keep proper reserves it might not have mattered, but they had lobbied to get the reserve requirements reduced, bringing a host of problems that were among the causes of the financial meltdowns. But that's for another day.

"When Derek was travelling in Europe he met a German woman named Margrit Kennedy who had done an analysis about the role interest plays as a cause of poverty, and the gap between the rich and the poor. She wrote a little book about it called *Occupy Money: Creating an Economy where Everyone Wins*. Imagine the world as two piles of sand: one larger, representing the wealth of those who have more than they need; and one smaller, representing those who have less. The people in Pile A want to invest their money and earn interest on it, while the people in Pile B need to borrow money, on which they must pay interest. If the interest is low, they can usually repay it without much damage. But when it is high, such as 20% for a credit card loan or as much as 4,000% for a short-term weekly loan, the wealth pours from Pile B into Pile A. The people in Pile A get richer and the people in Pile B get poorer. When Margrit calculated the impact of interest rates on the everyday cost of living she found that in a basket of everyday goods, 40% of the price came from interest payments."[435]

"But that's huge!" I said. "But is there a solution? Surely, people will always need to borrow."

"That's one of the questions the New Economy Team asked at the time of the OMEGA Days. One of the solutions is to turn the banks into regular utilities that are not allowed to engage in high risk lending. Some banks behaved very reasonably, building solid local relationships instead of indulging in risky global gambling, but most had caught the greed disease and needed to be inoculated against a future outbreak.[436]

"Another solution is public banking, but before we get to that there's the JAK Bank solution: if a thousand people pool their savings and lend to each other you've got a system than can operate without interest. There's a bank in Sweden called the JAK Bank that's been doing it for years, with very few problems." She walked over to the diagram and touched the circle at the top. It grew in size and a photo appeared of a group of people holding a JAK Bank sign. "This is one of the reasons I brought you here, since zero-interest mortgages are one of the services OUR Bank offers, using a fee to cover the cost of managing the transaction."[437]

"How does that work?"

"The people who join the JAK Bank forgo earning interest on their savings in exchange for being able to access an interest-free loan. They save some money every month and place it in a joint savings pool, in return for which they are able to access a zero-interest mortgage when they need it."

"That's really clever. But the big banks—are they still charging interest?"

"Yes, but they are more tightly regulated, as are the credit card companies. About 10% of the loans at OUR Bank are zero interest, with just a fee to cover costs. The rest are traditional loans on which people pay interest, but at a relatively low rate, since the risk is shared by The Bank of BC."

The Bank of BC? That didn't exist in my time. "What exactly *is* OUR Bank?" I asked.

"It's a branch of Vancity, Vancouver's member-owned credit union. It has been a strength of our local economy for years. This branch serves the Downtown Eastside."

"So this," she said, touching another circle, "is the second OUR Bank initiative, the Debt Circles." The circle expanded when she touched it to show a circle of people in a room. "It's a successful way to address chronic personal debt, modeled on Alcoholics Anonymous. It takes the shame out of being in debt and uses peer support to help people overcome it. With the Citizen's Income and student debt disappearing, personal debt is less of a problem than it used to be, but it's still a concern."

Back in my time, debt was a huge problem that often destroyed marriages and families, so this was a really valuable initiative, I thought to myself.

"Each Circle has a facilitator, paid by the bank, who uses very simple methods: you analyze your income and expenditures; find ways to consume less and earn more; and you share, borrow and make do instead of buying. You grow your own food, repair your own clothes, sell any stuff you don't need, consolidate your debts, pay off the smaller ones first to make you feel good, and find ways to celebrate that don't involve spending money. The groups often cook meals and organize children's parties together that don't require money."[438]

"That's very impressive," I said.

Naomi touched the third circle and an image appeared showing four currencies: the dollar and three others. "These are the co-operative currencies that we use alongside the regular dollar: Delta Dollars for the region as a whole, Eastside Hours here in the Downtown Eastside, and the micro-currencies that are used by residents of the same building or residents on the same street."

"What makes people believe in the new currencies? And are they backed by anything solid?"

"The Delta Dollars are backed by OUR Bank, which is backed by The Bank of BC, which is backed by the government of BC, so they're super-secure. The Eastside Hours and the smaller currencies are backed by simple trust and people's belief in the value of their community. They build social capital each time people use them. And they're 100% egalitarian: an hour is an hour, whether you're a consultant or a cook. After the crash there were so many people who had lost their jobs, but they still had talents and skills. The Eastside Hours allow them to continue to contribute to the community. It's like a parallel economy that operates alongside the mainstream economy."[439]

"Who organizes the currencies, and makes sure there's no cheating or forged money?"

"That's a good question, because there have been community currencies that failed. They need to be financially self-sustaining, so they use some of the currency for proper management. The Deltas are managed by a team that's funded by a small contribution from every bank in the region. The Eastside Hours are managed by volunteers from the Carnegie Centre, and the micro-currencies are run by part-timers who are paid by a small fee on the members' accounts."

Naomi touched the fourth circle, and the display showed five women sitting together. "Next we have our Micro Lending program for people who want to borrow small amounts up to $5,000 to launch or develop a business. It's based on the highly successful Grameen bank, which was pioneered in Bangladesh and is now all over the world. Compared to OUR Bank's main business, it's small, but for those who don't have a credit record to obtain a regular loan, it's everything."[440]

I thought of Emily, and her business selling flowers on the street. This must be how she got her loan.

"OUR Bank's main business is still regular banking, storing people's savings and lending them out for mortgages and business loans." Naomi touched the fifth circle, and an image appeared showing farmers and fishers, shop workers and software engineers. "The main difference compared to before the crash is that most local banks and credit unions have joined the Global Alliance for Banking on Values, and they take their commitment to social responsibility very seriously."[441]

She touched the circle again and it showed a map of the region, with a star for each investment the bank was making. She touched one of the stars and it showed a bakery in the Downtown Eastside.

"I feel really good banking here. We distribute our profits back to the community, and my savings support loans that I feel good about. There's also a mobile payment card that only charges 1% to participating businesses. It's part of the campaign to drive out credit cards. The income goes back to the Business Network and is used to promote local purchasing."

"How do you know if a business the bank makes a loan to is doing right by the environment, and operating in a socially just and sustainable manner?"

"Mostly by certification. That's what this is all about. She touched the sixth circle, and the display ran a slide show of businesses carrying the Oak Leaf logo in their windows, as I had seen at Musiceum. "I gather Laura's going to fill you in on this, so I'll move on." She touched the seventh circle, and a photo showed a group of business owners holding a sign saying 'WIR Banking with OUR Bank'.

"This is the WIR Bank, which operates independently, even though it's an OUR Bank initiative. It's a parallel electronic banking system run by businesses for their mutual benefit, which enables them to lend to each other when the regular banking sector seizes up for whatever reason. *Wir* is German for 'we'. It's based on the WIR Bank that was set up in Switzerland in the 1930s during the depression, when Swiss businesses realized that since they all had assets, they could use them to create their own money, and work around the financial collapse that had destroyed their normal sources of credit. It's one of the secrets that keeps the Swiss economy purring along, like their famous watches. It was people like Derek who helped spread awareness of the WIR bank, opening people's eyes to the fact that something like this was possible."[442]

"So finally there's the eighth circle." Naomi touched it, revealing a map of British Columbia with a circular logo announcing The Bank of BC, surrounded by a collage of people working: foresters, teachers, students, builders, nurses.

"This is the Bank of BC, that I mentioned earlier, which gives stability to the whole economy. It's a publicly owned bank, modeled on the Bank of North Dakota, which has been so successful in stabilizing North Dakota's economy and pulling it through both financial meltdowns.[443] It's equivalent to having a central bank for the province. It serves as a depository for the government's assets from taxes and fees, and it works in partnership with the banks and credit unions to support targeted economic development. It also provides subsidies for research and development, loan guarantees, and preferences for government procurement. And through its Earth Bank subsidiary, it provides loans and financing for local community economic planning, assisting each region of the province to create a blueprint for a green economy to guide development."[444]

"And just so that I'm clear," I responded, "it can create new money, the same way that the private banks do?"

"Yes. Its ability to create money is one of its most important functions. Since it was launched it has created and loaned out several billion dollars for targeted investments, including sustainable energy, building retrofits, farming, the electrification of transportation, and the knowledge economy. It also guarantees municipal bonds, which has made it cheaper for municipalities to borrow the funds needed to build bicycle routes and expand public transit. The interest on the loans returns to the public sector, where it helps with education and healthcare.[445] It is owned by the government, on behalf of the people of BC, but the management is strictly arm's length to prevent any possible corruption or conflicts of interest. Just because a bank is local doesn't mean its managers are immune to stupidity."

"So why were we told so often that governments would mess things up if they got involved in the economy?"

"A hundred and fifty years ago, when women were campaigning to get the vote, why did some men insist that they should be denied it? Was it really because women were unable to think clearly and incapable of making a political decision? It was pure prejudice in defense of private interests, that's all. The neo-liberals wanted government to be as small as possible so that they could be as large as possible and not have to bother with regulations, or even taxes. In other words, it was simply untrue. There's an Italian economist called Mariana Mazzucato who busted that myth wide open. She showed how some of the top companies such as Apple and Google were only successful because they received government research grants, and benefited from the pioneering work done in government research labs. Fundamental scientific research is probably the single most effective generator of progress, so whenever I hear of governments that don't support their scientists and don't believe they are worth investing in, I scratch my head and think 'what don't they understand?'

"But the private sector does not like to invest in the big breakthrough ideas where the chances of failure are high. You need an entrepreneurial state to do that.[446] That's one of the reasons why China has made such rapid breakthroughs in clean energy: its state banks have advanced trillions of dollars to solar, wind and electric vehicle companies. So now we can do the same. It's a great addition to our public portfolio, along with publicly owned power, insurance and health care."[447]

"What about the Bank of Canada? Has that changed, too?"

"Canada's fortunate, for its bank was always owned by the people through the government of Canada. Until 1974, the government paid for costs not covered by taxation through a mixture of issuing Canada Savings Bonds, and using the Bank of Canada to create money for big national investments, with the interest benefiting the public purse. In 1974, however, the newly formed private Bank of International Settlements persuaded or arm-twisted the government to stop creating money and to borrow from the private banking sector instead. The private sector created the money out of thin air, the way all bankers do, but the interest payments now went to the private banks, not back to the Canadian people. This led to public debt, as it did for governments all over the world. The bank has now reverted to the previous mixed approach, and the government's finances are far better balanced, thanks to this and other changes that have resulted from the new economy."[448]

"Is the new economy still capitalist?"

"You're full of pertinent questions, aren't you? You're just like I was when I went to Israel. Why this? Why that? It's a really good question. Some people say we still live in a capitalist economy, because they don't want to stir up unnecessary fears. But if the foundation of a capitalist economy is the primacy of competitive capital, and everything else—social costs, environmental costs, human costs—are an externality, then no, Dorothy, the economy is no longer capitalist. The fundamental principle that social, environmental and spiritual progress are just as important as economic progress has been accepted by the majority of Canadians, and government policy reflects the change. Most people now support the idea that cooperation works better than competition to build a strong economy. That's why there's support for the Business Networks. And that's why I tell people we no longer live in a capitalist economy: we live in a cooperative economy. It's still free market, the kind Adam Smith admired, but the primary impulse is cooperative instead of being about maximizing the growth of capital. And we still believe in sound monetary policy, alongside sound employment policy. It's still really important to take care of the needs of investors, who risk so much to create something new. They need the stability that sound monetary policy brings. In fact the cooperative approach brings *more* future stability, not less, so it creates a better scenario for investors."

Wow. Was capitalism really evolving into something new? That was a big idea to take on board.

"How does all this affect economic growth?" I asked. I had grown up to think that capitalism and economic growth were two sides of the same coin.

"That's another very pertinent question. Under capitalism, the fundamental goal was that capital should grow. If it didn't grow enough in beans, you took it out and put it into peas. That's why perfectly good companies were taken apart and destroyed, and why so much harm was done to the forests, the oceans and the habitat of so many species. It's also why workers were often treated like shit, because the lower their wages, the higher the return capital could get. The whole global economy was like a crazy machine stuck in top gear. Extract! Produce!

Consume! Destroy! Over and over, until we were dangerously close to there being nothing left.

"A cooperative economy is no longer driven to grow in purely financial terms. It seeks broader goals, using a wider and more balanced definition of growth. That's the essence of a B Corporation—it's no longer fixated on one goal to the exclusion of all else. So we no longer measure growth as GDP, but as Genuine Progress, and most businesses accept that green certification is necessary to show that the goods or services they sell are helping to restore the environment, not destroy it, whether deliberately or not. I hope there will come a day when a business is no longer allowed to operate if it has not achieved a certain level of green certification, and green auditing will be as normal as traditional auditing. We're not far off here in Vancouver and in many European countries, but globally there's a long way to go."

"But won't a cooperative economy still need to generate growth, even if it's not physical growth? How will people get their pensions, if they can no longer invest in stocks and bonds that grow by a certain percentage every year?"

"That's a big question. People will always want to borrow money to start new ventures, so banking and moneylending are going to be an important part of any economy, and I don't see any end to the charging of interest. It's the extortionate, rip-off interest rates of the past that needed to end. As long as humans have a spirit of adventure there's going to be growth: it's ecologically disruptive growth that needs to end, not other kinds of growth. But we need to be moving on, since we're only halfway through the tour. Wei-Ping said I should give you the works."

18
Earth's Furthest Dream

"RIGHT!" NAOMI SAID, taking a drink of water. "Everything we've looked at so far relates to the first and second of the five solutions in OMEGA M for Meaningful Work. So let's move on to the third and fourth, and Laura will fill you in on the fifth."

Walking out of the bank, I had an image of invisible threads connecting the hearts, minds and money of people throughout the region, and connecting them all to nature. For a few moments, I was transfixed. It was as if I had penetrated another realm and could see the connections that held everything together, building the new economy through the practical application of people's initiative, mutual care, dreams, good governance and care for the Earth. It was so *obvious*. And it was being made a reality.

"Are you okay? You look as if you've seen a ghost."

"I think I have," I said. "But not so much a ghost—more perhaps an angel. All this that you've been showing me, combining the ME economy, the WE economy and the Earth—it's so simple. Why did we have to go through such huge financial meltdowns before we could see it?"

"Ah," Naomi replied. "I used to feel the same way about Israel and Palestine. Why did it have to be so hard? Are the shells of our egos so tough, so determined, that they have to be smashed by something awful before we finally see the light? As soon as we start cooperating, life becomes so much easier. When you give, give, give, life showers you with happiness. When you take, take, take, it showers you with shit.

"Some people seemingly don't want to give: they *want* to fight it out as individuals, to try to dominate. And they group together to seize control. That's what happened in America, which ceased being a democracy because of it. Do you know the saying, 'the lunatics have taken over the asylum?' Well in America, the plutocrats took over democracy. At least, they did until the second American Revolution, but that's a whole other story."[449]

The second American Revolution? The *what*? Something told me to remain quiet, however, and hope to learn more later. I had lots of time.

"Where are we heading this time?" I asked, as we continued down the street.

"I'm taking you to the Community Economy Centre. But first we're going to drop in on Singh's Carpets. Not the factory—that's out in Surrey—but one of the local stores. It's a good example of the third OMEGA M solution, *Share the Work, Share the Wealth.*

Singh's Carpets was a small store with a window display that showed the interiors of rooms with digitally changing carpeting options. It had the same three stickers in the window as Musiceum and OUR Bank. The manager, Jumala Singh, wore a traditional red and blue Sikh dress with purple yoga pants and a blue headscarf over her blonde hair. Finnish mother, Indian father, Naomi told me afterwards. After introducing me, Naomi asked Jumala to tell me how they started their wage equity program.

"When Singh's Carpets joined the East Vancouver Business Network," Jumala said, "and went on to become a Benefit Corporation, one of the commitments the owner made concerned wage differentials. The managers used to get a bonus, and over the years this had increased the differential between the shop floor and store managers who were earning $50,000 and the CEO who was taking home $1.5 million. That's a 30:1 ratio. In other companies, some CEOs were earning 300 or even 3,000 times more than their employees. He was earning in ten days what it took me an entire year to earn. And that hurt.[450]

"We make five-star recycled carpet squares, and we have branches all over Western Canada. Gurmail Singh—the owner—invited the Green Economy Institute in to offer advice. When they interviewed us they found that some of the workers were sitting on ideas that could make the company a lot more profitable, but they were unwilling to share their ideas because they resented the big payouts to the bosses. As a result, Gurmail agreed to reduce the wage differential to 10:1 over a ten-year period. He brought in profit sharing and innovation sharing, and offered us share-ownership in the company. Our profits have increased every year since the changes, and we've become a poster-boy for social equity in the workplace, which is why people like Naomi keep bringing us visitors like you."[451]

We laughed, and then I asked, "Did he reduce the CEO's salary, or increase the workers' wages?"

"You haven't heard? We were expecting a bit of both—but then out of the blue he announced that he was cutting his personal salary to $100,000 and distributing the rest among the lower-paid members of the workforce. I feel so proud to be part of this company, and last year I received a $3,000 bonus through profit sharing. The company has also joined 2% for the Planet: every year we donate two percent of our profits to an environmental cause. Last year we supported women's environmental education in Pakistan."[452]

"Do you have a labor union?" I asked.

"Yes, and we also have union representation on the board of directors," Jumala replied. "It took a bit of a struggle, but we bombarded the board with arguments about the natural justice of allowing workers to organize together, and they finally relented. After all, Guru Nanak, the founder of Sikhism, was keen to eliminate hierarchy and inequality. So the board wasn't on strong ground, spiritually."

"The whole union movement went through a crisis as a result of the OMEGA

Days," Naomi told me. "Prior to the OMEGA Days we were seeing a decline in union membership, and in public awareness about what unions do. And to be fair, there were still some union organizers who defined 'success' as 'conflict,' and the mentality of some union leaders was quite antagonistic, which put people off. They were trapped in the dominance agenda, and not everyone got the importance of building a cooperative economy. That changed as the result of a grassroots drive within the labor movement that demanded more support for co-operatives, employee share-owning and reduced wage differentials. I'm sorry, Jumala—I got diverted."

"That's okay," Jumala replied. "So as I was saying, with all the changes, I'm really happy to come to work. Would you like to lease some recycled carpet squares for your home? They are made from 100% recycled biodegradable material and they are 100% recyclable in case they ever wear out. Which they never will, of course."

I chuckled, and explained that I was just visiting. "Can you still attract the senior managers you need at the lower salary level?" I asked.

"Well, I'd say $400,000 a year is pretty rich," she replied. "There's been no shortage of applicants for the senior jobs."

"It's fascinating," Naomi said, stretching and rubbing her hands over her face. "Studies show that on average, when people become rich they become less generous and less socially motivated than those who have to get by on a lower income. Becoming wealthy actually seems to make people more selfish.[453] It seems the higher their personal capital, the less they are willing to contribute to social capital. So reducing wage differentials is fundamental if we are to build a world that is fair and sustainable.

"The Business Network has a pledge that business owners can sign to reduce their wage differential to 10:1. It's not perfect, and it's twice the differential at Mondragon, but it's a big step forward. The government has agreed to reduce public sector salary differentials to the same ratio, increasing the wages of the lower paid and reducing the wages of the higher paid civil servants—and their pensions, too."

"I bet that ruffled some feathers."

"It did, but that's life. There was a book that inspired me twenty years ago when I was studying economics. It's called *The Spirit Level: Why Equality is Better for Everyone*. The authors used data from twenty-three countries and the fifty U.S. states to show that the more unequal a society is, the more it will suffer from a long list of social ills, including crime, imprisonment, anxiety, mental distress, lack of trust, eating disorders, high school drop-outs, teenage births, even lower math and literacy scores. It will also have lower rates of life expectancy, literacy and social mobility. It was a huge eye-opener."[454]

"I can see why that would be," Jumala said. "When our CEO was earning $1.5 million it was easy to want to steal something, or to cut corners in a way that would hurt the company. I remember the resentment I used to feel."

"Are you doing any other initiatives to reduce inequality?" I asked Naomi.

"Yes. We've set up the Businesses In Shared Ownership Network, BISON, which encourages businesses to support employee ownership. It's very motivational when employees become joint-owners, as Jumala is here. It creates trust, since they get to see the books and share in the important decisions."

"She's right," Jumala said. "As a shareholder, I see the annual accounts and other key documents. It's not just my job that depends on the success of the company; it's also my investment."

"They should put you on the board, Jumala. I always recommend that businesses appoint more women to the board."

"You really think it makes a difference?" I asked.

"Absolutely," Naomi replied. "Having more women on the board brings greater stability and better long-term returns. There was a lot of publicity about it recently when an older company director who had been replaced by a woman sued for wrongful dismissal, citing sex discrimination. The directors argued that they had appointed her to maximize the shareholders' returns, citing evidence that having women on the board brought higher returns on equity, sales and invested capital, as well as social and environmental gains. The court ruled in their favor, and ever since we've seen shareholders demanding more women on the board, and generally getting their way. But look, Jumala, we've taken enough of your time."[455]

We thanked her in the Indian style, hands held together, and stepped out onto the street. A long yellow bike passed being ridden by two women and eight smiling children. One of the women saw me smiling at them and gave me a wave. "Happy days!" she called out, as they rode on down the street.[456]

"A surprise a minute!" Naomi exclaimed. "Do you know what the science fiction writer H.G. Wells said about the bicycle? 'Whenever I see an adult on a bicycle,' he said, 'I have hope for the human race.' How can you not be happy, riding a bike like that? That gives me an idea. We can ride over to the Community Economy Centre; it'll save us ten minutes. There's a Bixi-Bike rack just around the corner."

A few minutes later we were cycling along on two bright red bicycles with sturdy upright handlebars. It felt good to be back on a bike—bubbles of happiness rose up my spine. With no cars on the route I was free to look around and enjoy the city going about its daily business. Everything seemed very normal. Compared to my time, there were simply more bicycles, more trees, more seats, more art, more flowers, and more planters filled with food. And more electric vehicles.

The Community Economy Centre was a new five-storey building sandwiched between two older buildings. The front was made from a glass-like material, decorated with curving lines of planted vegetables that climbed up the building in a repeating figure-of-eight. It looked stupendous, with bright salad greens against the shiny glass.

When we had parked our bikes I took a closer look: the plants were moving at a very slow speed, which would give someone on the ground time to pick the greens.

"Don't you love that?" Naomi said. "It's solar glass. It generates power for the building and helps keep it cool in summer and warm in winter."[457]

We entered through revolving glass doors and the inside was even more spectacular—an open atrium filled with light. And on the wall, just as you entered, another *Change the World* poster, this time with a photo of Wangari Matthai, the visionary and tree-planting activist from Kenya,[458] with these words of hers:

> *In the course of history, there comes a time when humanity is called*
> *to shift to a new level of consciousness, to reach a higher moral ground.*
> *A time when we have to shed our fear and give hope to each other.*
> *That time is now.*

That time is now. How much that was impressing me. Hanging down from somewhere far above there was a circular wheel with twelve spokes, each of which ended in a disc that carried a sculpture, some showing groups of people and some showing abstract shapes or buildings. In the center there was a larger sculpture with twelve people holding hands.

"What's this about?" I asked.

"It's the twelve dimensions of Vancouver's local economy, with the Community Economy Centre at the center." I looked more closely, and sure enough, each disc carried a label—the Benefit Corporations, the Business Network, the Co-operatives Network, Values-Based Banking, Green Business Certification, the Labor Movement, Community Land Trusts, The Nature Council, Community Welfare, Schools, Colleges, and Local Government.

"There's far more here than I can explain," Naomi said. "The Nature Council has reps from all the local environmental groups. The Green Business Certification Laura will explain. Each sector appoints a rep to the New Economy Circle, which guides the development of the economy."

"And the Community Economy Centre... where does it fit in?"

"Let's sit down, then I must be getting back to work. I'll call Laura to meet you here in half an hour. I believe she's going to take you on a tour of the Triko-Op."

We found ourselves a quiet corner with a couple of comfy armchairs, and Naomi continued. "The Community Economy Centre convenes the monthly meetings and keeps us all in touch. It also runs OUR Community Welfare, which I'll get to in a minute, and the Community Development Corporation, which takes on critical projects. It was instrumental in starting the Youth Enterprise Centre, and The Hub, the new center for farmers and food businesses. But let's get to the reason I brought you here, which is our new system of community welfare.

"The fourth OMEGA M solution addressed poverty and welfare: it's called *All for One and One for All.*

"It sounds like the Three Musketeers."

"It's the same idea of loyalty to the group, but it's about the transformation of poverty. When we look back in history, almost anywhere, if you didn't have a family or a tribe to support you when you were down on your luck, life was miserable. You had to swallow your pride and beg on the streets. It's one of the reasons people had children, to look after them in old age or when they got sick. The paradox is that it's economic development itself that's killing the extended family and the security it provides. It has given people independence and off they go to distant corners of the world. Look at me, an only child, leaving my parents in New York.

"So how should we respond when people are in distress?" she continued. "It's a question every culture has had to address. On a national level, the Citizen's Income has made a big difference, combined with the raised minimum wage, the subsidized daycare, the Living Wage Pledge that many employers have signed, the end of student debt, and all the tax and benefit changes. Together, they have taken the precariousness out of many people's lives. But we still have poverty, and we're working to tackle it in several additional ways. As well as the business development work, we've increased the availability of affordable housing; we're transforming the Food Banks; we've expanded community currencies; we're encouraging the sharing economy; and we've redesigned the way welfare works, changing it from a passive to an active process."

"How did the Citizen's Income come about? That must have been huge—and I would guess quite controversial?"

"Yes, it was. It took a lot of campaigning, getting the public to understand what it was and how it could be financed, and getting the politicians to embrace the idea. But the studies really helped. The Basic Income Earth Network persuaded a dozen countries to co-operate in a huge, networked global study in which people in a thousand different towns all received the Citizen's Income for a year, and then they analyzed the results to see what worked and what didn't.[459]

"It wasn't so much the Citizen's Income that was controversial; it was the fact that it's financed by a redistribution of wealth away from the wealthy, towards those who have less. So yes, it was definitely controversial. But as people realized how unequal Canadian society had become, and how difficult life was for a growing number of people, public sentiment changed, which made it possible for the political parties to champion the cause, and for the government to gather its courage and make it happen."[460]

"It's not just Citizen's Income," Naomi continued. "That's just one part of the National Strategy to Reduce Poverty and Inequality. After the OMEGA Days, which did so much to heighten public awareness, poverty and inequality became a national priority, and the political parties had to show they meant action. That in itself was valuable, because it convinced many people who didn't normally vote to do so, often for the first time."

"What else was in the National Strategy?" I asked. There was no such thing back in my time.

"It had five components: a strong foundation, and four strong pillars. The foundation contained the commitment, along with measures to remove the power of money from politics, and the formation of a broadly-based Genuine Progress Council to advise the government on the implementation of the strategy.

"The four strong pillars were Fair Economy, Fair Pay, Fair Taxes and Fair Benefits. The Fair Economy pillar covered a lot of the things I've just been showing you. Fair Pay included a raised minimum wage, going from $15 to $20, and the $30 living wage that some employers are now paying. It also included the restoration of legislated support for labor unions, and the guarantee of work in the community for anyone who has been unemployed for more than six months. Fair Taxes included a small increase in tax for the middle classes, raising the level of tax for people in the highest income bracket from 45% to 65%, and bringing in a lifetime tax on inheritances and gifts to tackle what we call the Picketty Gap."

"What's that?"

"It's the gap between income earned by means of economic growth, and income earned from the growth of capital, named after Thomas Piketty, the French economist I mentioned earlier.

"And finally there's Fair Benefits," she continued, "which includes the Citizen's Endowment for 21-year-olds, as well as the Citizen's Income, subsidized daycare and increased Child Benefit."[461]

"A Citizen's Endowment? What's that?"

"It's the new arrangement under which every resident Canadian receives $10,000 when he or she turns twenty-one. It's financed by the lifetime tax on inheritance and gifts, the purpose of which is to level the playing field a bit and reduce inequality, which has become so unfair and corrosive. It's only a fraction of the inheritance people typically receive if their parents own property, but it's a start. You've got to remember, many Canadians have parents who were never able to buy a house, so they don't get *any* inheritance to speak of. The *average* inheritance is $150,000, and when you exclude those who get almost nothing, those who *do* inherit are getting a lot more than that. So there is both justice and logic to the inheritance tax.[462]

"I've got a friend, Annie," Naomi continued, "who's a single mother with two teenage girls. They're both lovely kids. Before the changes, when the girls were young, she was forced to go out and find part-time jobs, and she really struggled. Since the changes, she has been able to devote herself to being a full-time mother, and now that the girls are at school she's going to college herself to improve her education. If she was still working at two part-time jobs those girls would be running wild and getting into all sorts of trouble. So I've seen the difference it makes."

"What do most young people do with the money they receive—the Citizen's Endowment?"

"According to a survey that was done last year, forty percent put it into some kind of investment; a quarter use it to support themselves while they're at college;

a fifth use it for various living expenses; eight percent use it to buy a car; and seven percent use it to start a small business."

These were big changes, I thought to myself. I had heard of Citizen's Income, or Basic Income as some called it, but it wasn't being discussed much back in my time. And the idea of a Citizen's Endowment was not on anyone's radar.

"When I first arrived," Naomi said, "Wei-Ping asked me to work on welfare. So I found a student who wanted to look into it and she gave us a really good background. Welfare 1.0 is simple mutual aid, she told us, the ancient tribal approach of caring for your family, neighbors and kin. It continues throughout the world and probably provides more support than any other system, even in the developed world. Have you come across the Canadian cultural anthropologist and explorer, Wade Davis? I attended a talk he gave and he was telling us that among the Penan people of Borneo, whose culture has been totally ravaged by illegal logging and the destruction of the forests they lived in, the greatest transgression among their people was something called *sihun*, which means failure to share. When one of the Penan people came to Vancouver, he just could not understand how we accepted homelessness, since to them, the existence of one poor person shames everyone. That's how deeply they took the importance of sharing. There's something similar among most indigenous people.[463]

"As tribes began to merge into larger human gatherings, however, mutual aid began to break down, and the new monotheistic religions had to address the suffering caused by poverty, exploitation and hardship, so they developed their own traditions, which became Welfare 2.0. The Jews embraced the importance of *tzedakah*, or charitable acts, putting a special emphasis on gifts that help people stand on their own two feet. The Christians urged people to love their neighbors as themselves. For Muslims, giving charity to those who deserve it is one of the five pillars of Islamic practice.

"The Byzantines, who ruled the eastern Mediterranean following the collapse of Rome, placed great importance on *agape*, the selfless love that reaches out to strangers who are not family or friends. They built a lot of hospitals and orphanages, and poorhouses for the lost and weary. In China's culture you owe a very strong obligation to your extended family, and maybe to some immediate neighbors, but after that the government is expected to pick up the pieces.

"In England, the monasteries took on the role of caring for the poor, but when King Henry—the one with all the wives—dissolved the monasteries, there were beggars and starving people all over the streets. So Queen Elizabeth's Parliament passed the first Poor Laws, making local parishes responsible for vagrants, beggars and people who had fallen on hard times. This is Welfare 3.0, where the state gets involved for the first time. Later, there were workhouses for the poor and unemployed, wretched places that were close to prisons where children were torn from parents and husbands from wives. It was either that, or starvation.

"The impulse for change arose during the Enlightenment, encouraged by people like Jean Jacques Rousseau, the French philosopher whose writings inspired the French Revolution. Then in the late 19th century Germany's chancellor Otto von Bismarck put the ideas into practice with Welfare 4.0, the world's first state welfare program, with universal old age pensions, health care and employment insurance. That's how the welfare state got started, ending the need to beg on the street or enter a workhouse if you lost your job, became ill or had an accident. During the 20th century most developed states adopted it, and Welfare 4.0 became a standard feature of the modern world.

"When the financial meltdowns happened, however, it was the people in the precariat who suffered most—the jobless, the homeless, and anyone struggling to get by on welfare, a minimum pension or part-time jobs. Welfare 4.0 was becoming harder and harder to hold together, with its complex rules and its poverty trap for jobseekers on welfare who had to pay an effective tax rate of 80% on any earned income. The shift to Citizen's Income, or Welfare 5.0, was the response.

"The Citizen's Income is not enough for everyone to survive on, however. It unifies a host of different programs and saves millions on bureaucracy, but there are still some who need a top-up. So we created Welfare 5.1 by adding reciprocity, restoring self-respect and mutual aid. The foundation of our approach is the fundamental importance of human dignity. It's right there at the top of the *Universal Declaration of Human Rights* that was written in 1948, following the horrors and sufferings of World War II: *All human beings are born free and equal in dignity and rights.* We adopted the premise that when people receive the gift of welfare from the community they should be encouraged to give something back to keep their dignity. Article 29 of the *Universal Declaration* states it well: *Everyone has duties to the community in which alone the free and full development of personality is possible.*[464]

"We wanted to restore the mutual aid relationship, so while the government still pays for welfare, its distribution is managed by the community centers, where relationships are local and people know each other. The staff are trained in Life Purpose Counseling, and everyone who receives welfare is encouraged to give back in a way that brings meaning and dignity, whether by volunteering in an urban farm or attending a training course."

"What's Life Purpose Counseling?"

"It helps you identify your natural skills and find your sense of purpose. Once your inner motivation takes over, everything becomes easier. Without it, people drift around, not knowing what their purpose in life is."[465]

We were sitting in the atrium close to the hanging wheel, with people coming and going. "How did people respond to the change?" I asked.

"Some of the welfare rights organizations were distrustful, but when they saw how people welcomed it, they came around. It makes people feel part of the community. Poverty experienced alone is far worse than poverty shared among friends and community. People are pleased to be able to give back, instead of feeling like the worthless recipients of charity. Are you familiar with the Lebanese poet Kahlil

Gibran, and his lovely book, *The Prophet*? Here, let me show you." Naomi said a few words to her phone and then projected Gibran's words onto a nearby wall:

> *You work that you may keep pace with the Earth and the soul of the Earth.*
> *For to be idle is to become a stranger unto the seasons,*
> *and to step out of life's procession,*
> *that marches in majesty and proud submission towards the infinite.*

> *When you work you fulfill a part of Earth's furthest dream,*
> *assigned to you when that dream was born...*
> *To love life through labour is to be intimate with life's inmost secret.*

"That's beautiful," I said. "How is the new system working?"

"People like volunteering, because it puts them in contact with others. It helps them build their skills and gives them a reason to get up in the morning. As a result, people spend less time on welfare and fewer get stuck in long-term unemployment. When you use your natural skills, you tap into something deep. I've seen people come into a community center with depression and defeat written all over their faces and six months later they are excited about the work they're doing, whether it's paid or voluntary. We run Encouragement Clubs where people help each other develop their dreams and embark on projects. It's like Kahlil Gibran says, work is so important. It's not just about earning an income. It's about expressing your purpose in life and contributing to the community."

"That's a very big vision."

"Yes—I suppose it is. I love the concept of Earth having a dream—it's like the Jewish and the Muslim dream of paradise, of a truly deep potential that's yearning to be realized.

"Another way we help people is through Time Banking, which is a form of community currency," she continued. "Many of our voluntary organizations are members of the Vancouver Time Bank. People are able to earn 'hours' by engaging in voluntary service, which they can spend on other services. Each person who receives welfare is given ten hours, which they can spend as they like, and they are encouraged to earn additional hours by providing a community service that reflects their skills and interests."[466]

"It must make a big difference for people to have their spirits lifted like that."

"Absolutely," Naomi replied. "Before the OMEGA Days most people never questioned the fact that many people went around carrying a heavy burden of despair. That's gradually changing."

"How has the new approach been received here in the Downtown Eastside?" I had already seen the difference, but I wanted to hear Naomi's take on it.

"That's a whole other story, because the sheer volume of problems makes it so much more complex. In 2020, when Vancouver missed out on being the greenest city in the world because of the poverty in the Downtown Eastside, there were over a thousand people living homeless, in spite of all the efforts the city had made.[467] It has taken a whole raft of initiatives to make an impact. Some of the people here have experienced a traumatic head injury; some have had seizures. Most have a mental health challenge of some kind, and the use of crack cocaine and crystal meth can cause people to have psychotic episodes. Many have experienced horrific abuse of one kind or another. Some are military veterans. Ten years ago most were using injection drugs and had a substance abuse disorder of some kind, and many had been exposed to hepatitis C or were HIV positive. It was a huge challenge."[468]

"How did you even begin?"

"Following the OMEGA Days there was a surge of determination to tackle the problems. A new healing initiative was launched and the network of mentors known as Judy's Angels was set up, named after Judy Graves, the Vancouver woman who had shown that if you have enough patience and determination you can cut through the barriers that keep people on the streets. Her Angels befriend and support homeless people for up to a year."[469]

So that was the origin of Judy's Angels, I thought to myself.

"The drugs problem has become a lot easier with the safest classes of recreational drugs being legalized. That has removed the criminal element and made it easier for people to tackle their addictions. We now have legal safe injection sites, detox and treatment on demand, and recovery coaching to help ex-addicts remain in their homes.[470]

"The sex-trade has also been transformed now that Canada has embraced the Scandinavian model, decriminalizing prostitution but making buying the services of a prostitute illegal, going after the johns instead of the women. The goal is to end prostitution entirely, since it is fundamentally abusive, but there's still a lot of debate, with some people saying we should adopt the Dutch model, which legalizes prostitution and enables the sex workers to run their own businesses and co-ops.[471]

"The housing revolution has also been important. The OMEGA Days brought a lot of changes, including the federal government's new national housing strategy and their commitment to the principles of Housing First.[472] That's how we got the funding to buy up the old motels along Kingsway and convert them into housing co-ops, with tax incentives for landlords who are willing to deconstruct their old rooming hotels and replace them with new ones. For every person who gets off the street, the city saves $16,000 a year: it's cheaper to provide a home and restore a person's ability to self-manage than to pay for shelters and deal with all the distress and disorder that results from homelessness."[473]

"That's a lot of change."

"Definitely, but there's still a long way to go. You can't heal such deep personal distress overnight. When I arrived in Vancouver there were twenty food banks and a hundred agencies providing meals to 15,000 people a week. Today, we are down to just two of the old-school food banks serving around a thousand people. There has been a huge drive to change things through a program called The Last Food Bank. Every food bank has been invited to join a five-step program of change: issuing vouchers that can be spent at a farmers' market instead of handing out actual food; organizing food-buying co-operatives where people pool their funds to buy nutritious food in bulk; setting up community kitchens where people learn how to cook nutritious meals together; organizing community gardens where people learn how to grow their own food; and hosting weekly community feasts, cooked and prepared by people who were previously food bank recipients. It's a complete turnaround, and now that people are used to it, it's become really popular. Previously, going to a food bank increased a person's sense of hopelessness, even if they didn't acknowledge it openly.[474] There is also more free food available, which is good because it's still needed. Ever since the law was passed banning supermarkets from throwing out waste food, there has been a reliable stream of food coming into the Community Food Centers.[475]

"Today, if you go into a Community Food Center you are likely to be invited to join a discussion group where people develop a critical consciousness of their reality, overcome their internalized feeling of oppression and learn how to participate in the transformation of their lives and the community: all through food. It's based on the work of Paolo Freire, the Brazilian education reformer from the last century."[476]

"My father used to talk about him."

"Yes. It's a shame his work is not better known. Speaking of liberation, there's another change that's also making a difference: the change in Canada's prisons policy. Prison used to be such a dysfunctional system, especially for people from Canada's First Nations. They were only four per cent of the population, but they were twenty-three per cent of the prison population.[477] Prison had become a revolving door for many of the Downtown Eastside's First Nations people, yet it cost $113,000 a year to keep someone in a federal prison, which was ridiculous.[478]

"The First Nations were very active during the OMEGA Days. They had a well-organized agenda for change with their own set of solutions. When we finally got the change of government in Ottawa, there was a complete shift in the relationship with the First Nations. The old *Indian Act* was scrapped, many treaties were completed with a waiver for the accumulated legal costs, and there was rapid movement to self-government, bringing ownership of the land. In the prisons, the changes brought an end to solitary confinement, more restorative justice, aboriginal healing circles instead of traditional sentencing, and funding for community-based healing lodges managed by the aboriginal communities.[479]

"When all these things came together—the affordable housing, the legalization of drugs, the restoration of the rooming hotels, Housing First, the prostitution changes, the funded recovery coaching, Judy's Angels, the prison changes, the sharing economy, Citizen's Income, community welfare, community economic development, the changes to transportation and street design, community arts, the community food movement—when all these came together, things finally began to change." Naomi paused. "It feels as if the air itself is finally being cleaned, after so many years of grief and suffering."

"So Vancouver is achieving a kind of spiritual healing, as well as physical healing?"

"It's a good way to put it. When Vancouver was declared The World's Greenest City, sharing the honors with Copenhagen and Portland, Mayor Reimer came to the Downtown Eastside to make the announcement since the judges had made a special point of praising the changes here."

"It's inspiring to see how much you have achieved."

"It takes a city to change a city. And yes, we've done quite a lot. When people woke up and realized that we were facing a global emergency, they knew they needed to do something. They couldn't sit by any more, watching like spectators.

"Vancouver, Berlin, Jerusalem, London, Cairo, Shanghai: we're all part of a huge wave of global awakening. If we hadn't organized, the initiative might have been seized by fascist groups like the Alphas and the Patriots. And that would have been grim.

"What we're doing here in Vancouver and elsewhere is not only a positive move towards paradise. It's also a defensive move against fascism, racism and narrow-minded intolerance. So much of the anger that fuels those things is displaced anger that picks on the easiest target—the new immigrants, the blacks, the Muslims, the Jews. The real anger is against personal failure, and the inability to overcome the forces that cause poverty and unemployment, debt and home foreclosure. That brings a feeling of emasculation and shame on top of the hardship. That's a really big burden to bear. In the past, you could align yourself with popular socialism, which made it easy to organize and to see the fault in the system, instead of carrying the burden of failure personally.

"From what I understand," Naomi continued, "the habit of displacement often starts in childhood if you have a dominating or oppressive mother or father who imposes irrational authority and control. The displacement of anger against your parents becomes an unconscious habit that's easy to apply elsewhere. You don't want to deal with your anger at your personal failure, so you take it out on someone else. That's how the fascists and the racists flourish.[480]

"When people are able live in harmony with their hearts, however, and to work together to create new forms of economic security, it's harder for the dark forces to take root. The bright rainbow colors of our cooperative economic efforts are the best defense against the black shadows of fascism and racism. Things could have gone so differently if the OMEGA Days had not been so committed to positive solutions. Globally, it's still a very open question. We're just one small corner of the Earth. If people fail to advance a similar agenda elsewhere, the millions who are struggling and frustrated in cities and villages around the world could yet turn to the dark side. If that happens, we'll see global entropy, not syntropy. And it will not be a pretty sight."

I sat there silently, absorbing Naomi's thoughts. I recalled my dream in

Dezzy's garden. A black hole, oozing danger. The black hole of fascism and hatred, sucking up everything light and beautiful because people saw no future security apart from reversion to clan loyalty, with its aggressive bonding and exclusion of strangers.

"You probably know that they've still not caught whoever it was who killed Derek Brooks," Naomi said. "I remind myself of it every day. It tells me that on some level they're still winning. Somewhere, his killers are laughing, mocking our efforts. There are plenty of people who dismiss Vancouver as west-coast woo-woo, where people dance in the street wearing yoga pants and the police ride bicycles and hand out donuts. Or was it the other way round, with the police wearing the yoga pants? I forget. At best, they dismiss us as irrelevant. At worst, they want to destroy what we're doing and re-assert the power of the wealthy, and private capital. So don't be fooled by all the wonderful things you see here."

It was a sobering thought, after all the positive things she had shown me.

"Hi Naomi! And you must be Patrick?"

My thoughts were interrupted by a youngish woman with curly black hair, a soft Scottish accent, a small blue Celtic tattoo on her arm, a green summer dress and a happy smile.

"Hi Laura," Naomi replied. "He's all yours! He's probably a bit shell-shocked, so treat him gently. It's been a pleasure showing you round, Patrick. I wish you luck in Sudan. Keep your eyes on the global stage: that's where things really matter."

19

Building a Circular Economy

"SHELL-SHOCKED, ARE WE?" Laura commented in a lyrical Scottish accent, as we walked out of the Community Economy Center. "That wouldna surprise me, knowing Naomi. She's quite the firecracker. How are ye liking Vancouver?"

She was pretty. *Very* pretty.

"I love it… it seems so alive. It's great that so many people are walking and cycling."

"Yes, Vancouver's good for that. Edinburgh's the same. There's been such a change over the last few years. I'm glad I don't live in Lagos, or Mumbai."

"You've been to Mumbai?" I asked.

"I spent a year there studying the Dalits. They're the lowest of the low in India's caste system, but now they're teaching in the schools and serving in high office. Some are even becoming chief ministers, in spite of hostility from people in the higher castes. I lived with a Dalit family and I saw their daily struggle, their humiliation and degradation, and how badly they want to escape from their poverty. The trouble is that sometimes when a family does make the break, they want to own as much stuff as possible to show the world they're no longer poor. If everyone who was once poor wants to do the same, I'm not sure our poor old world will be able to take it." She pronounced 'world' as 'wurruld' in her sing-song Scottish accent.

"But aren't there people in Vancouver too who like to show off their wealth?"

She turned and looked at me. "I've not been here long, but from what I've observed, no, there are not that many. People here make a big thing about how fortunate they are to live in such a beautiful city. From what I've seen, they find it rather embarrassing when someone shows off their wealth. Look—have ya seen this?"

On the side of a building there was a long wall painted black with the words *'Before I die I want to….'* in white with lines underneath, each completed by someone using chalk.

"Before I die I want to experience Heaven," Laura read out. "Isn't that just the greatest thing? *Before I die I want to ski naked down Grouse Mountain.* People say the greatest things!"

"Look at this one," I said. *"Before I die I want to leave the world a better place*. Or this one: *Before I die I want to enjoy at least one night's sleep without being woken by the pain.*"

"That's sad. How about this?" Laura said. "*Before I die I want to understand why I'm here*. It's like peering into someone's soul."

"What would you write?" Laura asked me with a smile.

"That's a tough one. Let me think. *Before I die I want to truly know what love is*. There: that's honest."

"You're so sweet!" she said. "You've got to write it on the wall—go on!"

I picked a piece of chalk and wrote my words. "How about you? What's your wish?"

She turned and stood with her back to me. Then she wrote, *Before I die I want to give as much back to the world as the world has given me.*

"That's beautiful." It was a powerful wall, opening hearts and filling them with warmth.

"There's one in Edinburgh too. But we'd best be getting on; I don't want to keep Merina waiting."[481]

"Where are we going?"

"To the Triko-Op. They make electric tricycles. 'Three wheels make a difference'—that's their byline."

We came to an old brick building bearing the Benefit Corporation logo and the Oak Leaf logo that proclaimed it a Five Star Sustainable Business. On entering, we found ourselves in a large open area with a central space, letting the light in and allowing trikes to be raised and lowered from one floor to another. The trikes were made from bamboo and teams of people were assembling them, finishing one set of tasks and then hooking a trike onto an overhead belt to go on to the next team.[482]

"The owner was making custom bikes," Laura explained. "Then as the boomers turned into zoomers and started the downhill progression to becoming tombers, he saw a growing market for tricycles."

"Did you say tombers?"

"Yes!" she laughed. "It's not a phrase I use when there are old folks around; for some reason they don't find it so amusing. The owner's son brought the company here to the Downtown Eastside Special Development Zone. He wanted to fulfill a social justice mission by hiring local people, and after a few years they became a co-operative, thanks to some Green Angel investors and a loan from OUR Bank.[483] I wonder if Merina's in—she's the Director of Operations. Then if you're free, we can find a café and I'll fill you in on Vancouver's green business program."

Laura disappeared and left me to watch a team turn a pile of polished bamboo lengths into a trike frame and send it on to the next station.

Merina Sam was a short woman in her thirties with copper skin and long black hair. She gave me a warm welcome and told me she was from the Squamish First Nation. She had an MBA in Sustainable Business Management from Royal Roads

University in Victoria, where she met her husband Martin. They had joined the Triko-Op together five years ago.

"We have a hundred people who work here," she told us as we toured the workshop. "We serve the market in British Columbia, and we have seven sister co-ops, including one in Cleveland, Ohio and one in North Korea that serves the Chinese market. China has over 200 million seniors, and the government is looking for ways to keep them active. They're building bike routes all over the country."

"You've got a co-op in North Korea? How are things going over there?" I asked, hoping not to display my ignorance. Back in my time, North Korea was a prison-camp state run by a weird dictator with probably the greatest concentration of human suffering anywhere on Earth. If you so much as criticized the government the regime could throw you and all your family into a prison camp. People had sometimes been so hungry they paid *not* to work so that they could spend a day gathering wild herbs to eat.[484]

"Following the coup by senior army officers that deposed Kim Jong-Un and sent him to stand trial in The Hague," she replied, "North Korea approached the South to start negotiations about re-unification. The problem is there's such a gap between the poorly educated, malnourished North Koreans and the wealthy, highly educated South Koreans. So they set up a bunch of cultural and sporting exchange programs. They're training the North Koreans in the skills needed to succeed in the global economy and South Korean investors are pouring billions into North Korea's Special Economic Development Zones. Their economy is so backward that they are among the lowest paid workers in the world, but that's an advantage when you're exporting. They learn fast, and their economy is growing by ten percent a year."

"Have they re-unified as a single country?"

"Not yet. They have created the United Federation of Korea, but they are continuing as separate states until they can reunite, and return to the way they had been for a thousand years until the Korean War tore them apart after World War II."

"How did the sister co-op come to be established?"

Merina paused beside a screen showing seven webcams, one for each sister co-op in Toronto, New York, Cleveland, Austin Texas, Brisbane, Jakarta and North Korea.

"As well as being a Benefit Co-op," she replied, "we're also a Social Justice Co-op. It's a relatively new movement. To become one our members made a commitment to three things: to include disadvantaged people in the workforce; to take on a specific social justice project; and to engage our members and customers in supporting a social justice campaign. We were on the lookout for a project when the North Korean coup happened, and one of our workers, Jiwoo Park, suggested that this might be our opportunity. So Jiwoo and my husband Martin went over and they came back with the sister co-op idea. Let me see if I can find Martin."

I asked Laura if she knew about this. "I knew they were politically active," she replied, "but I had no idea they were doing so much."

Martin was a tall man with short black hair, wearing blue dungarees. He wiped his hands before shaking ours.

"Merina tells me you're interested in our Korean sister co-op?"

"Yes," I replied. "I gather you were over there?"

"We went over six months after the coup. I remember how expensive the flight was, since the carbon cap was having an impact on airline prices. The capital, Pyongyang, was smart enough, though rather sterile, and most people seemed well fed, but as soon as we got out of the city it was awful to see the hunger and impoverishment. I was shocked at the complete loss of tree cover, and the barren farmland without any shrubs, hedgerows or signs of wildlife.[485]

"North Korea was in desperate need of economic development and China had a growing population of elderly people, so we suggested forming a sister co-op, using bamboo imported from China. Jiwoo spent a year there, and we found investors in South Korea. The idea of a co-operative comes naturally to the North Koreans, though they were very timid when we put them through the training course. They make electric trikes for the Chinese market and regular trikes for their own market, since they are cheaper and many of their homes still don't have electricity."

"It must be a huge step forward compared to the way their life used to be," I said.

"You wouldn't believe the stories that are emerging. They're heartbreaking. Do you know what their favorite Western movie is? It's *Les Miserables*. Some of them have seen it many times, and many people cry their eyes out during the final scene. We have a good relationship with the Indigo Global Humanities Project in Busan, South Korea, who do humanistic and global justice work with young people. They got us the invitation to visit the North. They have created an online course on sustainable village development that is being followed by tens of thousands of North Korean villagers, using $100 computers powered by solar panels. They have so much catch-up to do, after almost a century of oppression. They weren't only starved of food. They were also starved mentally, emotionally and spiritually."[486]

Martin made some gestures to the screen and a young Korean woman appeared wearing green dungarees. He introduced us and asked her how it was going. They chatted in fairly good English, then he asked if I wanted to say anything. I gathered my thoughts and said, "I am so happy to be part of our human family with you, and I wish you every good fortune." She smiled and bowed her head.

When they had finished chatting I plucked up my courage and asked a personal question.

"How did you two meet? You said it was at Royal Roads University?"

Martin looked at Merina and smiled.

"Yes," said Merina, "but we had known each other as children. We used to go fishing together as kids, and hunting for clams. But it wasn't until we were doing the HEART course at Royal Roads that we really connected. That stands for Healing Earth's Economy and Rediscovering Truth. It was a very intense time for Canada's First Nations, since the government had just scrapped the old *Indian Act* and was accelerating the treaty process, restoring self-governance and ownership of the land to First Nations, including guaranteed places at university for First Nations students.

"We were just friends, nothing more, but there was a three-day wilderness retreat during the course. We had to go in with a clear question, and without knowing it both Martin and I chose a question about finding our soul mate, the person we would fulfill our higher purpose with. On the third night I was sleeping on a bluff with a beautiful view of the ocean and the rising sun and I woke before dawn to see an eagle circling above me. As I was watching it I heard a voice behind me—a woman's voice—which said, 'You've already found him: it's Martin. But it's up to you to take the initiative.' I was really shocked, because I had never thought of him in that way."

"What happened for you, Martin?" Laura asked.

"Nothing like that. I just had a sense that I had to trust, that it would all be okay."

"I bet the first time you met after that was fun," Laura said. "Maybe even a wee bit embarrassing?"

"Yes, it was," Merina said. "I sat on it for a day or two, then I invited Martin for a walk on the beach down by Esquimalt Lagoon when I told him everything."

"And…?" Laura asked.

"Well, you tell it, Martin."

"I was completely bowled over," he said. "After she told me, I had the feeling that this was right. I took her hands in mine, knelt down in front of her, and started crying. She pulled me up, wiped my tears, kissed me, and we've been together ever since."

"Oh my god—that's so romantic!" Laura exclaimed. "Do things like that really happen?"

And there was me, equally impressed, and attracted to Laura knowing full well that nothing could come of it. Why did I fall for women who were so unavailable? Either they were probably dead, like the girl in Jerusalem, or not even in my time-reality, like Laura. Back in my time, she would be what, five years old?

"Yes," Martin said. "They do. We are the living evidence. Mind you, Merina's not always the easiest person to get along with…."

"It takes one to know one," Merina replied, wrapping her arm around Martin's back.

Back in my time, many of Canada's First Nations were trapped in poverty and hopelessness, with terrible housing and unemployment. They were also trapped in the colonial-era *Indian Act*, which deprived them of the power they needed to control their future and their land, and the treaty process was crawling along at the pace of a reluctant snail.

"What are the biggest changes that Canada's First Nations have seen over the last twenty years?" I asked. It was a big question so I was a bit nervous, but I was eager to learn.

Merina looked at me, said something to Martin, and invited us to follow her to her office. It was like entering a temple. The walls were covered with huge images of the forest; the ceiling was pale blue like the summer sky; and in the center there was a circular table with an I-Ball, the same as Wei-Ping's. Merina invited us to sit down and offered us some forest juice made from blueberries, Oregon grape berries, red huckleberries and pine mushrooms. Truly refreshing—and delicious. Then she turned and spoke in a deliberate, thoughtful manner.

"You ask about the progress we have made. Yes, we have made good progress. The *Indian Act* is no more. Most of Canada's First Nations have negotiated new treaties, and we have finally been granted the rights and freedoms other Canadians have enjoyed for centuries. We can now take out mortgages and own our own properties, with community covenants.

"Yes, many of our young people now get a good education, parts of which are taught in their native language. It's enabling them to rediscover their cultural history, and restoring their pride.[487]

"Yes, more of our young people are going on to become teachers and lawyers, doctors and engineers. And yes, I have seen reserves where the moldy houses have been torn down and new homes are being built to the highest green building standard, with a layout that recreates the way our villages were in the years before the white man arrived. And yes, it fills me with pride to see our young people building passive houses that need almost no heat, while earning a good living. And our new co-operatives and social enterprises are good, both for business and for the people who work in them.

"So yes, there is much we are proud of. But when I read the stories told by our elders long ago I am struck with a sense of how far we have to go before we regain the fortitude of purpose and sense of right conduct that was once our pride.

"It used to be the case, as one of our Squamish elders told it over a hundred years ago, that the very thought of owning more possessions than another when someone was in need was inconceivable. And yet today, many members of our people have lost respect for their elders, and follow the white man's ways.

"It used to be, in the years before the white man arrived, that if you were fortunate you would hold a great feast where you would give your possessions away. In this way, you would win the respect of your people and help those who were in need. We have a long way to go before those days return. Here, in our co-operative, we all earn the same income. We reward those who have to carry responsibilities home with a small bonus, and we reward anyone who makes a special contribution with a feast or a special gift of some kind. We are trying, in our way, to rebuild the spirit of solidarity that was once the norm among our people. And whenever there is hardship, when someone's child is sick or some other trouble strikes, we always help each other."

Merina was silent. She looked out of the window, then turned and scrutinized us carefully.

"I think you are good people, so I will tell you something else. It used also to be the case, in the years before the white man arrived, that our elders and wise women spoke a different language. They had not been taught to believe that the world was a material place where even the way we thought was governed by the white man's laws. They knew a different kind of power, a power that connects every living being to the Great Creator, not through muscle and bone but through spirit. They knew about the power of dreams, and of good and evil intentions. They knew that every living creature possesses those powers—the bear, the salmon, the beaver, the cedar trees. The words 'all our relations' were not just a pleasant greeting. They were a living reality, connected by threads that were invisible to the white man. They also saw visions. They paid attention to their dreams and learned how to guide their people, respecting the harmony the Creator has blessed us with. It was a great shock when they saw how the white man prayed to his God and sang hymns in his church but acted with so little respect for the works of their God and their fellow people.

"So, you may ask, how long do I think it will be before we return to the old ways? I do not know the answer, but I hope to see more progress in my lifetime."

I was silent before such moving words. It was Laura who came to my rescue.

"I am so appreciative, Merina. Back in Scotland, where I come from, there's a similar movement to restore the positive aspects of clan life, when people cared for each other. We have many young people who are finding ways to return to the land their ancestors were evicted from two hundred years ago, just as your people were driven from their traditional lands. My country has finally won its independence back after three hundred years, and we are relearning our Gaelic language. But there is much that we should consign to the past, for the clans were forever attacking each other and bludgeoning each other to death."[488]

"Yes, our people too used to fight and take slaves. So we too have much that is best left behind. Even our potlatches, when people gave away their possessions, were sometimes an exercise in power, when people forced obligations upon each other. But I have taken enough of your time. Shall we get back to the tour?"

I was sad to leave Merina's office. It was like another realm, a doorway to another way of being.

Back on the shop floor, I asked, "When it comes to sales, how do the trikes compare to bikes?"

"Trikes are five percent of the global bike market, or some eight million units a year," Merina replied. "We sell thirty thousand units a year, which is less than half a percent of the market, and we have to work hard to keep that share. We trade on our green certification, our willingness to take part payment for local sales in Delta Dollars, and our twenty-year warranties and service contracts. Our cargo-trikes come with a cargo-trike network membership. We have a customer loyalty program with riders' clubs, meet-ups and holidays; we invite our customers to join our social justice challenges; and we provide an urban trike-sharing service. We have fantastic customer loyalty: a third of our sales come through customer referral. A review recently said we were generating service envy, which is more important than product envy. "[489]

This was all new to me. "Do you manufacture all the parts yourselves?" I asked.

"The bamboo is grown locally. The aluminum parts are made at a company in Richmond, and we have an ultrafines-safe 3-D print shop, laser-cutter and CNC router in a building down the road for the polycarbonate parts. We contract out the electronics for the geesec."[490]

"What's geesec?" She pronounced it with a soft g, like 'gee whiz.' "And what are ultrafines-safe?"

"Geesec is GIS security. It tells you if someone's trying to steal your bike. And ultrafines-safe means the 3-D printing process recaptures the ultrafine contaminants before they get into your lungs. The early 3-D printers were terrible, spewing out billions of particles of ultrafines.[491]

"The most important part of the manufacturing process is the rigorous testing. We push every component to failure to make sure each trike is roadworthy in the toughest of conditions. We use a 3-D scanner to analyze for faults and we even burn a trike occasionally to test for air pollutants. We hand-weld all our joints to ensure quality. Do you want to see the design studio?"[492]

Merina led us to a room where designers were moving 3-D shapes on screens, using their hands to manipulate them while trying out different configurations with 3-D pens.[493]

"This is where we design the next generation of trikes. When we're happy with a design we send it out to our maker network for feedback."

"Your maker network...?"

"It's a nation-wide network of people who enjoy making, fixing and improving things. It was a maker who suggested adding gears to the trike wheelchairs, a maker who designed our airless smart-tires, and a maker who got us into the all-terrain electric sports trike market for people with disabilities. They're a really important part of our business."[494]

"What does it mean to be a certified green business? I saw your five stars sign at the entrance."

"It took us a good ten years to get that," Merina replied. "The first four stars were relatively easy. We had to show that we were making a useful product, educating our staff in the principles and practices of sustainability, recycling 95% of our wastes, using 90% recycled materials, and a few other things. It was the fifth star that was difficult: to be zero greenhouse gas emissions throughout the supply chain. We were using a small amount of steel for the gears, and it was impossible to find steel that didn't involve burning coke. Even recycled steel carries a carbon footprint. It wasn't until ThyssenKrupp Tata started to use molten oxide electrolysis that we were able to eliminate our carbon emissions entirely. They use a chromium-iron oxide to make the oxygen, which yields a finer steel with greater purity."[495]

"Is all steel-making done this way?"

"I wish. There are still countries that are dragging their feet, in spite of the tightening global carbon budget. The European Union is making a big effort to push the market. 40% of the steel sold there must be zero-carbon, rising to 100% by 2040. The ThyssenKrupp Tata steel works at Sault Ste. Marie uses hydrogen, which they source from wind energy from the deep water wind farm in Lake Superior, using the energy to split water."[496]

"Deep water turbines?"

"Yes. They float, supported by ballast and cabled to the lake or ocean floor."[497]

"What happens if there's a really big storm?"

"They go into automatic shutdown. There are also ships out there gathering wave energy, storing it in carbon nanotube batteries."[498]

This was all new to me—but so was everything.

"What about the power for the North Korean co-op? Is that green as well?"

"No. They get their power from China, some of which still comes from coal. But the carbon footprint of a trike is fifty times less than an electric car. We're on China's waiting list for green power, but it will be five years before they get to us, since the Chinese companies get priority."

"China has a waiting list for green energy?"

"Yes. Their businesses are rushing to acquire five-star status so that they can sell into the European market without incurring the eco-tax on non-certified products. Before the *Global Fair Trade Treaty*, China would have challenged it as an impediment to free trade, but now Chinese businesses are falling over each other to become certified. When I spoke to Zhejiang's Energy Minister in a teleconference last month he said they were aiming at 100% renewable energy by 2045."

"What about your electricity here—is that zero-carbon too?" I asked—while being amazed at what she'd just said.

"Yes. It has been 100% renewable since the last gas-fired power plant shut down. It comes from a mix of hydropower, solar, wind, geothermal and tidal. They've all got good EROI numbers."

"EROI?"

"Energy return on investment. It's a measure of how much energy you have to spend to obtain new energy. If it's one-for-one, why bother? In the early days, fossil fuels had a very high EROI. Coal used to have an EROI of 80, but it has fallen to 5, since all the easy stuff is gone. Oil was 40 back in the 1990s, but the bitumen they were squeezing out of the Alberta tar sands was as low as 2.4.[499] It's the same with natural gas. The easy stuff was 10; the shale gas as low as 3. And that's not counting its climate impact. Its ENVROI—its environmental return on investment, including all externalities—was hugely negative. The latest wind turbines have an EROI of 30. Regular solar is 8 and thin-film solar PV is coming in at 25, so they're all good numbers."[500]

"What about Canada's tar sands? Are they still operating?"

"Yes, but at only a quarter of a million barrels a day, a fraction of what they were producing forty years ago. They had planned on being at five million barrels a day by now."

"What stopped their expansion?"

"Where have you been, young man?" Merina said. "I thought everyone knew about the tar sands."

I coughed, and explained about living in Sudan.

"Hey, that's okay. The big squeeze started twenty years ago, with all the opposition to the pipelines. Then the carbon budget kicked in, the price of oil failed to rise because of the shift to electric vehicles, and Canada's new government started moving the whole country to 100% renewable energy. As their stock prices fell, companies shelved their expansion plans. Alberta's really moving on wind and geothermal power. They closed down their last coal-fired power plant ten years ago."

"What about coal in general?"

"All fossil fuels are being phased down to zero, driven by the global carbon budget. South of the border the Sierra Club ran a massive Stop Coal campaign, with its mountaintop arrests and its strippers against strip-mining where people would show up at a company AGM and strip off to reveal their coal-filthy naked bodies. When Chicago and New York rejected coal-fired power, there was a stampede to switch to renewable energy, which had become a good investment."

"I thought Canada's government supported the oil industry," I said.

"That was years ago, under the Conservatives. They only got elected because the opposition parties kept splitting the vote; they never had a popular majority. When the Liberals defeated the Conservatives, and Justin Trudeau became Prime Minister, they brought in proportional voting.[501] At the next election the Greens won fifty seats, and governed in coalition with the Liberals. There were fifteen First Nations MPs—we were very proud of that. Today there are twenty-five. The Green Party's Elizabeth May served as Minister of Environment and Climate Change, and she later became Minister of Finance and then Prime Minister, a position she held until three years ago when she stepped down at the grand old age of 78. What a gal! She was Canada's best Prime Minister since Lester Pearson."[502]

"That's really encouraging," I said. "What do you do for heat? I hope you don't mind all these questions."

"Not at all," Merina replied. "We're part of Vancouver's district heat network. We obtain our heat from a mixture of solar heat pumps, geothermal heat from under the railyards, water-source heat from the Burrard Inlet, sewage-source heat from the sewage pump stations, ice-source heat from the Nature's Path Arena where the Canucks play, and waste heat from the regional composting plant and several industrial operations, topped up with biogas when needed. There are super-insulated pipes that ship the heat around the neighborhood with almost no loss of temperature.[503]

"When we moved here," she continued, "we did a retrofit that cut our heating demand in half, with new windows, natural light tracking, under-floor insulation, radiant hydronic heating and sensors that deliver the heat where it's needed. We joined the Architecture 2030 Challenge, which has been a great source of support, and this part of the Downtown Eastside became a 2030 District, committed to accelerate the progress to zero-carbon sustainable buildings."[504]

Merina walked over to a screen on the wall of the design studio. "This is our energy display. It shows what's happening in every corner of the building, and what it's costing us. It creates an algorithm that optimizes our energy use."

"Do all companies have them?"

"Yes, it's pretty normal. We had a deal with BC Hydro—our utility—that if the display didn't pay for itself after two years they'd cover the cost. It paid for itself in eight months. It's great to have a public utility like that."

"What about garbage incineration? Do you get any heat energy that way?" I knew that several European cities had gone down this road, including Copenhagen and Stockholm.

"There was an attempt to do so, but it was killed off by public resistance. No-one wanted the air pollution, however many promises the experts made. There was always a concern about nanoparticles—fragments so small they can penetrate your lungs, brain and nervous system. There was also a concern that if Vancouver started incinerating, the public's recycling habit would weaken. I've been told that the engineers in Stockholm said that if they had to do it again they would not choose incineration, since it destroys too much valuable material."[505]

"So how do you handle your wastes?"

"We maximize source separation and recycling, including organic wastes, and the residue gets dumped in the landfill. The residue is getting smaller and smaller each year because of the recycling and the extended producer responsibility legislation, which requires us to recycle our products. Here in Vancouver we're recycling and composting 95% of our trash, and they say we should be able to cease dumping entirely soon.[506] San Francisco showed the way—they're the global leader."[507]

"What's stopping you from getting to 100%?"

"The non-recyclable imported products, the ones with composite parts that are glued together. The *Global Product Stewardship Treaty* is still quite new, and there are categories where the treaty doesn't kick in yet to give manufacturers time to redesign their products. Here at the Triko-Op we are legally responsible for the final disposal of every trike that leaves the factory, so we design for 100% recyclability. We attach a recovery fee to every trike and we pay a Recovery Co-op to either restore a dead trike or deconstruct it and return the parts for re-use. It creates a circular economy, which helps us manage our resources in harmony with nature."[508]

I was impressed. Why couldn't we do this back in my time?

"Do you think the whole global economy will be like this one day—a true circular economy?"

"We absolutely have to. For every one tonne of consumer products, seventy tonnes of raw materials are required to make them. Our only possible future is zero waste, using 100% recycling and resource recovery. The E-70 Group of Nations are making good progress. The horrendous garbage dumps in countries like Mexico and Brazil are being cleaned up and the garbage-picking jobs are being diverted to recycling. One of the new Universal Sustainable Development Goals is for the world to achieve a full global circular economy by 2055, so that gives us twenty-three years for every manufacturer, business, consumer and nation in the world to get on board. Some people say it's impossible, but if you don't set a goal you'll never begin."[509]

Once again, I was impressed. During my studies I had learned that at the end of the 19th century the people of Victoria, across the water from Vancouver, used to take their garbage down to the low-tide mark at night and let the tide carry it away. In 1908, they decided that was too gross, so they put it in barges and dumped it further out to sea. By 1958, that in turn was considered too gross, so they started dumping on the land, draining a lake to make a hole big enough. Then in the 1990s they finally said 'enough' and adopted the goal of zero waste. It was a complete evolution from being garbage scumbags to global citizens in 125 years.[510]

"About your heating—is everyone in Vancouver on a system like yours?" I asked.

"I'm not sure. Laura, do you know?"

"The district heat is mainly in the downtown, the industrial areas and the university, and new residential areas where development makes it financially viable."

"How much of it comes from renewable sources?"

"Almost all, I think," Laura replied. "Most homes use solar air-source heat-pumps, and most apartment blocks and condos use ground-source. In my apartment building our Green Team persuaded the landlord to join the Green Landlords program and she upgraded the building to ground-source heat, along with a bunch of other changes."[511]

"Green Landlords?" I asked.

"Yes. It's a program that encourages landlords to green up their properties. It helps with retrofits and encourages the tenants to set up Green Teams, which are a great way to meet your neighbors. In my block we're organizing a summer block party, we've joined Streetlife, and we've started our own internal currency. People are talking to each other and sharing interests. We've created a food garden and planted fruit bushes, and people are using balcony boxes and windowfarms."[512]

Windowfarms? Too many new things to keep up with.

"We have a similar program in our condo," Merina said. "We have a Green Team too, though I'm too busy to get involved. We took a dozen trikes to last year's Block Party when we closed the street and ran some races. People loved it.

It has completely changed the atmosphere in our building, creating a new sense of community."

"When you became a co-op, what difference did it make to your workforce?" I asked.

"Well, that's close to my heart," Merina replied with a smile. "When we moved to Vancouver and grew to a staff of fifty there were quite a few for whom it was just a job. But when we adopted employee share ownership and became a co-op, when we set up the daycare co-op and we put our workers through sustainability education, and then became a Social Justice Co-op, you could feel the difference. Absenteeism and sick days fell, and there was a steady increase in engagement. People began to take pride in their ownership and there was more joy in their work. The best motivator is the intrinsic motivation that comes from believing in your work and having the autonomy to get on with it, both individually and as a team. It enables people to be self-directed and to develop a sense of mastery. It fits well with syntropy theory, which tells us that consciousness is at the core of all reality. That's where it all begins and ends: right here in our inner being. That's where the motivation has to come from. A degree of autonomy is a pre-condition for self-motivation."[513]

I loved the way she thought, even if I didn't understand the syntropy connection.

"Are there other co-ops where the workers all earn the same wage?" Laura asked.

"Only a few: we're still a bit of a rarity," Merina replied. "But we're happy. When some people receive a lot more than others it weakens social trust. I was raised in a First Nations community and we have bonds of trust that go back thousands of years. I've seen First Nations chiefs who have broken that trust by awarding themselves big salaries, and I've seen the way it affects other people in the band. But enough with the talking. Would you like to try one of the trikes?"

"You mean, out on the street?"

"You could try the ceiling if you want to," Merina chuckled.

I laughed, and she brought me a neon blue recumbent tricycle. "Twist your right hand for the electric drive, and remember to lean into the corners."

Laura came and watched as I cycled off down the street. It was a delight to ride. At the corner, I leaned to the left and the trike corrected my balance. When I came to a speed bump it rolled over as smoothly as a soft massage. What a charmer!

"That was great!" I said, when I got back to the building.

"And in case someone tries to steal it," Merina responded, "the geesec safelock has a fingerprint sensor so only you or someone you authorize can ride it. If it's stolen, your personal device will tell you where it is. Our trikes have helped catch many thieves! But if you'll excuse me, I must be getting back to work. It's been a pleasure. There's a six-month waiting list if you want to order one!"[514]

20

Green Business and Love at the Tree Frog Hotel

STEPPING OUT INTO the late afternoon, Laura suggested that we find a café where she could tell me about the green business program. Some of the Triko-Op workers were loading a freight-pod with carefully packaged tricycles.

"They're off to Middle America," Laura said, using her phone to read the code on the side of the pod. "They'll be collected tonight and delivered to the Freight Centre outside Abbotsford for shipment on the west coast Hyperloop. Have you seen it? First in North America, Vancouver to San Diego. There's another being built from Quebec to Windsor, and two in the States. They say it goes like a bullet."[515]

Hyperloop? What was that?

"When did that get built?" I had a hundred questions.

"I dinna know. Hang on a tick." Laura got out her phone and said 'Vancouver Freight Tube.' "It was first proposed by someone called Daryl Oster in the 1980s, but there was no sense of urgency because the climate crisis hadna hit, so nothing came of it. But then the carbon cap arrived, carbon rationing put the squeeze on the trucking companies and the railways were full to the gills in spite of reduced coal shipments. So a group of company owners got together with the Hyperloop team at SpaceX and they formed a consortium."[516]

"How fast do they go?"

"Six hundred kilometres an hour. Vancouver to San Diego in four hours. So much more civilized than using a black oozy substance made from ancient sea creatures, don't ya think? They're solar-coated, with digital tracking on every pod."

"How wide are the tubes?" I was really quite astonished.

"Less than three metres, which makes them easy to build. In Chicago they're talking about tunneling under the city to bring the pods into the city center."

"What about ocean shipping?"

"What ocean shipping? Have ye nae been down to the harbor? Have ye nae seen all the empty ships sitting there going rusty?"

"No—I didn't know. So what's happening to all the global trade...?"

"It's doing wonders for local economies, that's what's happening. Instead of importing stuff from China we can make it ourselves! Just kiddin'. Tonne for tonne, container ships use far less fuel than trucks, but even they can't escape the global carbon cap,[517] so their owners have been scrambling to cover their ships with solar, install skysails,[518] and use underwater coatings that reduce the water friction along the hulls, copying the way the sharks do it. Anything to reduce the cost of bunker fuel.[519] There's no end of schemes to build zero emissions vessels,[520] giant airships,[521] and ships that run on biofuel. No end of schemes, but little reality so far. They're even making bioethanol from seaweed, but it's very expensive, and you need a vast area to farm the seaweed.[522] There's talk about building a freight-tube under the ocean floor from Vancouver to Shanghai, but I dinna know if it's for real. Here, let's have a coffee."[523]

My mind was racing as I realized that ocean shipping was failing to come up with zero-carbon alternatives. The same must apply to flying, which would be why Aliya mentioned going overland and sailing across the Atlantic as the best way to get to Syria.

The Tree Frog Hotel had outdoor seating and Laura took my arm, leading me over to a small circle of trees.

"Take a look at this—it's one of my favorite wee places."

Among the trees there was a small birch tree, its trunk separated in two and then woven into a figure of eight. There was a circle around it paved with green and blue mosaic tiles, with these words:

> We're all interconnected, the Earth and the Sun and the Moon.
> – Rosie Emery [524]

"It's one of the many heart-places that are being created around the city," she said. "They make me feel at home."

"A city can be full of black holes," she continued. "Ye dinna even know they're there until ye fall into one. Places where no-one cares, where there's nae love. They're really bad for your health. We need to remove the black holes and replace them with their opposites—heart-places that fill you with trust and generosity, immune-system boosters in the nervous system of the city. Urban acupuncture, some people call it—tiny wee spots scattered throughout the city where love is injected into our hearts and out into the wurruld."[525]

Laura was enchanting. I had never thought about energy that way. Back in my time there were so many places that were unloved and neglected where litter and graffiti accumulated. It didn't take much to fall into a black hole if you spent too much time around them.

A busker was playing a traditional Scottish melody, and Laura impulsively grabbed my hand and started to dance. Was she so spontaneous with everyone? I felt warm and connected.

"Let's order, then we can get doon to business," she said when the music ended. I ordered a cranberry coffee and a slice of fresh fava bean pie, while

Laura—I gasped—ordered smoked jellyfish and roasted locust burger.[526] Clearly not fasting.

"Eugh!"

"What's the problem, laddie?"

"It's, it's, it's nothing. Ow! I just got cramp in my foot. It's gone now. Do you often order the locust burger?"

From shock to admiration in three quick moves....

"They're only on the menu if there's been a plague somewhere. They're supposed to be very nutritious—high in protein, zinc and iron. There's other insects too, if you want. Here—take a look."

I examined the electronic table menu and sure enough I could have chosen buffalo worm soup, bug nuggets, crispy crickets, bugadilla falafel balls, a grasshopper burger or mealworm mash. Yikes.[527]

"How do they catch the locusts?" I asked. "Don't they swarm in ginormous numbers? And how did they get to be here in Vancouver if ocean shipping is so restricted?"

"I dinna know. Let's see what it says."

Laura clicked on the menu, which told us that the owner's son had been volunteering on a farm in Australia when the locusts swarmed, and when he sailed back he had brought a load with him. And yes, they swarm in the billions. The Aussies had developed a satellite system that predicts their direction and a pheromone amplifier the locusts apparently can't resist. Flavor of young virgin locust, or something like that. As soon as the farmers learn that a swarm is coming their way they drag a huge industrial crop-drying fan into the fields, set up the nets, switch on a really bright light, reverse the flow on the fans and haul in the harvest.[528]

"They're really quite tasty. They go well with the jellyfish. Here, have a bite." She passed me her burger and offered it directly to my mouth.

I crunched down on Laura's burger topped with jellyfish, and sure enough, it was definitely tasty, a bit like roasted sunflower seeds.[529] There was also something rather erotic about eating directly from her hand. Was she flirting with me, or did she act this way with everyone?

"Is jellyfish a common item on the menu, too?" I asked.[530]

"Unfortunately, yes. They're taking over wherever the fish stocks have been depleted, which is almost everywhere, alas. To think that in less than fifty years we have managed to deplete most of the ocean fish stocks—isn't that the most withering indictment of our so-called civilization? We're not creating new Marine Protected Areas anywhere near fast enough. They used to be less than one percent, and I believe they're up to fifteen percent but we need to protect fully forty percent of the world's oceans if the fish stocks are to recover. We urgently need a new approach. But Naomi said you wanted to learn about our Sustainable Business program—am I right?"[531]

"Yes. I was wondering how many businesses have signed up for certification?

And how do you create a single program when there are so many different kinds of business? A car repair shop is so different from a flower shop."

"To answer your first question, about 35% of Vancouver's small businesses have signed up, and 78% of the larger businesses with over ten million in revenue."

"How many is that in total?"

"For Vancouver as a whole, it's about nine thousand businesses."

"But that's enormous! However did you manage it?"

"When Vancouver decided to become the world's greenest city, creating a green economy was one of the goals, which included helping businesses to green up their operations. Whenever we're not doing business in a way that creates harmony with nature, we're doing it in a way that destroys it. So Wei-Ping and his team decided that our goal should be 100% sustainability for every business. It's a BHAG—a big, hairy, audacious goal." Laura flashed a grin at me. "A BHAG. I like that term."

"It's certainly audacious."

"They built on the pioneering work done by the Green Business Certification program on Vancouver Island[532] and a great wee program called Climate Smart that helps businesses reduce their carbon footprint.[533] They also had help from the Global Alliance of Benefit Corporations. They divided the economy into sectors and worked with teams of volunteers and business people to create a points system that recognizes the difference, as you say, between an auto repair shop and a flower shop. You can earn up to a hundred points if you do everything right. You need so many points for your first star, and you go on improving until you have the full five stars."[534]

"What were your biggest obstacles?"

"Verification, supply chain, corporate law, and the underground economy, where nothing is recorded or regulated. Imagine you're a major retail store. How do you track the supply chain for ten thousand different products? But without supply chain tracking, what use is certification? A company can be totally green in its operations but selling products made by destroying rainforests while pouring toxic chemicals into its workers' lungs."

"So how did they tackle that?"

"They went global, helped by the Global Initiative for Sustainability Rating.[535] The carbon rationing and the mandatory emissions reporting also helped. Unless they went global, they had no way to verify a company's claim that its suppliers were recycling or educating their staff. It's all done on the honor system, with random checks. You can lose your certification for up to ten years if you cheat."

"What motivates Vancouver's businesses to go to all the trouble?"

"We do, the customers. When a company gets its green logo, consumer demand increases by up to twelve percent, at the expense of its non-certified competitors. And many banks have taken to charging a higher rate of interest if a business isn't certified, because of the increased risk. Major institutional buyers like hospitals, universities and cities are requiring their purchases to come from certified businesses. The European Union is doing the same. There's a town in England called Todmorden where every single business has been certified, so we've still a way to go here in Vancouver."[536]

"What about all the different products, from shampoos to toaster ovens? Are they certified too?"

"Some are, but most are not. But everything is tagged, thanks to the Global Fair Trade Agreement. It's not as thorough as certification, but it's progress. When you scan a product's barcode you can see its source, its validations and supply chain tracking, and its green business certification, if it has it. So if I want to buy a chair I can use my device to read the code, which will tell me where the timber comes from, whether the forest has been managed sustainably, and if the company has been certified as a green business: things like that. Last week, I wanted to replace one of my artscreens with an organic LED screen that uses less energy and has greater pixellation. The tag said it was made in India, and the company was following the Fair Trade Fair Labour clause in the Dhaka Agreement. But it also told me that it used a rare earth from a mine in Inner Mongolia that had been flagged for labor code violations."

"So get this," Laura continued, waving her hands. "The app allowed me to send a message directly to the company telling them I was not buying their product because of the Mongolia violation, and why didn't they become a worker-owned co-operative?

"It also lets me send praise and encouragement. The combination of the bar-codes and the apps has done wonders for the drive to end child labor and other black-flagged labor conditions, and to eliminate toxic chemicals from imported products."

This was fascinating. All in the power of a well-designed app.

"You said corporate law was also a problem."

"Yes. That's a much deeper issue of legally enshrined selfishness. The best solution is the Benefit Corporation, where creating a community or environmental benefit becomes part of a B Corp director's duty. Lots of companies have made the change. There's an organized initiative in Vancouver and other cities to require businesses to either get certified, become a Benefit Corporation, or face lengthy delays in getting their business permits. The movement to end corporate abuse started in small towns in New England that were fed up with corporations coming in, trashing the environment and walking away scot-free, claiming they had the same rights as living, breathing people.[537] There's legislation being proposed that would require every business to become a Benefit Corporation: that will speed things up, and address the fundamental problem. After that we need legislation requiring every business to be certified."[538]

This was big. Corporate greed was one of the many problems we faced back in my time.

"What's to stop a corporation from simply switching its head office to avoid the new requirements?"

"Yes—there's been a bit of that. But we didn't end the African slave trade overnight; it took almost a hundred years. Things move a lot faster today, but it's probably going to be twenty or thirty years before the last rogue corporation either becomes a Benefit Corporation or shuts up shop. And you know what's so confusing about all this? It was the dreaded Walmart that helped get this going, more than twenty years ago. So there's a paradox for you. They sent a questionnaire to each of their hundred thousand suppliers requiring them to demonstrate what they were doing to improve sustainability.[539] That was the wake-up call that got the whole thing rolling. They also helped launch the Sustainability Consortium, which developed the scientific standards the program is based on. We'll get there. It's just a matter of time."[540]

"What gives you so much confidence that it will happen?"

"My study of history? My belief that in the end we humans will finally do the right thing? When it comes down to it, we're a pragmatic species. There's accumulating evidence that Benefit Corporations are outperforming traditional corporations in the things that count, like market share, profitability and employee loyalty. And then again, maybe I've fallen in love with syntropy theory, which says the drive to self-organize for the betterment of existence is intrinsic to the entire Universe. That's a powerful force to have on your side when you're trying to build a better wurruld."[541]

There it was again: syntropy.

"Or maybe it's simply that we Scots have a long tradition of wanting a better wurruld. Did ye know that some of the world's top universities were created by the Scots?" I shook my head. "Well, 'tis true! In the 18th century 'twas the Scots who led the wurruld in so many things. They called it the Scottish Renaissance. And we've been pretty persistent in getting what we wanted, when we weren't indulging in clan warfare. So how about you? What do *you* think will make for a better wurruld?"[542]

"Me?" I played for time, looking at my fingernails. "I think we'll just have to, because things will be so bad if we don't. Maybe I haven't got your confidence that it's inevitable. I see an awful lot of people making bad choices. Keep on shopping. Have another drink. Blame the politicians. There are too many people who don't feel any responsibility for making the world a better place. They just want to enjoy the ride while reserving the right to complain if the bus drops them off in a trash heap instead of Utopia."

"Is that why you've come here? Because you're so jaded with things back home?"

Laura was putting me on the spot, far more than she'd ever know. "Maybe. Or maybe it's because I can't cut the desire for a better world out of my heart."

"There you go," she replied. "Syntropy. What did I tell you? It's in all of us, whether we like it or not. Did ye ever meet someone who wished they had a lousy marriage? Or who wished they had hemorrhoids up the backside? I dinna think so. Everyone wants a better life. So we can either self-organize to make it happen, or sit back and complain. Take your pick—but I know which makes me happiest. Complaining and happiness are an incompatible couple, like eating jammy donuts and keeping fit."

I laughed. I was totally charmed by the girl—the gurrul. But I needed to get back on track. "What would you say are the biggest benefits of all the changes that are happening here?"

"Well, most of all it's the benefits for nature. The great thing about Vancouver is that so many people have embraced the idea that civilization *can* exist in harmony with nature. Can you imagine what it will be like when we finally get there? To know that every creature will be able to live its life peacefully without fear of assault by a human, and that every forest will be able to grow in peace and every human will be able to live with dignity and self-respect?"

"That would be amazing. I take it you're a vegan, and don't eat any meat—apart from locusts?"

"Veganish. I try to eat as little as possible because of the harm it does and yes, because I don't like the killing. That's been another huge change since I was a child. When I was at school we ate meat without thinking about it. But when I learned what the livestock industry was really like, how cruel it was, and how much damage it did to the world, I decided to quit. Do you know what that incredible woman Harriet Tubman said about dreams? She ought to know, since she did so much for humanity, after being born a slave...."

> *Every great dream begins with a dreamer.*
> *Always remember, you have within you the strength, the patience,*
> *and the passion to reach for the stars to change the world."* [543]

That must have been the trigger that led me to abandon all sense of caution. You only live once, I told myself.

"I don't know how to say this," I said. "I hardly know you, but I love the way you think and the song you carry in your heart. It makes me want to sing along with you." I'd never been that forward in my life. The future was having a very liberating effect on me.

Laura looked at me, then reached across the table, took my hands and kissed me on the cheek. "That's so sweet," she said. "I'll treasure your words forever. It's a shame that you're only here for a short while."

Only a short while. Was this a cruel trick my life was playing on me, or was it a beautiful opportunity to open myself to love and exorcise the ghost of the girl in Jerusalem from my heart? I decided to change the subject, before I became too mushy.

"Wei-Ping told me that your father's a good friend of a man called Liu Cheng, who was Chief Constable of the Vancouver Police at the time when Derek Brooks was assassinated."

I could swear Laura looked disappointed, but she quickly recovered and said, "Oh, you know about Derek? My da's been obsessing over his assassination ever since he and Liu Cheng worked on the case. He was invited over to help, since the VPD had no experience with assassinations and things like that. I expect Wei-Ping told you that my da used to work for MI5. He was always bothered by the fact that whenever he made enquiries to the Vancouver Police Department he felt that he was being blocked. His instincts told him there might be an inside connection, but Liu Cheng was very protective of his officers so he was unable to pursue that line of enquiry. When he returned to Edinburgh after the OMEGA Days he was very troubled that Derek's murder remained an open case. He often says it's his biggest regret."

"What do *you* think happened? You must have heard him talk."

"Four years ago when Liu retired he came to Scotland to do some fly-fishing in Scotland with my da. After sharing their frustrations, they decided to re-open the case in private. Two old detectives chasing an old case—they're like those old geezers in a TV show my parents used to watch. The only physical clues they had were two bullets, one bullet casing, and two seconds of video footage of when someone tried to drive Derek off the road. That led to a small-time criminal called Jerky who had been in Vancouver on the day of the incident, who later disappeared.

"Then a human foot washed up that was identified as belonging to Jerky, but the trail ran cold. They wondered if Jerky might somehow have been involved, so they redoubled their enquiries with the Washington State Police. Did Wei-Ping tell you my da's over here now, visiting Liu Cheng? They took a trip to Olympia last week to go over the files. And here's what's interesting. All the files relating to Jerky are missing. But they found an older file that refers to a crony of Jerky's called Buffalo John, a Canadian from Calgary who's said to have connections to the oil industry. That made them wonder if there might have been an insider who was part of a cover-up, so they used their contacts to dig up whatever they could about Buffalo John and they found that he ran a small energy trading company out of Calgary."

"It was just last week that they were doing this?"

"This week. They're driving to Calgary today to see if they can find him. His files include the interrogations by VPD Deputy Police Chief Ray Robinson, so they're wondering if he might have been working as an informer for the VPD, something Liu Cheng never knew about when he was Chief Constable."

"You've got quite the inside track."

"My da calls me every night, and we chat. But look, ye dinna want to listen ta me blethering on. I've got to go the washroom, then I must be getting off. My da's coming over for dinner and I haven't done any prep yet."

"More locust burgers?"

"No—freshly baked bread, a summer salad, an omelet and cranberry wine. Not everyone is so taken with the locusts."

While Laura left for the washroom I looked around the café. There were people of every skin-tone, from blonde to darkest black. I saw a screen on the wall and wandered over to look at the messages. Among the advertisements for solar repairs, solitude-enhancement, wilderness shuttle buses and something called the Drone Racing League, there was an invitation to a place called Joey's Farm offering a day of 'Farm, Friends and Syntrodance'—whatever that was—from noon till late for thirty dollars. It said the farm was two hours away in the Fraser Valley and a train left Vancouver at eleven in the morning. It would be great to get out of the city and see what the future looked like out on the land.

"What's that you're looking at?" Laura asked when she returned. "Oh—Joey's Farm! That's a wonderful place. Are ye thinking of going? I would if I were you. Their dances are amazing. Mina—she's Joey's wife—she's a good friend. I might even go myself if I'm free."

"You've been there?"

"Wouldna miss it for the wurruld. I met my first Canadian boyfriend there. Why do all the best men have to leave? My purse is full of sadness."

"Your purse?"

"It's where I carry my memories."

Memories. Is that what I will become—a memory?

"This has been great, Laura. I'll remember you."

"And me you. You have yourself a great time at Joey's Farm."

And with a not-so-quick kiss on the lips—yes, the lips, a lingering kiss on the lips—she danced off into the warm evening air, her summer dress floating as if she had just stepped out of a Renoir painting. *Pierre Auguste Renoir*—I love the way the words rolled off the tongue. *Laura MacGregor.*

Get a grip, Patrick. Get a grip!

 21

Song of the Universe

CYCLING HOME ON the Friday evening I saw people chatting on their porches, filling the sidewalk cafés, walking their dogs and playing with their children. Others were playing tennis, badminton, softball and soccer in parks surrounded by fruit trees. I passed a group playing bicycle polo, an outdoor gym where people were exercising on bars and swings, a parkour court with blocks, poles and jumps where young people were doing the most ridiculous things, a seniors exercise park, and a set of double-seated swings where a parent could sit opposite his or her child, swinging together.[544] In one park there was a kind of hammock-tree: twelve hammocks attached to a tree in a circle. An hour for five dollars, with one hammock free on the honor system for people who were broke. Even the joggers seemed to be having fun, with running shirts that said things on the back like SWEAT IS THE JUICE OF LIFE, and RUN-WORK-EAT-LOVE-SLEEP-HAPPY! There was such a pervasive air of happiness. Was this the future that people in my time were so unsure of?[545]

And on a wall, another *Change The World* poster. I never did learn who was responsible for putting them up. This one had a photograph of the Czech leader, Vaclav Havel,[546] and these words of his:

> *We must not be afraid of dreaming the seemingly impossible*
> *if we want it to become a reality.*

Then around a corner there suddenly came a group of people walking down the middle of the road with a police escort—old folks and young children, holding hands. GAGA—the Grandparents And Grandkids Alliance, the banner read, and their placards said things like *'Protect Our Future!'* and *'There Is No Planet B!'* I asked a grandmother where they were heading and she told me they were going to a rally for Climate Compassion down by the ocean.

"Come and join us!" the girl holding her hand said, but I declined, as I had a dinner appointment with Dezzy. And then I realized that I forgotten to ask Naomi to sign the petition—I'd forgotten to ask everyone, in fact.[547]

I was almost there when I passed a church and in front of it a small white circular building—The Urban Sanctuary. Curious, I approached the door and went in. A sign said it was always open and anyone was welcome to come in and pray, meditate, or just spend time in silence. There was just one man, sitting quietly.

The chairs were arranged in a circle around a large rock, on top of which there was a glass dome containing a holographic projection of a Jewish menorah. I sat down quietly and after a few minutes the hologram changed to the Taoist yin-tang symbol, then after a few more minutes to a Christian Celtic cross. The atmosphere was so quiet. What a blessing this place must be, I thought, and I wondered who had built it, but I couldn't linger since I'd promised Dezzie I wouldn't be late.[548]

"Come on in!" Dezzy said when I finally arrived, after returning Carl's bike. "We're just sitting down to dinner."

"Hi Patrick!" said young Jake, Dezzy's eight-year-old. "I've been making a salmon-run!"

"He means his class has been restoring a salmon creek," Dezzy explained. "It was buried underground and it's not seen salmon for over a hundred years."

"Yeah—and we're restoring it! There's going to be like millions of fish!" Jake added.

"Did you see the humpback whale feeding frenzy that happened in English Bay a few months ago?" Leo asked Jake.

"Yes, we did! We did!" Jake replied with super-enthusiasm. "That was like *so* wicked. There was like, these five humpback whales, and they were jumping and leaping and making these enormous splashes. Mum took me down to Kits beach and there were like, thousands of people watching."

"What were they feeding on?" I asked.

"Herring," Dezzie replied. "It's one of our big success stories. The herring used to spawn in incredible numbers until a hundred years ago, and then it all stopped. But over the past twenty years volunteers from the Squamish Streamkeepers have been systematically wrapping all the pilings in the harbour at False Creek and elsewhere to cover the creosote, which had been killing the herring spawn, and with the new regulations about oil spills and there being so many more electric boats that don't use oil the water's been clean enough for the herring to return."[549]

"Wow," I replied. That must have been amazing to watch.

"And like, the whales release all these bubbles," Jake enthused, "and they make a circular trap, and then, like, *whoom!* All the humpbacks burst out of the water at the same time, and they ate the herring!"

"It was really cool," Leo said. "It's one of the most amazing things I've ever seen. Rewilding—we're slowly rewilding the city."[550]

"So how's our intrepid traveler?" Lucas asked, as I took my place at the table for dinner. "Discovered any new secrets in our city?"

"Too many to mention. But I expect they're all normal to you."

"We've made a good beginning," Betska said. "But Vancouver is still only one small island of hope. Something's going to have to shift in the rest of the world, or we'll be dragged down by the larger failures."

"Oh, don't be so pessimistic, Baba," Leo said. "To you, it feels like a burden of worry. You've worked so hard and we're still not out of the woods. But to me and my friends it feels like a wonderful challenge. To transform the world, to bring fresh hope and spirit, to build the Age of Harmony. What a great adventure! It was you who encouraged me to not succumb to pessimism and despair. I feel that we're engaged in a wonderful journey."

"Why, are you planning to leave us?" his grandmother Betska asked nervously.

"Maybe. Who knows? There's lots to be done. Maybe I'll go to China."

"China? Is there something I don't know about?" Betska asked.

"You'd best get some make-up for that pasty white face of yours," Lucas said to Leo. "They'll think you're an alien from the Siberian landing."

"Aliens?" I said without thinking. "Siberia?"

"Didn't you hear?" Lucas responded. "They say they breathe methane, and this is the first time they've been able to land because of all the methane that's pouring out of the permafrost. Apparently they're being held captive in the Kremlin."

"The Kremlin?"

"Yes, they've got the mother-ship, too. They think it runs on a kind of solar essence that it extracts from passing stars. They're said to be amazingly beautiful. The Russians have had to lock up their sons and daughters to stop them from falling hopelessly in love."

"You're kidding...." and everyone burst into laughter, no-one louder than Jake.

"He believes it!" Jake exclaimed.

"You should see your face," Lucas added, unhelpfully.

"Patrick!" Jake exclaimed, his face full of exaggerated shock, doing service to eight-year-olds everywhere.

I felt foolish and scratched my head. Meanwhile, we were tucking into new potatoes, an omelet made from fresh eggs from a neighbor's chickens two doors down and a fresh salad with nuts, feta cheese and a variety of greens. A fresh, local, zero-mile diet, Dezzy said.[551]

"It's only a matter of time," Aliya said. "How can we live in such an incredible Universe filled with so many gazillion galaxies, each with so many gazillion stars, without there being other life?"

"What if other civilizations aren't limited by the space-time we live in?" Leo responded. "What if they've found a way to crack the gravity problem and understood what creates space-time? They might be able to organize it to suit their needs just as we've done for electromagnetism. Distance would be irrelevant, and time would be simply a dimension you chose to inhabit. They could be visiting us without leaving their dinner tables, just as we Skype our friends at the wink of a brainwave."

Wink of a brainwave?

"I don't think it'll happen in my lifetime," Lucas said. "We're still too dangerous. They've probably got a cosmic aggression scanner they apply to every planet they visit. If it reads too high they give it a pass."

"Why shouldn't it happen?" Leo asked. "Look what we've achieved in the past twenty years. If we keep the momentum up we should be able to abolish warfare soon, and consign it to a museum. We've already got the global ban on nuclear weapons and depleted uranium thanks to the Global Zero campaign, and it looks as if the ban on killer robots will come into effect soon."[552]

Killer robots? Another intriguing mystery, but I declined to ask.

"If we could just find a way to shut down the rest of the arms trade," he continued, "the remaining conflicts would fizzle out. We're at the birth of an incredible new era. Just think what might be possible when we finally end warfare, establish food security for all, phase out the final fossil fuels, protect the oceans and create safe habitat for all the world's species. It's all within reach. We've just got to keep working at it."[553]

"You sound just like my father," Dezzy said. "He used to talk like that."

"I'm with Leo," Aliya said. "I think they're waiting until we're civilized enough to welcome us into the Galactic Union, or whatever they call it. I think Lucas is just expressing his sense of realism, and maybe his frustration that he's not directly involved."

"That's so not true!" Lucas snapped back. "If you're not grounded in something real, who's to know you're not daydreaming or consuming too much homegrown bud? And besides, what do your Muslim sisters think about the Galactic Union? I don't think that's in the Koran."

"Now now, children," Dezzy said. "Remember your manners. Jake—pay no attention. We've a visitor present and we need to leave soon."

"I'm sorry, Aliya," Lucas said. "I didn't mean to be so sarcastic. I don't know what got into me."

"That's okay, honey. I forgive you," Aliya said. "The Qur'an actually says nothing against the idea of a Galactic Union. The Hadith says that seeking knowledge is obligatory upon every Muslim, both male and female. It says we are to seek truth, even if we have to go to China to find it. Going to China at the time of the Prophet—may Allah give Him peace—is like going to another planet today. Why should life be limited to one small planet? Science tells us we live in a single coherent Universe where everything is interconnected. If Allah is everywhere then He is also in the furthest galaxy, just as He was present at the Big Bang, and before it all started."

The table was silent before Aliya's passion.

"You are a blessing, Aliya," Betska said. "I hope, when we reach those galaxies, that they are filled with women as wise and loving as you. Lucas, you're a very lucky man."

"I know I am, Betska. She's my jewel, and I'm a stupid jerk." He reached out and took Aliya's hand.

"Okay, folks. Time we were leaving," Dezzy said. "Just clear the dishes into the kitchen. We can deal with them when we get back."

I had no idea what to expect from the evening. All I knew was that it somehow integrated science and spirit. It was a short walk from Dezzy's place, through sunny evening streets and winding footpaths cradled in the shadows of leafy trees. I walked in silence as thoughts of seductive aliens circled in my head. The city was magical. Dogs played freely in the parks, which Dezzy said was allowed before eight in the morning and after seven at night. A young boy ran past pushing a handle attached to a large yellow circular ball.

"What's that?" I asked.

"That's a washing machine," Jake answered. "We've got one, too. Mum sends me out to push it around and it's like the washing gets sloshed around. I prefer the pogo-washers—they're more fun. You just go out and bounce and, like, when you're tired, the washing's done!"[554]

"Jake's got the neighborhood record for the most pogo-jumps," Dezzy said.

"Three hundred and eighty-seven!" Jake said without hesitation. "The record for kids under ten is, like, more than a thousand, so I've a long way to go. The world record is, like, seventy thousand. Can you believe it?"[555]

My next surprise appeared around the corner where a stream flowed out of a stone wall into a small pool. The sign said it was the source of Brewery Creek, which had been lost to the world a century ago when development buried it underground. It had recently been restored by the Friends of Brewery Creek.[556]

"The goal is to restore it all the way down to the ocean," Leo said. "Imagine—a salmon creek running along Main Street, with birch trees and footpaths on either side. That would be quite something. They're cooling the water in summer using geothermal loops, so the salmon won't die before they can spawn. It's so wretched, that just when people are doing all these great things, the salmon themselves are in peril because of the warming rivers and the warming ocean."[557]

"That's so true," Aliya said. "But I love that the city is working with the neighborhood associations so that every street has a natural feature our souls can feast on. And I love all the tree planting, and the way they form a canopy in summer. That will help cool the creek as well."[558]

"I'd love if it there was more wildlife in the city," Lucas said.

"With wild cougars to take care of all the bunnies and deer?" Betska asked. "The sudden pounce of death on an innocent puppy?"

"The howl of the wolf," Lucas responded. "The silence of the deer as it stares death in the eye."

"Wolves in Stanley Park?" Dezzy suggested. "I'd prefer to stick to elves and fairies if we're going to populate the city with any more wild species."

"Look! It's the Poetry Walk," Aliya said. "Here's one of my favorites, by Leonard Cohen:

Ring the bells that still can ring
Forget your perfect offering
There is a crack in everything
That's how the light gets in.

Every five paces there was a screen in the sidewalk that contained a poem with words that glowed as you stepped on them.

"Here's one by Bud Osborne, from the Downtown Eastside," Lucas said.

Never believe
that the last word has been spoken
by those who could destroy things
rather than by those who could
create more life. [559]

"And here's one by Joni Mitchell, a classic from my day," Betska said.

We are stardust
Billion year old carbon
We are golden
Caught in the devil's bargain
And we've got to get ourselves
Back to the garden.

"Come on, you guys. If we read all the poems we'll be here till midnight," Leo said.

"Am I right that these can be changed?" I asked Betska.

"Yes. There's a website where people post poems they like and vote for their favorites. They change every month. There are Poetry Walks all around the city. You might enjoy the bench as well: if you've got earphones you can listen to stories from the neighborhood. Some go back over a hundred years to the legends of the Squamish people."[560]

"Have you seen the Art Walk on Robson Street, Gran?" Leo asked. "They've done the same for local artists, and they change every night. It's made it the most popular pedestrian street in the city. You should see the illusions! Some do a phase-change when you walk on them. It's very trippy. And there are holographic sculptures that change as you walk through them. It's wild, especially when they turn out the streetlights, and even more so when you're stoned. I've seen people playing holographic chess. The holographic advertising has been pretty wild of late as well. At least it's taxed."

"The advertising is taxed?" I asked.

"Yes. Every advertiser pays a five percent levy to the Public Wisdom Foundation, which uses the money to support public education and the arts."

So many possibilities! I loved the way art was being stitched into the fabric of the city, but my thoughts were cut short by our arrival at an old red brick church with large oak doors and stained glass windows.

As we entered the lobby people were busy greeting each other, and on the wall I saw a large framed poster with a photo of… Derek! The same Derek Brooks, Dezzie's brother. It was another *Change the World* poster, accompanied by these words:

> *It's a race. A race between the expanding reach of our empathy as it stretches across the world, bringing love and intelligent cooperation for the good of all, and the clutching fear of tribal distrust, made more powerful by modern technology. Which will win?*

It was a haunting question, lacking confidence about the outcome—and yet it had the effect of making me feel more determined to lend my weight to the side of the scales that would bring more love, and less fear and hate.

After a few minutes we entered a dark carpeted passage that curved around the building, lit by dim red floor lights. This led to the main body of the church, lit only by a model of the Earth hanging above us, glowing from within, and a sculptured lotus flower that glowed with a pale white light. There were three rows of seats in a circle around the edge, and cushions where people were sitting on the floor. Everyone was quiet.

When the shuffling ceased there was a period of silence, then a solo saxophone began a haunting melody. The light inside the lotus grew stronger and a choir entered, singing. We stood up, and a woman dressed in a white gown with an orange sash stepped into the center.

"Greetings, my friends. K'ayacht'n. Welcome. We are blessed to live on land that has been cherished and protected by the Squamish and the Coast Salish people for thousands of years, and we thank them for their good care and stewardship of the land. Bonsoir. Namaste, shalom, as-salaam 'alaykum. We gather here on this beautiful summer evening to celebrate the Universe that has created us, the love we share, and our hope for the future. Whether you are a regular friend or a visitor from afar, we welcome you. For some, this evening is a celebration of the wonder of the Universe. For some it is a celebration of God, the Great Creator. For some it is simply a celebration of existence. We are all created from the same reality, and we celebrate this in the Song of the Universe."[561]

Everyone rose, forming three circles. Three musicians played a violin, piano and saxophone and we linked arms, dancing slowly around the circle.

When the dance ended the woman invited us to find three other people and choose a place where we could talk. "This is your chance to share your personal journey," she said, "as you move along your path towards wholeness."

I was invited to join two older men and a younger woman, and for twenty minutes we shared something of our lives. I told them about my visit, and my purpose for being here. The first man introduced himself as a Persian-Canadian from Iran whose wife had died recently. He spoke of the comfort he found in the weekly gathering and how he used to be a professor of physics, but his main interests now were in poetry and music. He was a humanist and an atheist.

The second man was Jewish, here with his wife, children and grandchildren. They had shared their Shabbat family meal together before they came and he found it inspiring to see how reason and spirit could unite, bringing people together from religious and non-religious backgrounds.

The woman told us this was her first time here. Her parents were Kurdish Muslims who had come to Canada from Iraq in the 1990s. She told us of the difficulties her cousins were having in Iraq due to the country's rapidly declining oil revenues, and her worries for the planet's future. She hoped she might find understanding that would help her decide what to do with her life, so that she could make a contribution to the world. She liked the fact that the Song took place on a Friday night, since it gave her a sense of solidarity with Muslims around the world for whom Friday was a day of holiday, gathering and prayer.

Following the sharing a large screen descended. "Each week someone gives a presentation," Leo whispered. "Last week it was the oral history of the Coast Salish First Nations in the centuries before white settlement."

The speaker was a biologist, I'd guess in her fifties. I recall her telling us how our bodies contained seven thousand trillion trillion atoms, which had self-organized together to create us. There were a thousand times more atoms in our bodies than there were stars in the Universe, she said, and the hydrogen atoms in our bodies had self-created fourteen billion years ago as the first fundamental particles cooled and paired up. She showed how a simple bacterium stretched out a microbial nanowire to other bacteria, shuttling electrons back and forth to create a community of bacteria that then became a superorganism. Within our bodies and our brains, she said, the process of interconnection has been perfected to the gazillionth degree.[562]

The reality of consciousness, she said, had existed in raw form since the very beginning of the Universe; even the cells that constituted our bodies were conscious. The separation between mind and matter that had created so many problems for science and philosophy was finally being bridged. We possessed incredible potential, but the global challenges we faced were such that unless we speeded up our rate of global self-organization, our whole civilization might yet collapse.

She spoke far more than I can remember. She ended with a video that showed atoms self-organizing into molecules and then into the cells and organs of our bodies, all under the integrating influence of consciousness. I was mesmerized.

Afterwards, we were invited to find someone to share our thoughts with. Leo approached me and raised his eyebrows.

"It's incredible that there's so much complexity," I said, "just so that we can exist. All those atoms cooperating together."

"Think what we'll be able to do when we finally put our stupidity behind us," he replied. "All the wars, all the fighting over resources. I feel as if we are just beginning to live sensibly on the planet. It has taken the accumulated learning and wisdom of all previous civilizations to get us to the point where we can finally begin to cooperate. That's seventy thousand years, if you count from the time we left Africa. Imagine what the next seventy thousand years will bring, or the next million. The Sun doesn't begin to turn into a Red Giant for over a billion years. Can you *imagine* what that means? It's crazy. We're just beginning to be a mature planet. The way I see it, all of our history up until now has been a sorting out of the basics so that everyone can have a full stomach and a secure place to call home, without fear of death or hunger. Now we're getting ready to begin our real work, whatever that might be."[563]

"What do *you* think our real work is?"

"Well, that's the question. Can you come back in a million years?"

"No, seriously. Something tells me you've thought about it."

"Well, okay. My guess is that it has to do with exploring the potential of consciousness. I get the feeling that consciousness is like the deepest ocean, full of the most amazing places, and we—with the exception of a few great spirits—have been just paddling around in the shallow waters. Maybe if we explore it as thoroughly as we have explored the realm of matter we'll find that we can converse with a tree or an orangutan—or even a river."

I listened intently.

"Do you know the book by St. Augustine called *The City of God*?" he continued. "He was writing in the early 5th century, soon after Rome had fallen after being sacked and ravaged by the Visigoths. Many Romans were blaming Christianity for the collapse, saying it had enticed the Romans away from the old gods, and hence the downfall. Augustine wanted to contrast the City of Man, in which humans indulged their pride, greed and pagan beliefs, with the City of God, in which humans would live guided by the higher laws of Heaven. It strikes me that we are living through a very similar period of contrast. During the last two hundred years, humans have made unprecedented material progress, often motivated by the noblest of dreams, just as they did in Rome. But many rich people have become totally corrupt, also as they did in Rome, and embraced a selfish version of happiness that doesn't give a shit about other people. But something is happening in response. The OMEGA Days that happened here and around the world are part of it. There's a new vision arising, just as there was under Christianity. I feel as if we are creating a City of the Future, both here and in other places around the world, not just as a way to overcome our environmental problems but as a response to a higher set of values, a spiritual awakening. Maybe one day it will become corrupted too, but right now we are in its birth stages. Where it will take us, I don't know, but I'm loving the journey. It's so astonishing to be alive at this time. We hold everything in our hands: the birth of a new global civilization or collapse into another dark age, the way it happened to Rome. That's our choice."

I was so impressed by Leo's thinking. We were about the same age, but he had a vision of the future that was rare in my time.

"What makes you so confident?" I asked. "When I was with Li Wei-Ping this afternoon he was telling me how far we had to go before we would be living sustainably on the planet."

"But that's the whole point," Leo replied. "It's a journey. If you're an on-looker it may seem daunting, but when you're on the journey the only way is forward. It doesn't matter how many obstacles and difficulties there are. It's like she said—our consciousness affects even the way the cells operate in our bodies. The attitude we choose, whether positive or negative, affects everyone around us. We're all in this together. In the old capitalist world we were encouraged to think of ourselves as rugged individuals who happened to live on the same planet. But we're not. It's a lie. It always was a lie. Ask any First Nations person, or Buddhist. We're interconnected. We're an evolving superorganism of souls. Just think how much there is yet to discover."

When the discussion time was over we formed three large circles and danced again. I could tell by the smiles that people's spirits had been lifted.

After the dance there was a period of silence when all the lights were extinguished except for the Earth. In that silence, there was such a sense of unity and shared purpose. Then a man stepped into the center. "Every week," he said, "we take this moment to renew our Gratitude Pledge."

He asked us to think of one thing we were grateful for, and to tell the person next to us. Then he explained the importance of the pledge as a compass for our choices as we walked the path of life. Everyone then recited the Pledge:

> *In deepest gratitude for having been given the gift of Life, and the*
> *wonders that accompany it;*
> *In gratitude for the Earth our home, for Nature, and for all the*
> *incredible species we share it with;*
> *In gratitude for all who work to make this world a better place, and all*
> *who have done so in the past, stretching far back into the*
> *mists of time;*
> *In gratitude for the freedom I have to choose my path, as I take my turn*
> *carrying the baton of life;*
> *In gratitude for these gifts, I pledge that I will use my life to serve*

the betterment of all living beings, the healing of suffering, and the restoration of our beautiful planet the Earth.

Wow, I responded inwardly. That's powerful. That could make a difference.

There was a short period of silence for prayer and meditation, and then a glorious piece of music called the *Earth Anthem*, which began with a solo female and built into a full chorus. The final chord was followed by silence, then an explosion of celebration.

I was struck by the music, which remained with me after I returned home, so I wrote the words and music down and I contacted David Ballard, a student composer in Vancouver whose choral music I admired. He agreed to help and the choir at Canadian Memorial United Church on West 16th Avenue performed the World Premiere, which you can hear on SoundCloud.[564] These are the opening words:

On this planet that we live on, on this pearl among the heavens,
On this tiny floating sphere we call our home.
In this paradise of beauty, where all Nature has evolved,
From the elephants and tigers to the worms that grow the soil.
On this wondrous living miracle, this jewel in empty space,
On this planet where all humans live and die,
It is time for us to wake up, and give ourselves a shake up
And learn to live together in our home….

Chorus:

We are one, we are one, we are dancing in the one,
in the Universe that made us, we are one.
Ever since the great beginning,
in the love that gave us singing,
in the Universe that made us, we are one.

After the anthem there was more dancing, an earnest reminder about the Fast for Climate Compassion and a plea to sign the petition, and we ended by singing a version of *Hallelujah*, the song by Leonard Cohen. Then it was over. We formed two lines and linked hands, creating a tunnel and departing in pairs.

As we left the hall, the walls of the passage were lit with a sequence of images that told the story of the Universe, step by step. After the initial explosion of light we saw the first atoms, the first supernovae, the formation of the galaxies, the first planets, Earth's molten volcanic origins, Earth's all-encompassing super-ocean, the first primitive bacteria, all the way down to our appearance as humans in the last few images. The time scale was logarithmic, as the time since we left Africa is only the tiniest smidgeon of time since the beginning of the Universe.[565]

We stepped into the darkening dusk, with no streetlights. "It's Dark Sky Week," Dezzy explained. "It's an experiment the city's trying so that we can see the stars. It's also to give the city's wildlife a break—the frogs and insects, the moths, bats and snakes. They need the dark to reproduce—a lot like us."[566]

In the darkness, the crosswalks lit up when you stepped on them.[567] Some of the bike paths sparkled with light as if paved with diamonds,[568] and when a car passed, its lights made the white line down the middle of the road glow with photo-luminescence. It was quite unlike any city at night I'd ever seen.[569]

"Why don't they switch the lights off every night?" Lucas asked. "We no longer use the streetlights on our street."[570]

"How does that work?" I asked.

"The city wanted to retrofit the street with pole-less street lighting,[571] but we persuaded them to remove the overhead lights altogether and replace them with ground-lights that we click on when we need them. I love the darkness, and the old folks can use the clickers to see where they're going."[572]

"And they let you do that?"

"Yes. We registered it with the Green Streets Program. They support experimental ideas as long as you get enough people on board."

The night was not over yet. We walked through the dusk to a park where telescopes had been set up to view the heavens. The night sky was clear of clouds and the almost full moon was low on the western horizon, so the stars in the east could be seen in all their glory.[573]

"Just think," Dezzy said. "It was like this when our ancestors gazed at the sky fifty thousand years ago. Don't you find that incredible?"

"It is. I can never get over how stunning it is," I replied.

"Look! There's the Carl Sagan International Space Station," Lucas said, pointing to a moving point of light. Looking through one of the telescopes, I could make out its solar panels. The telescope also gave me immediate information, identifying various stars, and the craters of the moon.

"It was launched last year to replace the old one," Dezzy said. "You can pay them a visit if you've got enough money.[574] I have a Chinese friend working up there alongside the Russian, Canadian, American and Brazilian astronauts. It's a true work of global cooperation."

"You know one of the astronauts?" I said, with a degree of amazement.

"I do as well!" Jake shouted with excitement. "She's called Li Na, and she says, like, like, I can be an astronaut too if I work hard. Poke her! Send her a poke! Go on!"

"What, now? Well, I don't see why not." Dezzy got out her device and sent a quick message to her friend on the space station. Two minutes later she got a response.

"She says hi to you, Jake!" Jake did a sudden crazy dance, completely overcome with how cool this was. "She wonders why she can't see the lights of Vancouver. We met through the Perimeter Institute in Waterloo, where I met my husband. She's doing optics experiments in zero gravity—fascinating stuff."

I let that one pass. "Would you go into space yourself if you had the chance?"

"Not for a million dollars. Who would look after Jake if anything happened to me? How about you?"

"I've often thought that I'd like to. How can we not travel into the cosmos, to explore the great beyond?"

"To boldly go…?" Aliya asked.

"Yes," I replied with a laugh. "To boldly go."

"I hate to say this, but I don't think it'll happen," Aliya said. "The loss of bone density in zero gravity is too much. The astronauts lose one percent every month. If you did a quick six-month trip to Mars, spent a month there and came straight back you'd lose thirteen percent of your bone density. Anywhere further and you'd need bone replacement therapy for your entire body."[575]

"That might not be too bad," Leo chimed in, his eyes fixed to a telescope. "I've read that astronauts can have a genetic implant to change the nature of their bones. Or they could bioprint a set of replacements before they left. Welcome back to Earth! Here's your new skull. Can you just take your brain out?"

"That's wicked!" Jake exclaimed. "Can they really do that?"

"No Jake, Leo's joking," Dezzy said. "I must admit, losing that much bone density doesn't sound like fun. I think it more likely that we'll use some form of teleportation."

"Like, antimatter, and all that?" I replied.

"Yes," Dezzy said. "Maybe linked to the interface where consciousness forms the implicate order in pre-quantum reality."

"I'm afraid that doesn't mean much to me," I confessed.

"No matter. Sorry—that's a physics joke," Dezzy responded. "Teleportation would at least eliminate the problem of space junk. The exponential growth of the orbital shatterverse with its million pieces of microjunk zipping around at seven kilometres a second all smashing into each other has become quite the challenge. I volunteered as a NASA clickworker last year, identifying pieces of space-junk so that they could be targeted by lasers and pushed down to a lower orbit where they'd burn up in the Earth's atmosphere. Who would have thought we'd create an orbital garbage gyre in space as well as the garbage gyres in the ocean and the toxic fetal gyres in our own bodies? It's an awful legacy."[576]

———————✦———————

Walking home, I had a chance to talk to Leo. He impressed me, but I had no idea what he did for a living.

"I work in Tim Hortons," he said. [For non-Canadian readers, Tim Hortons is an iconic Canadian coffee and donut shop with locations everywhere.] "I love working there, and the restaurant has a great team. The owner encourages us to work cooperatively, with joint decision-making and profit sharing. What's also cool is that half our staff have a disability of some kind, but it makes no difference to their work. They are great people, and if anything, they are *more* motivated than some of the other workers."[577]

"How did that come about?"

"The owner, Sam, was so inspired by what was happening in Vancouver with all the co-operatives and employee-share owning that he wanted to do something himself. As well as including disabled workers, we practice open bookkeeping, so that we can participate in decision-making. We reinvest ten percent of the profits in the business, give ten percent to charity, and share the rest among ourselves."

"That's really impressive. What motivated him to share his profits?"

"He's a very special man. He lost his wife to Parkinson's and dementia a few years ago and he says the spiritual and emotional wealth he gets from sharing is far greater than any amount of monetary wealth. He comes and helps on the floor sometimes. We're one big family."

"How long do you plan to stay there?"

"I'm not sure. I've spent the last four years getting my philosophy degree, and I'm not sure what's next for me. I'm trying to live simply to save what I can. Five of us share a house, so my rent is low. I get Citizen's Income and I don't own a car. I pay three percent of my earnings for my student tuition and we grow a lot of our own food, so I am able to save quite a bit. I had an older cousin who couldn't take it. He was so down, what with his debt, his poverty and his general hopelessness that he committed suicide. Hanged himself in his basement suite. My aunt and uncle were devastated."

I was shocked. It brought home the stress and difficulty that many people had lived through.

"I'm so sorry," I said. "It seems the Twenties were hard on a lot of people. But what about *your* future?"

"I don't know. I did a session with a purpose counselor who told me I have a natural talent for enquiry, innovation and teamwork, but I'm no closer to knowing what my purpose is in life. I keep thinking that I need to get out of here, go traveling. Vancouver's too beautiful. I worry that it's going to stifle my soul, all this beauty and super-civilized behavior. It's too easy. I want to feel the quanta, mix it up with the strange attractors. Somewhere rough and raw. I'm thinking China, maybe. Maybe I could teach a course in philosophy there. From Socrates to Syntropy. Sound enticing?"

"I'd sign up. But what do you mean when you say 'feel the quanta' and 'mix it up with the strange attractors'?"

"It's where quantum theory flirts with chaos theory. The quanta never quite know what they're going to be. They like to hang loose, unfolding in the wave before deciding to become a particle—or not. The in-between, the unknowing: that's where the possibilities hang out. One becomes a wage-slave, the next a revolutionary. What decides? The strange attractors. They float through the air, invisible organizing forces that influence your path, one way or the other. If you don't know which way to go, where better to hang out? Someone should open a café to attract the strange attractors, a place where everyone knows that anything is possible."

"Wouldn't it lead to rampant sex?" I joked.

"I doubt it. We don't exactly live in a sex-deprived world. I'm thinking something with a higher wyrd factor—that's 'wyrd' with a 'y.'"

"What's 'wyrd' with a 'y,' when it's at home?"

"It's never at home: that's the whole point. It's the word the Anglo-Saxons used when they spoke about the strange attractors, long before there was such a thing as science. It's the spirit of possibility that hangs in the air when you either choose your destiny or let fate choose it for you."[578]

"Why don't *you* open a café and call it *The Strange Attractors* to help people find their direction?"

"That's a great idea! But I have to find my own direction first. I keep thinking how dull most of my philosophy studies were at university. Thank heaven for that book, *Sophie's World*, which laid out the history of philosophy as a story for children.[579] There's got to be a better way to bring critical questions to light so that they have relevance for ordinary people. There's got to be a way to portray the evolution of philosophy as a syntropic emergence, the instinctive quest of the human mind to find answers to the questions that concern us all. We all die. We all have our hearts broken sooner or later. We all wonder why we're here. There's got to be a way to show how Socrates, Ibn al-Haytham, Kant and Mukherjee are different expressions of the same syntropic quest for wholeness. There's a unity that holds it all together, and that's so much better than the doctrinal squabbling and hatreds that have caused so many wars, inquisitions and burnings. Maybe I could contribute something useful."

Suddenly a flash of color zipped by, missing us by inches.

"What was that?" I yelled, startled.

"Whoa, where've you been, dude? You've not seen one of those before? It's a bicycle. What did you think it was, a teleporter?"

"No, uh, er, I mean the color." As I spoke another bicycle zipped past us, its wheels a circular whirr of pale blue.

"They're Revolights. All the fancy bikes have them. You wait till you see a proper Night-Bike. They're really something."[580]

Sure enough, two minutes later a bike came by with its entire frame lit up, changing color as it passed.[581] And the night was not yet done with its surprises. As we walked home I saw cyclists with jacket sleeves that contained turn-lights and cyclists with lights in the handlebars. Then a man walked towards us wearing a T-shirt with a brightly lit display that said:

MAKE LOVE NOW

As he got close, the words changed….

TOO LATE TTFN

We'd all seen him, and when we read the words on his back we burst out laughing....

GLYDE VEGAN CONDOMS

⁂

When we got home I told Dezzy about my plans for the next day and she lined me up with a tent, a sleeping bag, a travel card and a picnic lunch. She said she'd called a friend who used to be a city councilor and she'd agreed to meet me for breakfast to tell me more about Vancouver's Greenest City initiative, but I'd have to leave early in the morning.

"You look as if you've had a great day," Dezzy said.

"I have," I grinned. "I even met a girl I really like."

"You're a fast mover! You have a great time at the farm tomorrow, and we'll see you on Sunday evening. I have two very special friends coming for dinner at seven, one of whom is Thaba, my ex-husband. I think you'll enjoy their company. So don't be late!"

Later that night, Leo and I sat up in Dezzy's living room watching television on the large screen that faded into the wall when it was not being used. Traditional television had merged with the on-line world and everything was either free, free with advertising, or pay-as-you-go. As well as channels I had never heard of, I was happy to see that Al-Jazeera was still there, along with the CBC (Canadian Broadcasting Corporation), the Knowledge Network from British Columbia, the BBC from England, and PBS from America. When Leo wanted to change the channel he used a hand-motion that he had tailored to his favorite channels through a tiny device that he wore under his turtleneck. A flick for this, a wave for that, and there was the channel.[582]

To access a pay channel, he held his wrist to his forehead and concentrated for five seconds. When I asked what he was doing he said nonchalantly that he was giving his password. When I asked again, he realized that I didn't understand. He explained that the black band on his wrist was his phone, and his password was encoded into his brainwave response to a particular song he had recalled. Only he knew what the song was, and only his brain could recall its unique brainwave configuration, which a sensor on his phone picked up and translated into a unique digital password. He said he could use this one password for every use, instead of having to remember a host of different passwords, and he could sign into anything in the world as long as he had his phone.[583] He could also use his heartbeat, but the song method was more cool.[584] He asked if I wanted to watch a particular movie, and I thought of *Hope's Sister* that Wei-Ping had praised, but time lag was catching up on me so I told him I needed to hit the sack.

But Leo had other plans in mind. "Have you tried the latest Oculus Rift body-suit?" he asked. "It combines Oculus Touch, Oculus Smell and Oculus Feel."

"No," I replied, not having a clue what Oculus Rift was.[585]

"Oh, have you got a treat in store." He opened a chest and pulled out two bodysuits complete with gloves and helmets. He helped me put mine on and gave me a pair of hand controls. He put his own on, switched them on, and whoa! I was immersed in the full dimensional world of virtual reality. "I'll show you my favorites to give you a sense of what's possible."

What followed was incredible. First he took me into the underwater world of microscopic phytoplankton, swimming among the geometric shapes of living plankton as if we were the same size, inhabiting a world I scarcely knew. I could even reach out and touch them, giving me every reason to believe I was really there.

Next he took us into what felt like a real-life medieval village, where we watched a drama unfold as a young woman was distracted from milking a cow by a knight on horseback who scooped her onto his horse and rode away with her, chased by a man on foot, her apparent lover. Unlike a traditional movie, the activity surrounded me in all directions. I could smell the horse manure, feel the wind on my face and feel the earth vibrate as the horse galloped away. We could watch the whole movie if we wanted, Leo said, but he had other things he wanted to show me.[586]

My next jaunt was—wow—way up into the sky. He had made me a bird! "Move your arms!" he called out, and as I did my wings beat up and down. I was soaring over a beautiful landscape of farms and forests, able to control whether I hovered, climbed or dived.[587]

He loaded another sequence and we were on the surface of the Moon, the Earth a distant blue-green planet floating far away in the blackness of space. I walked across the surface of the moon, experiencing the absence of gravity with little weight and a strange ease of movement. A woman approached me in a space suit and held out her hand to shake—which I did, experiencing it as solid and real. She then asked me a question—"How did I like it on the Moon?"—and when I answered that it was amazing she nodded her head and told me she had been here five years ago, and this was her recorded message. Wow.

Then Leo put me inside a caribou in a boreal forest, aided, he told me, by tiny microcams carried on the rack of an actual caribou. My caribou was in the forest at that very moment, quietly grazing on lichens and vegetation; even the mosquitos and other insects felt real. It was dark in Vancouver but still light in the forest, since it was midsummer in the far north. Then he switched to some archived footage and wow! My caribou was fighting with another caribou bull, our horns smashing and clashing as we fought for dominance. It was so real, so utterly vivid. If people wanted to get a sense of the caribou's life, I thought, this had to be an amazing way to do it.[588]

For his next surprise Leo took me into the sky above a city and showed me great bubbles of carbon dioxide arising from trucks and buildings, circles of blue floating upwards. Then we flew inside one of the bubbles and I smelt the diesel and heard the noise of a truck. If this was how people were learning about the climate crisis, I could see how powerful it could be.

"Ready for a rollercoaster?" he asked.

"For sure," I said, assuming he meant another adventure, but no, he meant the real thing, and ohmygod I screamed, squealed and almost fell over as I experienced a real 2030s rollercoaster, all stomach-heaving drops and lurches. This was w-a-a-a-y too real.

"How about some overseas travel?" he asked, when I had recovered. "Which country have you always wanted to visit?"

"Tibet," I said. And suddenly there I was, walking into a monastery, all red and orange colours, surrounded by local Tibetan people, fully immersed in their world. This was astonishing. I couldn't interact with them as I had with the woman on the moon, so it was in that sense a passive experience, but my point of view was central to everything, and when "I" sat down to meditate, I felt as if I was totally there meditating among the other monks. I smelled the incense and heard the shuffle of the monks next to me. I could have stayed there forever.

"One more?" he asked.

"You mean I can go anywhere, literally anywhere I like?"

"Not quite—only where Google Oculus has been. But the choices are enormous, and growing every year. They've just added parts of Antarctica. "How about something completely different to end the night? How's your embarrassment factor?" he enquired, setting me up for what was to follow.

What followed was shocking, amazing and exciting—virtual reality sex. I pondered censoring this part of my story, but it was part of my journey, and it shows one aspect of what people were doing, so here it is.

Instead of looking at a photo of couples having sex or watching the things people do in porn movies, I was present myself, and a sexy brunette woman with curly brown hair was approaching to make love to me with nothing veiled or hidden, as if she was really there. When I reached out my virtual hands I could touch her virtual breasts and other places, just as I would a real lover, and I had real sensations in my fingers. When she sat astride me it was a total turn-on—my bodysuit had sensors in all the right places. It was totally arousing, and made all the porn I had ever watched seem like nothing.

Then Leo said, "Do you want to switch it up?" and before I could answer I was in a woman's body and a man was approaching to penetrate me and make love to me. "No way!" I yelled instinctively but Leo shouted, "Hang in there! Be cool!" and I experienced the very strange sight—and feeling—yes, feeling—of a man making love to me, thrusting in and touching my breasts. It was too weird, but as I relaxed and accepted what was happening it was as if I had actually become a woman, experiencing sex as a woman might experience it. And then—I've no idea how—my body experienced what I can only imagine was supposed to be a female orgasm through goodness knows what sensors embedded in my bodysuit, though I'm pretty sure it didn't do a very good job of it.[589]

"That was crazy," I said to Leo, when we had taken our bodysuits off. "Is this replacing real sex?" I really didn't know what to think.

"Not at all," he said. "They call it full-dimensional but it's actually very one-dimensional. It's exciting on a physical level, but it doesn't bring the love and intimacy or the closeness you enjoy with a real person."

"Do lovers program it so that they can make love when they're away from each other?"

"Yes," he replied, "I've done it myself. We coordinated our lovemaking and shared a voice connection, talking to each other as we did it. It was definitely a turn-on, but it left me feeling empty afterwards, aching for her real touch, her real kisses, and her real body next to mine. So no, I don't think it's going to replace real sex. Not to worry!"

———✦———

Later, lying awake in bed, I cast my mind back over the day. There was so much to ponder, from my discussions with Aliya about food and healthcare to the morning bike ride, the Climate Compassion fast, the Future Café, talking to Emily in the Downtown Eastside, meeting Wei-Ping, Naomi's time in Israel, our tour of Vancouver's new economy, my visit to Triko-Op, my discussions with Laura, and then the Song of the Universe, our conversations walking home and the virtual reality experience with Leo. I loved this future, and I already knew it was going to be hard to return. There was so much to digest, and I was only halfway through my journey.

Which made me think of Laura. Laura MacGregor. Was she thinking of me? I doubted it. But you never know....

22
Renovating Democracy

I WOKE EARLY, crept downstairs and activated Dezzy's screen, touching 'City' and then 'Transport'. A map appeared with buttons for the various modes of transportation and local destinations.

I touched my destination, the Three Moons Café east of False Creek, and then touched 'Transit.' The bus route showed up with the time of the next bus. I touched 'Carsharing' and a series of stars indicated the nearest vehicles. I touched 'Ridesharing' and a screen appeared where I could register to give or receive a ride. I touched 'bike' and it showed the bike route to the café and places where I could rent a City-Bike. This set-up had everything! I jotted down a few notes and went to the kitchen in search of a coffee.

Outside, the quiet early morning air felt liberating. I walked four blocks to the bus shelter, a curvaceous structure built from a glass-like material, and when I used my card on the scanner, the time of the next bus appeared in the glass, accompanied by words from a poem by Kate Braid:

> These trees worked hard to get up here
> one ring at a time. The prize is sky
> and the freedom of birds. [590]

My early-morning mind was captivated, soaring up to the treetops and into the sky. Did the poems change? I pressed a button and an abstract painting appeared full of flowing colors. 'By Nitya Mohan, age 14,' it said. 'To submit a poem or work of art, go to vancouverart.com/bus-shelter.'

The bus was full of people chatting or quietly reading their screens. I waved my card across the scanner and found a seat. A screen showed a map of the area and the next stops in English, Chinese, Punjabi and Urdu. I contemplated the day ahead. First I was going to the Agora in Vancouver's Urban Farm District to meet Alice Woodsworth, who Dezzy said had been very involved with the Greenest City initiative, and then I was off by train to Joey's Farm.

The bus seemed different, but I couldn't discern how until I saw a display that said it weighed half of what buses used to weigh, and the seats were made with compressed air bubbles. As a result, it needed half the energy and had twice the range of older, more traditional electric buses. It had a 150 kW battery and could operate eighteen hours a day, catching a partial overhead recharge while the driver took a break and a full recharge at night.[591]

I got off in the Urban Farm District behind the bus and railway terminal. In my time, it was a fenced wasteland of derelict industrial sites, car parks and non-descript commercial sheds. Today, it was a magical world of waterways, food gardens, green spaces, bike trails and businesses, focused around an urban village. The concrete overhead viaducts—part of a never-realized plan to carry a freeway into the heart of Vancouver—had gone.[592]

At the bus stop there was another *Change The World* poster, this one carrying a photograph of the famous primatologist Jane Goodall, who had done so much to build the world's understanding of chimpanzees and how their habitats needed to be protected,[593] with these words of hers:

What you do makes a difference,
and you have to decide what kind of difference you want to make.

They got you thinking, these posters. It was only 7:30 a.m. but the place was bustling with activity. The village was designed around a lagoon, giving it the air of a miniature Venice. The sidewalk cafés were full and the shops were open, in spite of the early hour. At the Cottonwood Gardens Urban Farm people were cleaning the animal stalls and exercising the horses.[594]

My attention was drawn to a large circular building about thirty metres high with windows and balconies that rose towards a glass pinnacle. It reminded me of images I'd seen of Les Galeries Lafayette in Paris, and then of something else: a kind of broccoli that grows in fractal spirals. The ground floor was occupied by various food businesses and the upper floors were residential. I entered through an arch and found myself in an enormous hall with a high vaulted ceiling. People were busy buying and selling, with cargo-bikes and small electric trucks coming and going.[595]

Light poured in through the roof, so I climbed the stairs to an upper walkway and gazed down on the activity. Another set of stairs, and I was out on the roof. The view was spectacular, north to the mountains, south to the tree-filled streets of Vancouver and west to the shining towers of the downtown, catching the morning sun. The air was crystal clear, with no hint of pollution. Below me, the apartment roofs had small minarets tiled with solar shingles. Winding around the building, spiraling downwards, there was a... slide! I watched a woman leave an apartment, get on and disappear from view. These Vancouverites knew how to enjoy themselves. It felt as if the Viennese architect Friedensreich Hundertwasser had reincarnated and come to Vancouver.[596]

Alice Woodsworth was a middle-aged woman with pale green hair bound up in a braided plait. She greeted me warmly in the Three Moons Café, inside the

Agora. We ordered breakfast from the electronic table menu—scrambled ginger tofu on sourdough toast with chives, roasted hazelnuts, red pepper jelly and solar hothouse tomatoes: perfect for a bright Saturday morning. I ordered a mug of Goldie's Dandychic coffee, made from roasted dandelion and chicory roots, roasted pearl barley, cinnamon, carob powder and maple syrup, harvested at local farms and blended here in the Agora. How can I describe such a delicious taste?[597]

Alice was an old school-friend of Dezzy's mother. When they were at college together they had cycled across Canada with several other women, camping under the stars. She had gone on to teach anthropology at UBC before entering politics.

"What made you stop being a professor and become a city councilor?" I asked.

"I was inspired by the city's goal to become the greenest city in the world," she replied. "I wanted to do more than academic work. Some of my students were engaged, and they encouraged me to run for office. When the city started out on its Greenest City path there were ten thousand people who played some part in creating the first Greenest City Action Plan. There were camps, conferences and a citywide brainstorm that generated hundreds of ideas.[598]

"Their goals were admirable. They wanted a green economy, climate leadership, green buildings and green transportation. They also wanted zero waste, access to nature, a lighter footprint, clear water, clean air and local food. It was all laid out in a very comprehensive manner, and integrated into the city's work plans and budgets.

"By the time I became a councilor, however, the public interest had begun to fade. Even though the number of climate disasters was increasing every year, and Calgary had just had its second thousand-year flood in ten years, the unemployment and the ridiculous price of housing were more immediate concerns. There was steady progress on transit, bike lanes and local food production, but there wasn't widespread public engagement."

"But I thought you said ten thousand people had been involved."

"Yes, but that's less than two percent of the population, so even in the beginning ninety-eight percent of the people were not engaged. We didn't have the level of engagement we have today.[599] We didn't have any number of things, such as the 'Improve My City' community participation app, or PlaceSpeak, which makes it so easy for neighbors to share in a public consultation.[600]

"Let me be clear," she said, in a tone that brooked no nonsense. "If you want to change the world at the community level you need to satisfy at least seven conditions. I wish it wasn't so many, but that's the way it is."

I focused my attention on taking mental notes.

"*First*, you need a lot of people who are really pissed off with the status quo, who have a burning passion to change things. It took the OMEGA Days to get us to the tipping point, and then over.

"*Next*, you need a positive vision of a desirable future that will make people want to get involved. It took us a while to achieve this, but when someone produced a video that painted a really bright, positive picture of what the future could look like, people began to get engaged.

"*The third thing* you need is cohesive organization. We had good internal management at City Hall but we suffered from departmental silos, and there wasn't cohesive organization for the city as a whole. It was only later that we established better organizational roots with the Village Assemblies and the Business Networks."

"What do you mean by departmental silos?"

"Oh, the water people not talking to the buildings people, the transport people not talking to the civic engagement people, all the normal stuff. After I became a councilor we went through a big upgrade with the goal of becoming an integral city. The consultant encouraged us to think of the city as a beehive or a forest, not just an assemblage of buildings. We learned to think of the city as a soulscape, a mindscape and a heartscape, as well as a streetscape, a landscape, a foodscape, a historyscape, a moneyscape and an ecoscape. It was quite exhilarating, integrating all those scapes into a single vision.[601]

"A few years ago we needed to upgrade the sewage treatment plant, which was more than sixty years old and right on the ocean. If you think of sewage as waste, it's natural to want to dispose of it. Using integral city thinking, however, we understood that sewage contains nutrients and energy, and you can grow biodegradable plastic in sewage bacteria. So we asked the engineers to look at it from an industrial ecology perspective and they came back with a proposal to build five decentralized units, some as small as two city lots, linked to the city's district heat network. So now we are selling heating and cooling to nearby developments. We've started the bioplastic experiment and we're producing water that's clean enough to discharge into a constructed wetland, providing green space and wildlife habitat complete with herons and tree-frogs."[602]

"What a great story!"

"Whenever an upgrade of some kind is due you can either approach it the old way and just replace it, or approach it in an integrated manner and discover all sorts of new opportunities. But you need skills to make that happen, which brings me to the *fourth* condition, which is community training. We persuaded our community centers to run courses on things like group facilitation, and how to start a social enterprise, so that we could find people with leadership skills and get them working.[603]

"Then we ran into another problem. It seems there are some people for whom a little bit of power goes to the head. They become self-important, and they drive other volunteers away. It's a narcissistic pathology that's probably as old as the hills. By naming it, *cratophilia*, we were at least able to train people to be aware of it, and address it sensitively."[604]

"My grandfather used to be in the army," I replied. "I remember him telling me about the self-important behavior of some of the newly promoted corporals."

"I put it down to self-acceptance deficit," Alice said. "The status some people gain from thinking they are important makes up for their lack of self-worth. There's another aspect of social change that it's good to know about, too. The conservative brain is wired to threats and dangers, rather than innovation and change, so conservative people tend to be more reactive. Once you know that, it's possible to frame innovative ideas in ways that are less threatening. If you're hot-headed, many conservatives will dig their heels in and resist.[605]

"The *fifth* condition for successful change is committed leadership, and I must say we have been blessed in that department: we've had some great mayors, and great leaders in the non-profit sector, the universities and the business community—and the schools and places of worship. It used to be unfashionable to talk about leadership when I was at college, but without it, nothing gets done.

"The *sixth* condition is pretty obvious: you need a majority on council to vote for the changes you want. We always had to keep an eye on the next election, reminding people that running for office is just as important as running an urban farm.

"And *finally*, you need a well-organized system of public engagement that's strong enough to cope with the inevitable backlash. In the early days we just didn't have enough people."

"So what happened to turn things around?"

"The OMEGA Days. They shook things up, and integrated all the various concerns. When the greenest city work started very few people linked corruption on Wall Street with climate change, for instance, or homelessness with marine fish stocks depletion. It took the OMEGA Days to pull all the pieces together.

"I'd been studying this stuff for years," Alice continued. "The OMEGA Days were the first time I'd seen a movement integrate environmental action with action on housing, poverty, economic justice and democracy. They lifted everything to a higher level, and generated a new consensus. And then syntropy theory came along, claiming to provide a new scientific foundation. Whether it's true or not, time will tell. But those who accept it seem to have a new moral clarity, similar to that which inspired the abolition of the slave trade in the late 1700s. Take the United Nations Global Declaration of Interdependence, and the Global Constitution Clauses. I doubt they would have been possible if the people behind them had not had the confidence that comes from really clear vision."[606]

"The Global Constitution Clauses—what are they?"

"They are a set of clauses that nations are being encouraged to add to their constitutions. On the environmental side they include granting legal rights to nature, recognizing the human right to live in a healthy environment,[607] and recognizing ecocide as a crime against peace, subject to the International Criminal Court.[608] On the social side, they enshrine the fundamental human right to affordable housing; and on the economic side they clarify that governments have the right to operate public banks and to create their own money, a duty to end tax-evasion and to work with other nations to close down the offshore tax havens, and that serving the public benefit is a legitimate purpose for a private corporation.

Have you read that series of history books by the husband and wife team, Will and Ariel Durant? I learned from them that in the years following the French Revolution, whenever Napoleon invaded a new country he brought with him a new code of laws, known as the Code Napoleon. As well as the end of feudalism, religious freedom and freedom for the Jews, the new Code brought equality before the law, a competent judiciary, the jury system, uniform taxation, the end of internal tolls, uniform weights and measures, responsibility for looking after the poor, support for science, literature and the arts, and a host of other social and economic reforms. Napoleon tore up the old traditions that protected the aristocracies and feudalism, and in their place he imposed laws and institutions that laid the foundation for the modern world. I say 'imposed,' because he used his armies to do so, but he was strongly supported by liberals and progressives in the lands he occupied, who rejoiced at the end of the old ways. His Code still allowed the oppression of women, slavery and hard labor, but it was a big advance.[609]

"My point is that the constitution of a nation expresses its cultural and political DNA. So when the principles of a cooperative economy in harmony with nature are enshrined into a constitution, it speaks volumes. It signals the end of the era of global capitalist expansion that Napoleon did so much to launch."

"That's a fascinating perspective."

"If we don't learn to cooperate together we're going to die together, and take nature down with us."

"How many nations have adopted the new clauses in their constitutions?"

"Very few so far—it's a big deal to open up a constitution. But many countries have signed Statements of Intent to do so."

"Were you implying that this has something to do with syntropy theory?"

"Some people say so. I'm an anthropologist, not a physicist. It's clear that species self-organize for their own survival. But do atoms and molecules? Is consciousness the self-organizing matrix? Did consciousness exist before the first material form? I'm not qualified to answer those questions. But I do notice that whenever people claim that syntropy suggests a fundamental directionality in the Universe, I want to argue with them. I know it's not a rational response, but there it is. If syntropy theory does turn out to be valid, I'm going to have to widen my understanding to include trans-conscious communication and the potential entanglement of everything in the Universe. Maybe I resist because I was taught so firmly that things like telepathy and spiritual healing have no scientific foundation, and I've done enough anthropology to know how deluded people can become under the influence of warm fuzzy ideas that discourage rational thought. There's a great book that a friend of mine wrote called *Kults, Konspiracies and Kool-Aid*. It shows the vulnerability of the rational mind to seductive ideas. Have you read it?"

"No. But weren't you saying that syntropy was one of the factors behind the success of the OMEGA Days?"

"No. Satyanendra Mukherjee's famous paper on syntropy wasn't published until after the OMEGA Days. But don't get me wrong: I'm a huge believer in the need for social and environmental progress. But as an anthropologist I recognize that when people become insecure and afraid of growing disorder, the absence of rational explanations creates a mental vacuum that sucks in unifying theories, including the weird and dangerous. It happened with early socialism and communism, when industrialization caused such suffering and poverty. It happened under the Nazis, when Germany experienced a period of economic chaos. And it happened in the US with the Tea Party, when many white working and middle class people felt that their familiar world was crumbling around them. It's a natural biological response when under threat to want to latch onto something that promises to restore order. And that presents me with a paradox, for it's precisely what syntropy theory predicts should happen."

Alice twitched her fingers and scratched the back of her neck. Paradox and conflict clearly did not sit easily with her.

"Do you think that one day things will swing back towards individualism, away from the emphasis on community and neighborhood?"

"Quite possibly," she replied. "Nothing lasts forever. The new becomes old and the old becomes boring. The danger I see is that the new synthesis will create so much harmony and greenery that it will blind people to the larger forces that threaten, while giving the young and restless nothing to sharpen their claws on. Forgive me for going on. I have always been interested in big picture thinking, and the bumpy road to progress."

"So you believe in progress?" I asked.

"Well, that's another difficult question," she replied with a strained expression. "As an anthropologist, my answer has to be no. All change is relative; no one culture is better than another. That's perhaps postmodernism's greatest contribution. On the other hand, my heart would love to believe there's an inherent self-organizing tendency within existence that's as fundamental as gravity or space-time. If that turns out to be true, it might justify the idea that history is guided by an invisible inanimate presence and pre-determined to move in a purposeful direction. After all, humans are purposeful, and humans make history. But maybe we should get back to more practical things?"

I was floored. There was a dimension of thought here that I was not used to. But my time with Alice was limited if I was going to catch the train to Joey's Farm.

"Did you get arrested during the OMEGA Days? And did you take part in the street protests?"

"You bet we did!" she replied with a laugh, lightening up. "Me, my partner Aisha and our two children. We took it in turns. Aisha had come to Canada as a refugee from Ethiopia, and she knew all about the need for political action. The kids loved it. They said it was the most fun they'd ever had. I was a city councilor, so getting arrested was a high stakes game, especially with the municipal elections coming up, and I couldn't vote on anything if I was in jail. But there was so much at stake, and the OMEGA solutions were so well developed. I only spent two weeks in prison with our daughter Amina, because they didn't want to make a hero out of me. But that was long enough to build a relationship with women who had put their lives on the line to create a better world for their children and for the Earth."

"How did the OMEGA Days affect the greenest city movement?"

"They gave everything a tremendous boost. Remember how I told you there were ten thousand people involved at the beginning? Well, there were a hundred thousand people on the streets that summer, all eager for change, wanting to make a difference. There was a surge of interest in street greening, home retrofits and growing food. Someone created the Change the World Community Portal as a doorway to the movement, which has been adopted by cities all over the world. If your interest is transportation, it shows how you can get involved with transit, cycling or ridesharing, etcetera, and so on for every area of change. [609B]

"It wasn't just us, remember. There was a similar movement happening all around the world. They used the same OMEGA framework, but they crafted their own local solutions. The Green Cities Alliance captured the best ideas and made them easy to adopt. That's where we learned about the value of visible indicators, for instance, which inspired us to install the Green City Progress sign outside the new Art Gallery. And that's how the Humane Society came to us with their proposal that Vancouver should join the Humane Cities Alliance and work to improve animal care in the city, everything from improving the way pets are cared for to ending the sale of endangered fish, and boycotting farms that didn't have the organic star for kindness to farm animals. We had already ended the sale of eggs from caged chickens and meat from factory farms, banned the sale of shark fin soup and created off-leash areas where dogs can run free in most city parks in the early mornings and evenings. It's amazing how much support we received over that one. People here really love their animals.[610]

"The social capital movement took off that summer too, enabling people to invest in local business ventures. And we set up the Green City Assembly, which led to the Village Assemblies in each neighborhood. So many people found new ways to contribute. We had a record turnout at the next municipal election, and a strong vote of confidence for the improved Greenest City Action Plan, which had been extended to include affordable housing, community planning, poverty and homelessness—social issues that had been neglected the first time round. There was a very strong coherence between the OMEGA Days and the greenest city movement."

"And were you re-elected?"

"Yes, I'm happy to say I was," she replied with a smile. "It was the first time sixteen-year-olds were allowed to vote, and the first time that we used a transferable vote with run-offs for the Mayor and council, which made voting more democratic. It was also the first time people could vote by smartphone, which increased voter turnout. How was your breakfast, by the way?"

"It was delicious; the Dandychic coffee is amazing."

"Mine too. I was fasting yesterday in support of Climate Compassion, so I expect it sharpened my taste buds. Would you excuse me for a moment while I take this call?"

While Alice turned away, I looked around at the busy market with its open displays of fruits, flowers, vegetables, meats, cheeses and fish. It seemed so normal, almost timeless. It was great to know that the food was all organic, and as a consequence every farm it came from was being restored ecologically.

"I'm sorry about that. That was Aisha, about our son Amara. A few years ago when he was at high school he took a lot of heat because Aisha and I are gay. The Ethiopian community is really intolerant when it comes to homosexuality, even here in Vancouver, which has to be one of the most gay-friendly cities in the world. So when Amara left school he went to Toronto, hoping to avoid the stigma, but he got in with a bad crowd."

"Is everything okay?"

"It's more than okay! He's been in jail for three years for being in a gang that beat up gay men; he was leaning over backwards to distance himself from having been raised by two gay women. His parole board has agreed that he can be transferred back to Vancouver to enter a restorative justice program. So we'll be able to see him regularly, which will be wonderful. Hopefully he'll come to terms with his past."

"That must have been really hard for you."

"Yes. It's been very difficult. Aisha comes from Ethiopia. We met while I was volunteering with Amnesty International. We had adopted her as a prisoner of conscience and I went to Ethiopia to try to get her freed.

"She was a teacher, and she had been jailed for protesting the forced eviction of villagers from their ancestral villages to meet the demands of Chinese and Indian investors who were buying up huge tracts of farmland.[611] We knew she was also gay, which increased the need to get her out of Ethiopia since she faced extreme intimidation, and she had been badly beaten while she was in jail."[612]

Alice paused and stared out across the market.

"I had no intimation that I was going to fall in love, or how profound it would be. I'd been in love before, but this was different. When we were apart I felt all wrong, sometimes even physically nauseous, and when we were together, and especially when we woke up together in the mornings, it was just so wonderful. It felt as if heaven was growing inside our souls. I hope you don't mind my speaking so personally. When people fear homosexuality they don't understand that. They just think about the sexual goings on, maybe because of their own sexual repression. But it has nothing to do with that. Sex is just one of the many ways that people express their love. As soon as we got Aisha out of Ethiopia and back to Vancouver, she and I got married. We adopted Amara when he was six from an orphanage in Addis Ababa. He has always been a difficult child; we knew that when we took him on. He experienced some terrible things in the first few years of his life."

"You said you have a daughter too?"

"Yes, Amina. She is Aisha's by birth; we're both her mother. We're very close. She's training to be a teacher."

"I'm very happy for you. Getting back to what we were talking about, if that's okay, do you recall what the OMEGA solutions were for democracy? I've been wondering."

"The O for Open Democracy?"

"Yes. Did that have five solutions too?"

"Yes, everything came in fives. Let me see… the first solution was to make provincial and federal voting proportional, so that every voter would be fairly represented. Under the old 'first past the post' system you could have five people seeking to become a Member of Parliament and they could each win about 20% of the vote, but the one who got ten votes more would become the MP even though almost 80% of the people had not voted for him or her.[613] We also wanted voting to be electronic."

"You weren't worried about privacy, or vote-hacking?"

"No. There's software that can identify fraud and people have been making on-line financial transactions for years without any worries. There was an attempt to corrupt the electoral process before the OMEGA Days, but Canada has never suffered the corruption of democracy America has known. We have always trusted our electoral process to be non-partisan."

"Did you consider making voting compulsory, the way they do in Australia and various other countries?"[614]

"Yes, but we went with automatic voter registration at the age of sixteen[615] and a $100 voter's tax credit instead. If you're not earning enough to pay tax you accumulate the credit against future taxation. If you choose not to vote, you get nothing. It's very popular.[616]

"Our second solution was to lower the voting age to 16," she continued, "so now most young people experience their first election while they're still at school where they can discuss the issues, meet the candidates and get in the habit of voting. When it was 18, most young people treated voting as something adults do, and therefore not cool. Since the changes, the turnout in our municipal elections has increased to 65%, and in the last provincial election it reached 85%."[617]

"That must be very satisfying."

"Yes, it is. Our third solution was for every party to receive a public subsidy based on the number of votes it got at the last election, with a limit on campaign fundraising, and no-one being allowed to donate more than $1,000 over the course of a year—all of which has been successfully implemented. This was one of the things that made Canada so different from the States. They've had a huge struggle to amend their constitution to end the power of money in politics, and to end the legal fallacy that a corporation was a person, entitled to spend as much money as it wanted to influence or buy an election. I'm so glad the OMEGA movement in America is finally having success on that front."[618]

I was excited by the talk of change in America, but I didn't want to interrupt Alice's flow.

"The fourth solution was to create more transparency and accountability among the politicians, lobbyists and civil servants, with disclosure rules designed to end all the closed doors meetings and secret deals."

"That's important. And the fifth?"

"The fifth was more amorphous, but it goes to the core of what democracy is about. It was a call for political parties and governments to support the process of cooperative self-organization, which is the foundation of our efforts to build a better world. Self-organization is a fundamental principle of nature, but the major parties had cut themselves off from its life-giving current of change, and in so doing they had lost their vitality and become inherently boring, especially to young people."

"Why did that happen?" It was a big problem back in my time.

"Over the years, I think the impulse by the party leaders to control suppressed the impulse to self-organize and innovate, except at election time, and it deprived the parties of the life-giving process of rejuvenation. In response, many people no longer felt a pulse of vitality in the parties—so why should they participate? The party insiders never felt that, for they were the ones exercising the control, but to outsiders it was a fact of life: except at election time, political parties were boring. At the community level there have been great examples of people coming together to tackle a challenge such as homelessness, or to create a new urban farm, initiatives that are packed with vitality. We wanted to restore the impulse for self-organization to the political realm. To use a medical analogy, the political parties, which are the heart of democracy, needed an emergency infusion."

"Were you successful?"

"I think so. A team of sociologists created a questionnaire that measures each party's commitment to self-organization as an essential principle of democracy and they published the results, which proved embarrassing. In response, the politicians really opened up. These days, most parties encourage their members to participate in local community initiatives and they have opened lines of communication to thousands of non-profits and charities, actively seeking their ideas. They have also opened up the policy-making process, instead of keeping it close to their chests until it was time for an election. Syntropic democracy, they call it. The New Democrats, the Greens, the Vancouver Island Alliance and the Northern Alliance are more like open-door parties these days, where new people and new ideas are welcome. They now form the provincial government, working in coalition, and the Green Party's Charlene Jack has become BC's Premier—it's the first time a First Nations person has ever held the post."

"That sounds like a really big change."

"It is! In the last election the rallies and meet-ups were so engaging, with their visuals, music and instant audience polling. And the young people organized Youth Vote Pledge Circles, where you earn points for different ways of participating.[619]

"Do you do electronic ballots on key issues, the way they do in the States?"

"Yes. We liked what we saw in America but we didn't like all the money that was involved, so we brought in rules that require signatures from five percent of the electorate, and neither side can spend more than $100,000. We wanted more democracy, but not the kind money can buy. Can you guess what our first local ballot was about?"

I was stumped.

"It was about Vancouver's municipal golf courses. There were large areas of Vancouver that lacked any decent green space, yet we had these three large publicly owned golf courses on the south side of the city. They were being managed sustainably, but the land was being used exclusively by the golfers, who most people saw as a privileged minority. So the nature-lovers and the foodies got together with a ballot proposal to convert one of the golf courses into a mixture of community gardens, a food forest and a native plant restoration area, with picnic areas and public trails. With only seven percent of the voters being golfers they never stood a chance, and the ballot passed easily. The neighborhood forest school is using the space for its outdoor classroom, instead of having to bus the kids to Stanley Park. It may even help reduce the number of people suffering from allergies, since the more people are exposed to forests and green space the less they suffer.[620] At the last election there was a successful petition to reclaim the second golf course too, so now we've got two new parks and community gardens on the south side of the city. I doubt we would have had the courage to take that on as a council. It needed a citizens' ballot."

"How is the local food movement progressing?"

"I believe there are some fifty farmers markets in the city, and twenty thousand community garden plots. We still call them farmers markets, but not many farmers come any more. They don't have the time, and their food distribution is handled so much more effectively here in the Agora and by The Hub, the farmer's co-operative. So it's mostly neighbors selling their surplus produce."[621]

Alice's phone buzzed again. She took the call and then returned her attention to me. "I hate these things. Three of my students have died from phone-related brain tumors, and twelve have been through brain surgery or tumor treatment regimes. They say it takes ten years of using the phone for an hour a day, but I know lots of young people who are on the phone far more than that. I'm always going on at Amina to reduce her phone-time."[622]

I agreed that it was troubling; I had heard the same from Aliya.

"When you were working to make Vancouver the greenest city, what were your biggest frustrations?"

"That's a good question. As a councilor, one of my frustrations was our own bureaucracy. It sometimes took forever to get a simple thing done, with all the forms and paperwork. There was an episode a few years ago when one of the Village Assemblies wanted to create a nature trail through their neighborhood. They ran into so many obstacles that they erected a sign across from City Hall with a green leafy image labeled PARADISE and a sticker pasted across it saying CANCELLED DUE TO BUREAUCRATIC CONSTIPATION. That was tough.

"It's easy to criticize, but whenever we cut corners and something goes wrong there's always someone trying to sue, demanding that we do a better job. That's why the lawyers insist that we stick to the book and do everything by the letter of the law. The green lawyers are no different. When we revised the City Charter we enshrined the rights of nature, reflecting the worldwide movement. So now there's a wildlife group demanding that Dark Sky Week happen every week, because the streetlights deprive the snakes and bats of the darkness they need, and claiming that they have a legal right to darkness.[623]

"Whatever next?" she continued. "Will they be demanding that we bring back the wolf? After all, they were here before we were. Maybe it's good that I've gone back to anthropology. It's bad enough with the deer—and the raccoons, thanks to all the food people are growing and people switching off the streetlights at night. Was there anything else you wanted to ask?"

"I can't imagine what it's like dealing with all the conflicting interests. It's amazing what you have achieved."

"It's the people of Vancouver who did the work, who made it possible. Without them we couldn't have done a thing."

"I do have another question, if you've got the time. I gather Vancouver's close to being a 100% renewable energy city. I know that most cars, trucks, buses and trains are electric, and I've been told about the freight-tubes. But what about long-distance trucking? And what about the ferries, and all the ships and airplanes? What do they use if they're no longer using fossil fuels?"[624]

Alice sat up and paid fresh attention.

"That has been one of our greatest technical challenges. When the city established its goal to go 100% renewable energy back in 2015 they didn't have a clue how we were going to get there—and trade goes far beyond our borders. It made us very aware how dependent we were on fossil fuels when we tried to plan for a future without them.[625]

"Take a typical 18-wheeler with a 500 horsepower engine, burning 40,000 litres of diesel a year. Without the diesel you'd need 500 horses to pull it, or 5,000 humans. Can you picture that? No wonder so many pre-industrial societies practiced slavery. You might think that cynical, but so many of those grand country houses in Britain and estates in the Carolinas were based directly or indirectly on slavery. As we developed coal-fired steam we replaced the slaves and servants with fossil fuels, and now we need to replace the fossil fuels because of all the dangers they've created."[626]

"So how are you solving the trucking problem?"

"Half the problem is cultural. We would never have come to expect oranges from Florida and computers from Korea if it wasn't for cheap fossil fuels. So part of the solution has been reducing our wants and meeting more of our needs locally, using the sharing economy, local businesses and advanced 3-D printing."

I knew a little about 3-D printing, but back in my time it was more in the blogs than the marketplace.

"What kind of things can 3-D printing produce?"

"What can't it produce is more the question. There's an entire micro-village of 3-D printed homes in Burnaby, complete with printed toilets, sinks and doors. There's a company in Richmond printing electric cars, using liquid carbon fiber for the bodywork, and another printing computers under license—all they have to ship in is the motherboard. There's furniture, office supplies, medical supplies, prosthetic limbs, machine tools, play equipment, boats, shoes—basically anything that can be made from plastic. Over in Victoria, Viking Air is printing aircraft wings, propellers and small fuselages."[627]

"That's very different from life in Sudan," I said, covering for my ignorance.

"But it's just as important to reduce the need in the first place. Making things last longer. Repairing them locally. Lending and borrowing instead of buying. When I needed a ladder to fix a broken window last weekend I borrowed it from my neighborhood tool-sharing co-operative. I found the glass in the Building ReStore and was able to fix the window myself. Aisha was most impressed.[628] The sharing economy is reducing the cost of living, thanks to the shared vehicles, shared tools, shared living, shared meals, shared child-minding, shared urban farming and so on. Consumerism used to be such a Ponzi scheme, persuading people that they had to keep on buying if they wanted to feel alright.

"During the Transformative Twenties it would have been far worse if people hadn't been supporting each other. Even a small thing like toy sharing became important. We got Amina her first bike as a trade for coaching a neighbor's daughter. We had people offering to cook for us most nights, so Aisha and I could come home after a day's work and know that a neighbor had supper ready for us."

"And the trucking…?"

"Right. The truckers are freight-sharing to eliminate empty return trips,[629] but the real problem is technological, replacing diesel with a renewable fuel. On the ferries, the shorter routes are all electric now, and for the longer routes BC Ferries has just commissioned a new fleet that will be half the weight, powered by a hybrid electric-hydrogen drive.[630]

"So anyway, a number of big trucking companies got together and launched the Sustainable Fuel Challenge.[631] There was a proposal for airborne blimps, but the weather was too unpredictable over the mountains and the accident risk was too high. There was one to use wind energy to make green methanol, but the process needed carbon dioxide from fracked shale gas, which made it a non-starter. They claimed they could do aerial capture of the CO_2, but the technology is nowhere near ready.[632]

"There was another proposal to make green methanol from forest wastes, but you had to use energy to truck the wastes from the forest to the methanol plant and if the forest floor isn't renewed biologically you end up with barren land. Finland is building a methane economy using biogas from its forest wastes, but their distances are smaller.[633]

"There was also a proposal to make biogas from farm wastes, urban organic wastes and gasified wood wastes, but the volume was nowhere near sufficient. It's the same with biodiesel from food and animal wastes: there's not enough supply. In the end, only two technologies made sense: hydrogen and electrification.

"British Columbia has plenty of wind energy that we could use to make hydrogen, but without a continent-wide refueling network no-one wanted to build a hydrogen-fuelled truck. And there was never the will to invest, because truck electrification was advancing so fast, and a straight electric drive is three times more efficient than a fuel-cell electric drive. So the question came down to which electrification approach would it be? Would it be plug-in fast charging, battery switching, overhead charging through wires, charging from the road, or stationary inductive charging from above? It's amazing how you learn these terms—it makes me sound like an engineer."[634]

"And…?"

"The winner seems to be overhead stationary fast-inductive charging. The new electric trucks have recharging capacity on the roof that they use at weigh-stations, truck-stops and other locations. There has also been a lot more use of the railways, which had spare capacity following the phasing out of coal. The railways are all being electrified with solar panels along the tracks, and Canada's power will be 100% renewable as soon as the last gas-fired power plant closes down."[635]

"How far can an electric truck travel on a single charge?"

"About four hundred kilometres, or a four-hour drive on a good highway. Speaking of transport, have you seen the Solar Hyperloop, that uses air pressure tubes to carry people at a thousand kilometres an hour?"[636]

"No. Are they for real?" I was surprised that Laura hadn't mentioned them when she told me about the freight-tubes.

"For sure they're for real. They're similar to the freight-tubes, but designed for people. I travelled on one last year when I went from San Francisco to Los Angeles for a conference. The whole trip took under an hour. It uses compressed air and very low air pressure to propel the capsules and it gets all its power from solar on the exterior of the tube, combined with battery storage. There are several tube technologies around; the one I rode in was very comfortable."

"What about flying?" I was intrigued by the solar tubes, but I needed to press on.

"That's more difficult. The *Global Aviation Solutions Treaty* requires airlines to phase out kerosene by 2045, and they've formed a consortium to ensure that there'll be enough biofuel and hydrogen, but they're not sure where the biofuel's going to come from, since the algae plans have not panned out."[637]

"No?"

"You need a constant stream of CO_2 to fertilize the algae, and without fossil fuels there's no obvious source. They hoped to extract it straight from the atmosphere, but as I said, the technology's still developing."

"How can you be sure that biofuel won't compete with growing food?"

"That's the challenge: the certification has to be very rigorous. There's also investment going into hydrogen planes, and maybe there'll be a breakthrough in ultra-lightweight batteries that will make electric planes viable."[638]

"Changing the subject, how did you find the funds to build all the new bike lanes, and buy all the new electric buses?"

"My, you're a sucker for punishment. My brain's getting exhausted. The buses were self-financed by the savings on energy over the twelve-year life of a bus, and the bike-lanes were financed by carbon taxes, regular taxes and road tolls. But what do you think of this building, the Agora?"[639]

"It's amazing," I replied, feeling embarrassed for pushing too far. "Am I right in thinking it's designed to look like a variety of broccoli?"

"Yes, indeed—Romanesco! It was designed by a student of Zaha Hadid, the Iraqi architect who's so famous for her curvaceous buildings. It won the full five petals in the Living Building Challenge.[640] It's a passive house design, so it hardly needs any heat, and it has self-cleaning carbon nanotube solar windows. It's timber-frame construction throughout. Did you see the wonderful slide?"[641]

"Yes, I did!"

"They say it represents the best of classic green funkytecture. I use it every Thursday evening to celebrate the end of the working week."

I *loved* this future!

"Can I ask about something else I've been wondering about?"

"Fire away—then I really must be getting on."

"It's about sea-level rise. The land here seems to be the same level as the ocean at False Creek, and Wei-Ping from the Green Economy Institute was telling me that the sea will rise by more than a metre before the century's out, and anywhere from five to twenty metres in the centuries to follow."[642]

"Yes. We had a very heated discussion about that. Some people said these lands should be returned to wetlands, and we should locate the Agora on higher land in Burnaby. But the planners had fallen in love with the idea of the Urban Farm Zone, the Director of Finance at City Hall said we needed the revenues, and the foodies had fallen in love with the site. It's not just here that sea-level rise threatens: it's all the way up the Fraser River, as far as Hope. Huge areas of Metro Vancouver are at sea level. If you combine a big high tide with a sudden spring melt and a sea-level rise of only ten centimetres you've got flooding everywhere."[643]

"So what is the city planning to do about it?"

"The region hired a team of engineers to come up with estimates for different scenarios. We're talking billions of dollars to raise the sea walls along a hundred and fifty kilometres of riverbank, to raise all the bridges and re-engineer all the sewer drains and wastewater treatment plants. We know we're in trouble. The last time the world was 3°C warmer, the sea level was twenty-five metres higher, and it looks as if it's going to pass 2°C in spite of all the efforts people are making. It would take several thousand years for the ocean to rise by twenty-five metres, but it's our responsibility to make sure it doesn't happen."

"So...?"

"We're collecting an annual levy on every property to pay for a future two-metre seawall when it's needed, spreading the cost among everyone, not just those whose homes and businesses are at risk. There's no way we can keep out a twenty-five metre rise, so we urgently need the planet as a whole to phase out the final fossil fuels and make the transition to 100% renewable energy. But fundamentally, we're making a statement of commitment by building it here that we're not going to allow the climate crisis to defeat us. Our daughter Amina is extremely worried. She wants to have children too, and they'll be alive well into the next century unless they do something stupid like join the McJunkies. Can you believe that the city just gave heritage status to a McDonald's store? Whatever next?" Alice snorted in disgust.[644]

"It's the new reality," she continued. "Nature's going to throw everything she's got at us, and we're going to have to do things differently if we want to survive. My hope is that it'll make us draw closer, and we'll learn from our mistakes. It's the third stage of civilization, learning to cooperate. Stage One is the Age of Discovery, when the first humans enter an unoccupied land and enjoy nature's incredible abundance. Stage Two is the Age of Empire, when people fight among themselves to control the diminishing resources. Stage Three is the Age of Cooperation when people finally learn to share, just as we do in a healthy family. I think of it the way a mother does. In Stage One, Mother Earth gives us everything. In Stage Two, we throw temper tantrums to try to get our own way, and in Stage Three we finally learn to cooperate and discover how great life can be when you work and play together as a family."

I paused, taking in her words. There seemed to be a lot of wisdom behind such a simple way of describing the evolution of civilization.

"It's so exciting to be alive at such a time," I said.

"You've got that right! But we're only half way there. The shadow of the dying Age of Empire is still enormous."

"On that topic, can I ask you about Derek Brooks?"

"You mean the Derek Brooks who was shot during the OMEGA Days? That was a terrible affair. They still don't know who killed him. What do you want to know?"

"Do you have any theories about who might have done it?"

"No more than anyone. I've always assumed it was an extreme fundamental-ist; someone from the Texas Taliban, perhaps. Or maybe someone connected with the banks—they really didn't like the fact that Derek understood how banking worked, and wanted to end the banks' monopoly over the creation of money and debt. But there *is* one thing that has always left me wondering. When I was in jail I had a conversation with a fellow prisoner that left me with a strange feeling. She was boasting that her man had done something really, really big, but then he'd gone and left her without so much as a note. He was such a jerk, she kept saying; that's why they called him Jerky, she said."

"Isn't that the name of the man whose foot washed up nine months after Derek's death?"

"I don't know, but for some reason the conversation has never left me. Joan: that was her name. Joan Sadulski. But look, I've got to be going. I've promised Amina we'll do the Grouse Grind this afternoon. It's only three kilometres to the top, but it's almost three thousand steps, so it's quite the workout. Wish me luck!"[645]

And with that, I thanked Alice profusely and headed off to the railway station.

23

Taking Down the Tax Havens

THE PATH TO the Pacific Central railway station passed through the village center, a five-sided pedestrian plaza facing the new canal. It felt timeless, as if from an old European town. Even the architecture was traditional, with window balconies and an arch under which pedestrians came and went. Only the roof showed the difference, with its undulating solar array.

The station had a huge bicycle facility with parking for four thousand bikes. The first thing I needed was a way to call Laura. I had a card from Dezzy but no phone, but I found a place where I could use it. She wasn't in, which was disappointing, so I left a message telling Laura about my meeting with Alice and her jail conversation with the woman called Joan; could she pass it on to her father?

I used my card to buy a ticket and found a seat on the train opposite a man with long blond hair and a heavily lined face, with sores on his cheeks and lips. His hands were also inflamed.

He nodded to me, but I felt repulsed and looked away. The train's window had a button that said 'Heritage Tour' so I pressed it, and words appeared in the glass with information about the view I was seeing on the North Shore, across the Burrard Inlet. In June 1886, it said, a fire had destroyed almost the entire early city of Vancouver, but many residents had been saved when people from the Squamish First Nation paddled their canoes across the water to rescue them. The window also told me about Grouse Mountain, with its cluster of ten wind turbines high above North Vancouver, and the Second Narrows Bridge, which was so badly built in 1925 that a large ocean-going barge got wedged under it. When the tide rose, the barge took out the whole central span, tumbling it into the water.

"What a mess! Why couldn't use simple measuring tape?" the man opposite said stiffly, through pursed lips.

"Yes," I replied. "We humans so often screw up in the most obvious of ways."

"And most devious. Most people no idea. No idea at all."

I introduced myself, and encouraged him to say more.

"Brad. Brad Walfisch," he said, speaking slowly, as if the sores on his face made it difficult to form the words. "Used to be big-shot dealer, finance world. Condo Vancouver, hideout Switzerland, penthouse Shanghai. Saw all the stupid things, year after year. I know. Did a lot of them myself."

I was taken aback and didn't know how to respond, so I nodded.

"Worked in money business. High risk, high adrenalin. Hobnobbing with big boys. Part of invisible cabal. Our own cult *omerta*: be silent or die. Thirteen years old when started buying, selling. My dad encouraged—said could go far. Never made much of his life. Civil servant, worked for government. Transferred expectations to me. When fifteen, developed an app, easy to speculate. Had schoolmates doing it. First million before left school. By time thirty, two hundred million in offshore accounts, sitting pretty, no interference taxman. Wife and two girls Vancouver. Girlfriend Shanghai. Speed boat, holidays Kamchatka, fly-fishing. Top of game. And then this...." He showed me his hands and pointed to his face.

"What is it?" I asked, feeling shocked, and still wondering why he had such a strange curtailed way of speaking.

"Gonorrhea," he said, looking me straight in the eye. He held my gaze to see if I'd flinch. I was feeling nervous, but I held my ground.

"HR73a multiple-drug-resistant gonorrhea," he said quietly, after a long pause. "No remaining antibiotics. No wonder drugs from Amazon. No ancient Tibetan herbal remedies. Only because eat well, train every day, painkillers. Keeps at bay. Sorry about strange speech. Easier talk."[646]

"Gee, that's terrible," I replied awkwardly.

"Don't mind talking about it. Least can do. Maybe warn others. I'm healthy vegan," he continued. "No trouble performing, but hurts like hell. No woman this side Mars wants come near me. Divorced from wife. Daughter with Chinese girlfriend born blind. Died meningitis, age two." Brad's expression crumbled before my eyes.

"Broke my heart. Far more than divorce. Can't explain why. Really loved that girl. Mother jumped. Suicide. Twenty-fifth floor, high-rise Shanghai." Brad paused, his lips tightening.

"Left note. Couldn't go on." He paused, and I could tell he was hurting.

"Wasn't her fault baby born blind, got sick. Who I, tell her not other boyfriends? How we to know gonorrhea gone MDR? No symptoms. No way knowing."

"Your ex-wife, is she okay?"

"Yea, but could easily have passed it on. She suspected me seeing someone, so not sleeping together. Only fair we divorced. Who'd want live someone like me? Still see kids, but breaks my heart every time, see me like this."

"Is it becoming widespread, the gonorrhea?"

"Becoming so. People use condoms, but spreading by oral sex." He lowered his voice so others on the train wouldn't hear. "They say avoid if cling-film," he whispered, "but what kind of thing is that? I'm pariah. Red-carded."

"Red-carded?"

"In ID. Shows up biotest screening, border. They say developing new antibiotic, mimics chemistry soil fungus. Might help, but seen too many promises. HR73a chat room helpful. Least know not alone."[647]

Wishing desperately to change the subject, I said the first thing that popped into my head. "So what keeps you busy these days? Are you still in the money business?"

"No. Whole other story," he replied, sitting back in his seat. "You got time?"

"Sure, if you're willing."

"Used to be hotshot, übershot. Million dollar deals. When got clap, went into shock. Whole year doing nothing. Anger, blame, self-pity. Ex-wife social worker. Yes, incongruous. Girlfriend Shanghai. Her friends worked twelve hours day, knick-knacks, sale to west. Miserable conditions. So yes, I knew real world, real people suffering."

The window told me we were passing the Belcarra Regional Park across the water to the north, where the Tsleil-waututh First Nation had had a traditional summer village since 2,750 BC—4,600 years of habitation. Ever since the time of the pyramids, I thought to myself. Then the white settlers arrived, bringing smallpox. By the 1860s the camp had been abandoned. Would we ever truly comprehend how much misery our world has seen—and how much was still going on?

"Did lot of thinking that year," Brad went on. "Went right down. Heart of darkness. Had my Bucky moment."

"Your Bucky moment?"

"Buckminster Fuller. You heard of him?"

"Yes, of course. The man who invented the geodesic dome."

"And so much more. Inventor, architect, futurist. Had phase when deeply depressed. Daughter died. Jobless. Drinking. Family near bankruptcy. Decided to drown himself. Went to edge of lake. But then big realization, deeply selfish. Life not his to throw away. Didn't belong to him. Belonged to Universe. So he thought, if willing to end, why not live rest of life experiment, see what one could do change world, benefit all humanity? Live for others, not solely self.[648]

"Did lot of reflecting, way I'd been living," he continued. "Not pleasant. Friend of ex told me about Tax Justice Network, office in Vancouver.[649] Got in touch. Went to conferences, did lot of reading. Became obsessed. I was insider. Trust accounts, anonymous ownership, flee clauses, offshore marketing—knew all the tricks. Transfer pricing, shell banks, re-invoicing. Checked conscience at door. Everyone did. Conscience for lefties and sissies: not playing real game. That's way we saw it."

By now, you—the reader—should have a good idea of how difficult Brad found it to speak. So from now on, to make things easier and to ensure that I communicate his ideas the way he would have wanted, I will report his words the way I think he would have wanted them, had he not been so afflicted.

"This was all some time ago," he said, "when governments were struggling with their debts and wanting to close the tax havens, repatriate the money, and stop the freewheeling flight capital from playing hell with the currency markets. It was all so deeply embedded—and not just the small players. The big players too: Switzerland, Britain, Holland, and several states in the US. Some of their tax regimes were so liberal you could set up a head office and hide your operations in five minutes. There were huge global outfits, gaming the system, making profits in the billions and paying no taxes. It was all by the book, they said—they had professional tax accountants. Exxon, Bank of America, General Electric, Chevron, Microsoft, Google, Walmart, Starbucks, Coca Cola, Amazon—they were all doing it. They used the magic of transfer pricing, moving their incomes around. They would overstuff their offshore accounts with equity capital and then demand tax holidays to bring the money home. Half of all world trade was passing through the tax havens, a third of all global wealth. Can you believe it? It wasn't until I was out of the game that I saw whole picture—and how by being part of it I had killed my own daughter, and my girlfriend." Brad bit his lip.[650]

"How can you say that?"

"It was the way I lived—it was like an infection. Gonorrhea of the mind. It made us believe we could do no wrong. The world's problems? They were all the fault of governments. That's what we told ourselves. They were the ones wasting the public's money with their handouts to the poor, not supporting the real entrepreneurs, the people like me. If I hadn't been so full of shit, and so blind to the truth, none of this would have happened. I've got to live with that till I die."

I was silent, looking at Brad as the tears welled up in his eyes.

"It's so fucking stupid," he said. "So fucking stupid."

I had no response, but I reached across the table and placed my hand on his. He quickly brushed it off, indicating that he didn't want to talk about it anymore.

"How much money are we talking about in the tax havens?" I asked, hoping to get back onto safer ground.

"As much as $50 trillion, before the crash. But you've got to understand—the tax havens were not really about the money. The real game was a more pure form of evil. That's strong language, but that's the way I see it. Pure individualism. If I showed any weakness, I'd be on the way out. The way we saw it, governments were out to spy on us, take our money and put it in their own corrupt coffers. They were building an all-intrusive world government with twenty-four-hour electronic surveillance. We, the entrepreneurs, the risk-takers—it was *we* who made the economy work, and created the jobs. Not them. That's the way we saw it. They should have *celebrated* us, not try to punish us."

Brad was on a roll....

"Freedom. That was our flag. Freedom to keep the wealth we had earned, and to enjoy the rewards. Tuscan villas. Palaces on Lake Como. Private islands. Private art collections stored in tax-free warehouses.[651] *No heaven above us, no hell below:* John Lennon. Do you know that song from *Les Mis*, where the greedy innkeeper celebrates his spoils? *I raise my eyes to see the heavens, and only the moon looks down.* No God! Only us! We could choose to live any way we wanted. There was no judge at pearly gates. Girls, drugs, booze, private jets. We used to hobnob with the intellectuals at fancy conferences, and make seem we were one with them. When you were at school, did they teach you about the political philosopher, Edmund Burke? *When bad men combine,* he wrote, *the good must associate; else they will fall, one by one, an unpitied sacrifice in a contemptible struggle.*[652] That was us, all over. We had found out how to combine and make the world work for us. The rest of world, they were too busy arguing, watching sports and reality TV. We, we took it all. We were good at what we did. Paying taxes, giving to charity—that was for losers."[653]

"But now you've changed sides?"

"Yes. I've got to earn my redemption. So I do what I can. I used to joke about how we should establish a separate sovereign state for people like me, with no taxes and no intrusive regulations. Free and proud, we'd be, serving humankind by our entrepreneurial zeal. You've got to understand—the most important thing about a tax haven is that its government must be weak and malleable. If you bring legislation to remove the requirement that auditors must be accountable, or to reduce banking reserve requirements, you want the legislature to rubber-stamp it into law, and ask no questions. That's what the tax havens really were—assault weapons of deregulation, a power base for the plutocrats, the royalists, the banksters—call us what you want. We would establish a bulkhead in Jersey or Delaware and then blackmail the other states, demanding similar legislation. Do as we want, we'd say, or the whole financial sector would up and leave. That was our threat. Some of our accountants—they had the Jerseys and the Cayman Islands wrapped entirely around their little fingers.[654]

"Our philosophy was pure Ayn Rand. Neo-liberalism," Brad continued. "Free markets, free capital movement, freedom from regulation. The entrepreneur as the hero. We wanted sovereign control over global financial trading, just as previous empires had controlled Africa, India and the Middle East. Why did we need governments? Why did we even need countries? We had our own private paradises, and our money was safe in shell banks and trust funds with no government control. We were a free-reigning oligarchy, safe in our secrecy jurisdictions. If other people chose not to work and become rich, that was their fault. That was our view. It was they who caused the big government debts, not us, because governments had to support them with welfare. Look in a mirror? That's something we never did."

"You're being very honest," I said. I couldn't believe I'd struck such a rich vein of pained reality.

"I've lived so many lies. From now on, I'm trying to be nakedly honest. It's the only way to remain hopeful, and it's part of my commitment to repay. I was deep inside it. I saw what was happening, and I did nothing. It was a deliberate set of actions, a shared belief that if you were clever enough to become wealthy you were entitled to keep it all. Bugger the poor and homeless—let them go fuck themselves. Wasn't that what they did anyway—overbreeding, causing all the world's problems? If they failed, that was their fault. That was our view. Stop whining. Do something about it! The philosopher Ayn Rand, she was our inspiration. She said it all. *I swear, by my life and my love of it,* she wrote, *that I will never live for the sake of another man, nor ask another man to live for mine.*[655]

"I used be like that," he continued. "It troubled me sometimes, sharing my yacht with big-shot drug dealers and dodgy accountants. But I always found a way to hide it. Any hint of weakness, and I'd have lost their respect, and then I'd have lost out on the next big play."

"So what changed?"

"This." He showed me his hands and lips. "I became one of the fallen."

He peered into my eyes and leant forward.

"I spent three days in a hotel in Barbados, all alone. Anger, denial, contempt, self-pity—I indulged them all. I was contemplating suicide, just like Bucky did. I really hit rock bottom. Then I went right back to my childhood in Belize. I had a favorite nanny, a Creole woman. Alejandra, her name was. She had taught me how to pray, when I was six. I went right back, and in my mind, I asked her for help.

"Then something incredible happened. A cleaning lady came in, Filipino. I never knew her name. She just sat by me, and listened. I'd never had anyone listen to me like that before. Not the way she did. She sat next to me, her arm around me. I was so full of anger and fear. I was all mixed up. She put her arm around me and brushed my forehead. And then I collapsed. Cried like a baby, sobbing in her lap. That's when I discovered what human connection means, what it really means. To be so helpless, and yet still to be loved by another."

Brad paused, clearly still moved by the memory.

"So, long story short, I came out of my funk. I knew I had to change. So I re-invented myself. I found new friends, and I adopted new attitudes. I discovered what it means to be part of the human race, where people are not such assholes, where they actually help each other, instead of exploiting each other. I bought a house in great cohousing project here in Abbotsford. It's as green as they come. Car-free, great neighbors. It's great for the kids on the weekend. We eat together three nights week, and in the evenings we play games together, or work on various projects. It's far better than eating alone, watching TV alone. I've got a neighbor who's teaching me how to use tempera paints, how to open up creatively. I'm discovering a whole new side of myself that I never knew existed before."

"That sounds great!" I said with enthusiasm. "And you were also working to close down the world's tax havens?"

"Yes. I offered my services to the Tax Justice Network. At their peak, as I said, the tax havens were holding as much as \$50 trillion[656] and the annual tax loss to the world's governments was around \$400 billion. Global government debt was \$62 trillion.[657] Now, with the havens gone, and with the taxes on repatriated assets, it has fallen to \$44 trillion. It's all part of the Dhaka Agreement. A share of the taxes on returned flight capital must be used to pay down government debt. Are you really interested in this stuff? Most people find it boring."[658]

"Yes, I am—please go on."

"The Dhaka Agreement came up with a formula for international corporate taxation. There was to be no more transfer pricing, claiming expenses in countries with highest tax regimes and income in countries with lowest. Here in Canada there's a fifty percent tax on untaxed offshore profit from earlier years unless it's repatriated by certain date, in which case it falls to thirty-five percent. And corporations have to pay out their repatriated profits as dividends, not as bonuses, to get the money circulating.[659]

"So, I did an economics degree at Oxford that covered flight capital and repatriation. By time I was through, however, it was not just the tax havens that I was into—it was everything, including global debt. Working with others, we gradually figured out the whole picture. How had the world gotten into such mess? And how to make it sustainable, both financially and ecologically? That was the big question. It had to be both."

"And you got it figured out?"

"Just the debt part. We published a paper that laid it all out: *Earth Harmony: The Road to Financial and Ecological Balance.* We analyzed the potential to pay down government debts, using income from closing the tax havens, and various new taxes. We supported public banking, with government reclaiming the power to create new money for the public good. Globally, the global financial transactions tax, or the Robin Hood tax, as some people call it, is bringing in $2 trillion a year.[660] The sale of permits under the global carbon cap, half in tax breaks, half in revenue: $1.1 trillion.[661] Methane permit fees: $10 billion.[662] Junk food taxes, like Europe's: $250 billion a year, declining as sales fell.[663] Add the taxes on marijuana, and the savings on prisons and law enforcement, along with many tax changes and closing the loopholes."[664]

Brad picked up his screen and showed me a chart. In total, globally, the new revenues came to almost four trillion dollars a year.[665]

"It's a good result!" he said. "This had been defeating the economists for years. Some even thought that democracy and long-term balanced budgets were inherently incompatible, since governments would always feel the public pressure to spend more and borrow more, deferring payment to the next generation.

"Here's how it works for the US," he continued. "Total debt: $20 trillion. New annual income: $1 trillion a year. Closing the tax havens: $50 billion. Fair taxes on the super-rich: $50 billion.[666] A five percent federal sales tax, same as in Canada: $180 billion.[667] Reducing the military budget, and closing most US military bases overseas: saves $200 billion.[668] The financial transactions tax: $300 billion. Legalizing and taxing drugs: $20 billion. Carbon permit fees: $110 billion.[669] Methane permit fees: $2 billion. The junk food tax: $25 billion. Half of the new income to eliminate the deficit, half to pay down the debt. Clear in forty years."

"But that's stupendous!" I replied, knowing what an enormous weight America's debt was on its hopes for the future. Had these people in the future really solved the government debt problem?

The train stopped in Port Moody, where a smart new station had been built surrounded by new buildings, with people walking and cycling around. It was a far cry from back in my time, when it was a dull, characterless affair surrounded by equally characterless commercial buildings and parking lots.

"But that's not all," Brad continued. "Until they started to take the climate crisis seriously, most nations were spending huge sums of money to import and subsidize fossil fuels. With renewable energy, most of the money stays at home, where it supports local business and generates taxes. And on top of that, the air pollution disappears, so health care costs fall. What's not to love?"[670]

"It sounds like a very sane way of managing the world."

Brad looked at me, nodding his head as I stated the obvious.

"So how many of the ideas that you proposed in your paper have been adopted?" I asked.

"Quite a few, especially in Europe. The global economy is much healthier now, and debts are shrinking. It's all part of the New Track commitment."

"What's that?"

"It's the global agreement to harmonize debt repayments and increase banking transparency—part of the Dhaka Agreement. The banks now have to be totally upfront about who they lend to, following an automatic exchange protocol. Interbank lending used to be a terrible a black box, creating every reason for caution, since it created billions in toxic debts. That was the main cause of the meltdowns. Without good information, the governments and the banks couldn't analyze the risks they were exposed to. How could they? There were so many shell banks and offshore accounts, and they were all covering their tracks."

"What about the people who benefited from the old ways, the super-rich with their luxury yachts, and the corporations that had been avoiding taxes for years? I can't believe they just rolled over."

"They didn't. There were some huge struggles, most of which were never reported. I knew of one scheme to assassinate five of the major bankers who were supporting the plan to close the havens. They got three: one in a car crash, one in the trunk of a burning car and one by a sudden heart attack. It's still not clear who was behind it. I received thousands of threats when my name came out supporting the changes, so I had to lay low. British Columbia has some wonderful fly-fishing. I needed the downtime, anyway."

"Speaking of assassinations, do you know about Derek Brooks, who was assassinated during the OMEGA Days?"

"Yea. What of it?"

"I'm staying with his sister Dezzy. She was telling me that they still haven't caught whoever was responsible. Do you have any idea who might have been behind it?"

"I didn't follow it very closely. It was probably someone with something to hide, who wanted to stop the campaign for transparency regulations. It wouldn't surprise me if they had a list, and Derek was on it."

Another conversation, another theory.

"How much do you know about the OMEGA Days?" I asked. "I'm told that the E in OMEGA stood for a New Economy. Do you know what kind of solutions were involved?"

"I wasn't paying much attention—I was too busy enjoying the good life in Hong Kong. But I can look it up if you want." Brad lifted his screen and said, "'OMEGA Days, New Economy.' Here they are. Okay if I read them?"

"For sure—that would be great." Another piece in the puzzle was about to slot into place.

"OMEGA E for a New Economy. Five Solutions. *Solution One: Shift to real world definitions of wealth and growth such as the Genuine Progress Indicator (GPI). Redefine wealth and growth to include their social, environmental and cultural dimensions, using new indicators and systems of analysis.*

"That's pretty good," he added. "My divorce cost me a bundle. By the old way of measuring things I should have gone Wahoo!—I've just increased Canada's wealth. Maybe I should have crashed my car as well to increase the GDP further. It was same with the environment. If you left a forest standing it did nothing for GDP even though it was home for the salmon, the bears, and all the other critters. But cut it down, and Bingo! You've just increased the GDP. Take a bonus and retire to the Cayman Islands. It was like counting your gross income but leaving out all your expenses. It was totally dumb."[671]

"Do you know how the OMEGA solution is being realized in practice?"

"It says here that seventy-four countries have adopted the GPI, and a further thirty-six are considering it. That's pretty good. Let me click on Canada. It says that Canada's GPI fell steadily from 1970 to 2023, but it has been rising steadily ever since.[672]

"It says company reporting is changing, too. Every company in Canada is now using the Global Reporting Initiative framework, and they must file an annual report showing workplace, social and environmental progress as well as financial progress.[673] And not just businesses: every public body, hospital, school and prison must do so too."[674]

"That's really encouraging. What's the OMEGA Solution Two?"

"*Solution Two: Shift to real world sustainable business. Reframe corporate law to require the pursuit of social and environmental benefit as well as profit.* That's pretty good."

"What does it mean in practice?"

"It says that Canadian companies have been given five years to become a certified B Corporation, embracing a higher standard of purpose, accountability and transparency. That's what B stands for. Be the change. After 2035, those that don't will have to pay a higher rate of corporate taxation. It also says that the government is giving most of its supply contracts to B Corporations—that's got to be a big incentive. Overall in Canada, thirty-nine percent of businesses have become B Corps. In US, forty states have enacted B Corp legislation, and fifteen percent of businesses have made the change. Changing the legal definition of a corporation is a very big deal—it ends the legal requirement that company directors must behave in effect like assholes and psychopaths. Ready for Solution Three?"[675]

"For sure."

"*Solution Three: Shift to real world sustainable banking and money-creation, including publicly-owned state and federal banks. Reframe corporate banking law to require the pursuit of social and environmental benefit, as well as profit.*

"This is the same as Solution Two—the B Corporation thing applied to the banks. When we wanted a loan to replace the solar on our Common House with solar shingles, we had to show the environmental gains."

"That's something to smile at," I said. "How it has affected the banks in general?"

Brad went on to tell me how, following the Dhaka Agreement, all of the world's major banks were now required to be fully transparent about their loans and transactions, to reduce the risk of another crash.[676] He told me that thirty-six of the states in the US had passed legislation permitting the formation of a state-owned public bank, that eighteen states had done so, and that the new Greenpeace Gaia Bank would soon be as big as Barclays, it had attracted so many investors."[677]

"Are the banks still being allowed to create new money the way they used to, simply by pressing a key on a computer?" I asked.

He replied that they were, but the reserve requirements were now much larger, especially for the more exposed banks. Before Dhaka, he said, some countries had been allowing their banks to create money with almost no reserves at all, and it was no wonder the money supply had gone crazy, creating so much debt and inflation. It had been the same in Spain when they imported all that stolen gold from the Americas, he said, causing massive inflation and eventually the end of the Spanish Empire.

When I asked him what he thought of the idea that governments should take control of the money supply and create it in the public interest, he replied that as long as there are limits and controls, and third-party oversight, it was a good idea that should have happened years ago. The banks should also be allowed to create money, he said, as they had been for years, as long as they kept good reserves. When banks had persistently lobbied to lower their reserve requirements, he said, alarm bells should have gone off all over world.[678]

That's when they really abused the public trust. What they really wanted was the freedom to create as much money as they wanted, he said, since they made such big profits by gambling with it. That's why we had the meltdowns, followed by the multi-trillion dollar bank bailouts.[679] The current approach was much better, he thought, with public banking, the closure of the tax havens, tightened capital reserve requirements, the new transparency requirements, and the re-imposition of the legal separation between Main Street and investment banking that had protected America from major banking crises for fifty years from the

1930s to the 1980s, when the banks persuaded Ronald Reagan to cut them some slack and get rid of the regulations. Even after the 2008 meltdown, Brad said, the financial lobbyists were still pushing their pet senators and congressmen—the ones they had bought with campaign donations—to block the re-instatement of the separation that was so essential for a safe, healthy banking system. The corruption had been very deep before the Dhaka Agreement, he said. It had corrupted the whole world, and undermined our ability to act together to stop it.[680]

"Many banks are now becoming B Corporations—so that's a good step forward," he continued.

"How many of them have done it so far?" I asked.

"In Canada, it says here that four of five major banks have made the change. In the States, it's five hundred out of ten thousand, so there's still a long way to go. Ready for Solution Four? I'm enjoying this."

"For sure." I felt excited, which was weird for something as complex as financial reform.

"Solution Four: Shift to real world taxation and regulation. End tax avoidance; capture external costs through appropriate taxation; re-impose the necessary regulations including capital controls with strict standards of transparency and public scrutiny; end the corporate capture of the regulators.[681]

"That's a big one," he said. "There was an almighty struggle before the world's nations finally signed the Dhaka Agreement, shutting down the tax havens and ending the shadow banking sector. But we're finally winning. It's the same with the global carbon cap and the financial transactions tax. It was painfully slow, but we finally got there."

"But how can you stop collaboration between governments, big banks and the corporations," I asked, "which has so often resulted in corruption of the regulators?"

"That's a good question," he replied. "I remember being on a skiing holiday in Aspen, Colorado, where I heard some Wall Street brokers boasting that they controlled the US Federal Reserve, the Securities and Exchange Commission and even the US Department Treasury. One even claimed that he controlled the World Bank. These days, if you're a politician or a civil servant, you have to publish a full record of all your meetings on the public purse, and there are firewalls to prevent you from moving from industry to the government positions that regulate that industry, and back again, the way they used to. Solution Five?"[682]

I indicated yes.

"Solution Five: Shift to real world global financial architecture. Reframe the charters and goals of the Free Trade Agreements and the Bretton Woods Institutions—the World Bank, the IMF and the Regional Development Banks—so that they support global social, environmental and cultural progress as well as economic progress.

"There's some stuff here about the Terra, too—the new global reserve currency based on basket of currencies and commodities. There's also stuff about the

Global Green Fund, that has replaced the IMF, and the Global Justice Settlements Bank that has replaced the Bank of International Settlements, based in Basel, Switzerland.[683]

"There's also stuff about the fair trade thresholds that are being allowed for some developing nations, the importance of economic self-reliance and self-determination, and the value of trade barriers and capital controls in countries that need time to build their economies. How else can they compete? It's a big change from the days when people were told to accept free trade as a blessing regardless of whether your economy was strong or weak, or how many jobs were lost or forests destroyed.[684]

"But you know what?" Brad continued, with a change of emphasis and a seeming lightening of his heart. "I've moved on. I find my inner life much more interesting these days. I've come to see a lot of things differently. Does that kind of thing interest you? You never know who you'll bump into on the train."

"Yes, it does," I replied. "I'm glad I bumped into you."

"Since I moved to Abbotsford I've been doing a type of Jungian therapy called psychosynthesis.[685] We're all on a journey to wholeness, it says, guided by our higher selves, but our egos don't know the game plan, so they're constantly getting in the way, creating sub-personalities that try to run our lives. Some people have a sub-personality that's a judge, that's always telling you you're not good enough, for instance, encouraging you to judge others to make yourself feel better. Some have a sub-personality that's a child, who acts irresponsibly and wants to be looked after. Some have a spoiled prince or a princess who demands to be the center of attention and is very narcissistic. I had one who was a selfish king, always trying to accumulate treasure.[686]

"I find it very helpful, this way of seeing things," he continued. "It has given me the tools to discover my sub-personalities and understand their scripts, so that they no longer create so many dramas. Once they're under control, they slowly fade away, making it easier to remember who you are and to get on with your journey. I'm no longer a set of competing sub-personalities. I'm becoming a more whole person; at least, I'm trying to be. I have to keep reminding myself that I'm more than my mind, more than my feelings, and more than my poor crippled body."

Brad became calm as he spoke, and his hands, in spite of the lesions, became quite expressive.

"When my king sub-personality was ruling he never allowed me the space to question, or to ask if this was what I really wanted. He just assumed the right to rule, for his own selfish benefit. You see it a lot in some sports champions and musical stars. He also gave me permission to take as many lovers as I wanted, justifying it by telling me that I worked hard, so I deserved them.

"There's a new school of thinking that interests me called *global psychosynthesis*. It sees each civilization as a being that's striving to achieve wholeness, just as we do. It's fascinating stuff. It suggests that every civilization has an ego as well as a soul, and that its ego has similar sub-personalities. Globally, they express themselves through all eight billion of us: we each carry fragments, who live out their struggles through us.[687]

"One of our civilization's sub-personalities is that same selfish king who wants to accumulate money and power. When becoming rich and powerful becomes a civilizational goal in its own right, things really begin to go wrong. Non-corrupted governments represent the attempt by the higher self to keep the selfish king under control.

"When governments allowed money to go wherever it wanted, to hang out offshore, corrupting individuals and corporations, they allowed this sub-personality to sabotage the whole journey of civilization towards cooperation and harmony. It became an evil entity, dedicated to the corruption of civilization's higher purpose. Pyramid schemes and Ponzi schemes: they all appeal to the selfish king. Understanding this, my work to close down the offshore empire and to end the dominance of capital became far more than a personal interest—it became a driving goal. I want to heal far more than my own soul; I want to contribute to the healing of the global soul."

"Wow. That's a pretty big goal you're taking on," I said with a smile.

"I know that it seems ridiculously grandiose—maybe that's what happens when you go cold turkey on an extreme way of being. But I'm not completely whacko. I've read lot of history since I retired from the money-games. When the British Member of Parliament William Wilberforce was campaigning to end the slave trade, two hundred years ago, one of his concerns was that the trade didn't just cause unimaginable cruelty and suffering to the slaves. It also corrupted the souls of the people involved, making them more violent and brutal."[688]

I was listening intently. Global psychosynthesis was a very insightful concept, and Brad was a fascinating man. And to think that I had met him by chance.

"I remember learning about Wilberforce," I said. "It's very heartening when an individual stands up for humanity and succeeds against all the odds."

The train was crossing the Fraser River on a new bridge, and we'd soon be in Abbotsford, where a bus would take me to the farm.

"And here *you* are," I added, "fighting to stop the corruption. I'm really glad I met you."

"I've been fighter all my life," he said.

I nodded.

"I have one more question, if you'll indulge me," I said. "Can you explain how the tax havens were finally shut down? Did someone invade them and take them over?"

"Ha! That's a nice thought. You've got to understand—those digital deceptions moved at the speed of light. One whiff of the cavalry and poof! A secret flee-clause would be activated and the money would be off in Uruguay or the Seychelles.[689] The Dhaka Agreement finally drew the line in sand. It established a new set of global financial and regulatory standards, and Britain, Switzerland

and America finally signed on, thanks to the persistent detective work of the International Consortium of Investigative Journalists, combined with public pressure.[690]

"Those three nations were the biggest culprits, along with Luxembourg, Hong Kong, Singapore and Japan. The British ultimately signed because their public debt was so overwhelming. Switzerland signed because their people demanded it in a public referendum. They were so embarrassed that their bankers were causing them such shame. It was the same in Japan: public shame. America was the big holdout, since it needed the support of all fifty states, including some pretty corrupt governments. There was a lot of arm-twisting needed to bring Nevada on board."[691]

"What exactly did the nations agree to regarding the tax havens? I haven't been following it very closely."

"Most people don't—that's part of the problem. First, they agreed to a 'no place to hide' transparency agreement. So today, the banks are required to submit automatic reports on their foreign account-holders' earnings, with proof that they are being taxed appropriately.

"The *Global Tax Evasion Complicity Treaty* has made it a jailable offense to knowingly serve as a fund manager, accountant, trustee, lawyer or corporate nominee, or to knowingly help a tax evader. It hit Nevada badly, that one. There's a whole new crowd of people pacing up and down in their jail cells who never saw it coming.[692]

"The treaty also established a unitary approach to corporate taxation, divvying up a corporation's profits, apportioning them to the jurisdictions where it has its activities, ending the transfer pricing game that allowed a corporation to file its income in one place and its expenses in another, playing the high tax rates off against the low. Now that the loophole's been closed, governments are receiving billions in previously evaded taxes.

"Also, the British government has finally agreed to abolish the Corporation of the City of London, the square mile of the financial district that lies at the heart of London, which was the head of the offshore octopus. Barclays Bank used to have more than three hundred offshore tax haven subsidiaries, if you can believe it. Very few people, even in Britain, knew that the City of London was not governed by Parliament, and that it routinely ignored anything the British government told it to do. It was a quaint, corrupt, medieval plutocracy that allowed the bankers to do whatever they wanted, with no government oversight."[693]

"I had no idea. I always assumed that London was London, just like any other city."

"No. The City of London was a financial fiefdom within the larger city, with its own rules, some of which were over a thousand years old. Even old King Charles couldn't enter without their permission: that's how independent they were.

"The end of the City was brought about by a combination things—global pressure to close the tax havens, a rare consensus among Britain's political parties, and campaigning pressure from groups like the Tax Justice Network, Platform London, and Londoners for Sustainable Democracy. I was there when Parliament debated the motion to strip the City of its privileges. There were fifty thousand of us, and we formed a circle three deep all around the square mile of the City, along the Thames and the streets north of the river. What a day that was! I was sitting down sharing some food when the first tweets came through that Parliament had passed the motion. What a moment! There was whooping and hollering, and I joined a massive conga line, dancing all around the now-abolished City limits.

"Before I finish, let me tell you one last story. The State of Delaware, in America, was a soft, easily corrupted jurisdiction, where any corporation could ride roughshod over its other stakeholders. There was an anonymous two-storey yellow building at 1209 North Orange Street, Wilmington, where *three hundred thousand* corporations had their legally registered head offices. Tax evasion? Tut-tut. Of course not. It was all perfectly legal in Delaware.[694]

"Well, around the time that things were getting hot, the billionaire Warren Buffett set up Buffett's Bounty, a five billion dollar bounty fund, as an incentive for people working in the tax shelters to break their contractual secrecy and spill the beans.

"It was a little old lady, Minny Minkelbaum, who finally stripped the fig leaf off Delaware's dainties. Wonderful character, that Minny. She had worked for forty years and she was coming up for retirement, but underneath her pinned-up hair and her widow's clothing she was a Raging Granny and a member of Hell's Hackers. She was so pissed off at all the greed and selfishness when so many were struggling to feed their kids and put a roof over their heads. So she scooped all the data, loaded it onto a two-terabyte USB key and dropped it off at the International Consortium of Investigative Journalists, before heading for the hills in her beat-up old Chevy.[695]

"When the media published the data all hell broke loose. In Congress, it was revealed that more than three hundred congressmen and women were receiving pay-offs from Orange Street corporations. Delawhores, the media called them. One senator was on seventy-three different boards. The public outrage was incredible. The Wilmington Police had to call the National Guard.

"And Minny? She picked up $50 million from Buffet's Bounty and got herself a new life, a new lover and a lifetime supply of daiquiris on some island off Cuba. I can picture her now with her feet up, cigar in hand, receiving a foot pedicure from some lush dude.[696]

"In the mid-term elections the public was so angry that all the Delawhores were thrown out, and the new Congress passed the Financial Transparency Act, putting an end to the tricks. I'm pleased to have been able to play some small part in it. I will never make full amends, but it's good beginning."

"I'm very grateful to you, Brad. You've taught me a lot. That Minny—she sounds like quite the lady."

"She is! She was very prim and proper. You'd never guess that she played a critical role in bringing down the global offshore empire."

"Abbotsford: it seems we're arriving," I said. "It's been great talking to you."

"That's what I like about the train," he replied. "I've never had a decent conversation in a traffic jam, except with myself. *'Abbotsford: Green by Nature'*—ain't that true! It's a great place live. We've got a basketball game this afternoon with the Firehall Brawlers. They're busting to win, since we beat them last time. I'm allowed to wear gloves. It gets a bit painful, but I love it. And that farm you're going to? They're doing great things from what I hear. Sayonara, kiddo!"

And with that, he disappeared into the crowd.

24

Worms and Horses

THE ABBOTSFORD TRAVEL Hub had a shimmering solar roof with hanging flower baskets that looked like floral UFOs. There was a 3-D map of the region in the floor that showed the roads, mountains and farmlands, with the Fraser River flowing through them, and a touch-screen where you could explore your travel options, including bike sharing and walking.[697]

And what do you know—there was another *Change The World* poster, this one with a photo of the famous American feminist Gloria Steinem,[698] and these words:

Without leaps of imagination, or dreaming, we lose the excitement
of possibilities. Dreaming, after all, is a form of planning.

I had a while before the bus arrived so I sat down to collect my thoughts, and ponder the power of dreaming. My time with Brad had given me a lot to digest, and it had been hard work following what he said with his strange way of speaking. Did the changes he told me about amount to a shift away from capitalism, as Naomi had suggested? It was hard to tell when I didn't have a global grasp of what was happening.

Or was capitalism inherently flexible, able to bend like a Tai Chi master and adjust its bow-tie after each crisis? But if it could embrace changes as progressive as values-based banking and B Corporations, should it be called capitalism at all? Was it evolving into economic democracy, just as the feudal states evolved into political democracies? And if it was called capitalism because the power of capital was its dominant feature, what should it be called when cooperation and mutual support were dominant? A cooperative economy, as Naomi suggested?

Or are there two tendencies that are constantly in tension with each other: the impulse of individualistic people to band together to maximize their selfish interests, and the impulse of ordinary people to cooperate and live together harmoniously? Is that a fundamental duality, me versus we, individualism versus community? Or do we each have a bullying dominator and a kind cooperator within us, and it's up to each of us which we feed, like the two wolves that live inside each of us in the native Cherokee legend?[699]

I also wondered what was happening in America. That was a big blank in the picture I was building. Canada had its problems, but however much Canadians back in my time like to complain that their democracy was messed up, with too many parties splitting the vote, it had not been corrupted the way America's had. Were the changes I was learning about in Vancouver even possible south of the border?[700]

———✦———

There were twenty of us on the bus to Joey's Farm, and I made friends with two young women, Sarah and Pelly, I'd guess in their late twenties with blue eyes and reddish wavy hair. Twins! Sarah was about to start farming and her sister was a neuroscientist.

Abbotsford's streets had been planted with fruit trees and they had generous bike lanes, with the same variety of bicycles I'd seen in Vancouver. It was Saturday morning, and everyone seemed to be going somewhere. I couldn't pay much attention, however, since Sarah and Pelly were arguing.

"I never said I supported it," Pelly was saying. "I just said it needs to be given a fair hearing. If there's even a *possibility* that genetically modified organic wheat can produce a higher yield, and the studies show there's no risk, surely we owe it to the millions who go to bed hungry to try to make it work?"

"I can't believe you are saying that, Pelly," Sarah insisted. "You, of all people, should know that rogue genetically modified proteins can go anywhere and can't be tightly controlled. Are you willing to risk brain interference or the release of genetically modified pollen for a possible five percent yield increase? People just need to eat less meat if we want to feed the world. It's as simple as that."

"I don't dispute that, Sarah, but meanwhile people are still starving. You go and tell your neighbors to stop eating meat. I'm just saying that if it's shown to be safe, why would we *not* want to do it? If people had taken your approach a hundred and fifty years ago Gregor Mendel would never have been allowed to work on plant hybridization."

"Oh, don't be so stupid. Hybrids result from natural variation, whereas genetic modification is gross interference with the biology of a species."

"Like the heart transplants that have been happening for the past sixty years? Or the brain cell replacement therapy and ultrasound treatments that we're using to fight Alzheimer's? Aren't they also 'gross interference'? Come on, Sarah. You're turning into quite the squinch. Lighten up."[701]

A squinch? A scientific grinch, I later found out.

"I'm sorry. And I apologize that you should see us fighting like this, Patrick," Sarah said. "We're really entangled. When Pelly thinks something it's as if some of her thoughts drift into my head, which gets me rattled. And I'm still freaked out by Brazil's experience with the Terminator seeds."

"You mean the seeds that were genetically modified to ensure that the plants are sterile?" I asked. "I thought they'd been banned years ago."

"The very same. And yes, they were banned—but oh no, Brazil's politicians thought they knew better. They are GMO's equivalent of a nuclear meltdown."

"What happened?"

"A group of corporate farmers persuaded the Brazilian Congress to allow them to use the Terminator seeds, and just as predicted, the pollen began to make other plants sterile. If Greenpeace and Via Campesina hadn't organized a global protest and got the legislation withdrawn, it could have been far worse."[702]

—————✦—————

The landscape outside the city was glorious, with the coast mountains to the north, the dark forest of Sumas Mountain rising over the plain to the east and the snow-covered peak of Mount Baker towering over the south like an enchanted kingdom. Snow on the peak... but seemingly a lot less snow than I was used to seeing in summer.

The road to the farm took us down an avenue of maples and at the entrance we were greeted by a sign that said *'Joey's Farm—Working with Nature'*, with hand-painted images of animals and food. Past the gate the lane led to a shaded courtyard with a pond and a gentle fountain. Clematis and wisteria covered the buildings and petunias and geraniums hung from the eaves.

A short blonde woman wearing jeans and a pale blue blouse greeted us. "My name is Mina," she said. "Welcome to the farm. Joey's my husband. We've been farming here with our daughters for twenty-five years. Let's start with introductions, then we'll do a tour of the farm and the village and put you to work. The feast is at eight, there's community slug-picking at nine and then the dance."

Slug-picking! It reminded me of a holiday on my grandma's farm in Ireland.

The group was mostly young. As well as Sarah and Pelly there was a black woman named Isabella from Mozambique, who introduced herself through a device that translated her words into passable English; a young man called José from the Philippines; a black African teacher named Serge; and an older white man, Ryan.

"Before we start," Mina said, "let me explain our philosophy. We've got eighty hectares here and there's a lot going on." She walked us over to a diagram with seven concentric circles.

"At the center of our work is love—love for nature, with all its plants and animals, love for our fellow humans, and love for God, or the spirit of the Universe, however you like to think of it.

"The circle around the love represents the land—the soil, the worms, the plants, the wildflowers, the trees and hedgerows, the ducks and horses, the food we grow and the seeds we save. Our goal is to produce food that's healthy and delicious: the kind we like to eat ourselves.

"The kind of farming we practice here is quite a mouthful. Are you ready? It's five-star organic, zero-till, agro-ecological, horse-and-worm cultivated, bio-intensive, zero net energy, community-supported polyculture. Otherwise known as 'the new farming.'"

"Is that the same as permaculture?" someone asked.

"We don't follow the permaculture books, and most of our crops are annuals not perennials, but apart from that, yes, it's similar. When you look at the evolution of farming, starting in the Neolithic, our way of farming represents the fifth agricultural revolution. I can explain that later if anyone's interested."

"What did you have to do to earn the five stars in your organic status?" I asked.

"The first star is for being a traditional organic farm with no chemical inputs. The second is for meeting the Kindness to Animals code, which means we are Animal Welfare Approved.[703] The third is for encouraging biodiversity through agro-ecology. The fourth is for soil restoration, and the fifth is for embracing fair trade and fair wages. The system's about five years old now—it was introduced to emphasize that going organic is only the beginning, not the end of the journey."

This was the same that Aliya had told me about. It was impressive that organic farming had come so far.

"The third circle is our connection to the global movement for change, and Via Campesina, which I'll come to. Here in the Fraser Valley, compared to most parts of the world, we have an easy time of it. We have a mild climate, and great soil dropped by numerous glaciations. Humans have been farming here for less than a hundred years, so we haven't caused the soil erosion yet that's common elsewhere in the world.

"It's tempting to sit back and enjoy this paradise," she continued, "but we're struggling to feed the eight and a half billion people who want to enjoy life on our planet. Since humans started farming ten thousand years ago we have stripped eighty percent of the ancient forest cover and lost seventy-five percent of the top-soil we need to grow food. Ever since the Mesopotamians, one of the reasons why civilizations collapsed was the loss of topsoil and the collapse of agriculture. So one of our most important goals is to demonstrate how we can turn this around. If we don't change globally, the hunger, the suffering and the frustration will only get worse. That's why our work here is so much more than our labor on this small patch of land, and why we do so much educational activity." Mina had a passion that was infectious. It wasn't just me; everyone else was giving her full attention.

"The fourth circle represents our educational programs. We have a long-standing relationship with Thunderbird Elementary School in Vancouver, and a dormitory that sleeps thirty. Every month one of their classes comes here for a week, learning where their food comes from and joining in the life of the farm.

"We have a year-long young farmers training program with interns from all across Canada, a mid-week seniors program, a city teens program, a mental health recovery program, and pregnancy retreats for women who want to give their babies the best possible start in life, with fresh air, clean water and healthy food, away from any contaminants that might harm the baby.

"We also have weekend work parties like the one you are on now. We have Monk-for-a-Day Silent Sundays and singles weekends for people who like to get their hands in the dirt. We've had several couples who met here who have become part of our extended family, along with their children. Who knows, maybe some of you will too?"

There was a ripple of laughter, and several people looked around the circle with curiosity.

"We run courses on everything from seed saving and worm propagation to draft-horse farming, zero-till and food preservation. Everyone is put to work, young or old, so you can all expect to get dirt under your fingernails. We even had a group of nudists once, from Seattle. I think they got dirt under more than their fingernails, but we'd prefer it if you left your clothes on, as this is a fairly conservative part of the province." Mina smiled, and there was more laughter.

"Which reminds me," she said, "can you switch off your devices?"

Several people reached into their pockets and three put up their hands.

"Yes?"

"I'm asthmatic," a man said. "I wear a brain sensor that warns me when an attack might be coming. Is it okay if I leave it on?"

"For sure. There's a lot of pollen around. If you need a clean space the cob sanctuary in the village is very calming."

"I'm on a neurofeedback program to unwind my ADHD," a young woman said. "My brain-sensor tells me when I need to take five to work on new connections. Is it okay…?"

"Yes, for sure. And yes?"

"I'm recovering from cocaine addiction and my brain-sensor warns me when my craving threshold is close. I may need to take five for brainwork too. It seems there are a lot of us."

"That's fine. How many of you are on a brainwork program of some kind?"

Fifteen hands went up. I was astonished.

"Is that normal?" I whispered to Pelly.

"Yes," she replied. "But most people are probably on simple learning programs, not the remedial ones that require sensors. They've been proliferating ever since we realized that the brain changes its structure in response to conscious thought, and there's no need to remain trapped in negative patterns. People are using them for all sorts of things, from improving their eyesight to overcoming their anger. The doctors are prescribing brainwork as well as daily exercise."

Someone shushed us with an irritated look. "We can talk about it later—it's fascinating," Pelly whispered.

"I'd love to," I replied.

"The fifth circle," Mina continued, "represents our commitment to be self-sufficient in energy and water. You'll see this with the draft horses, the solar panels on the buildings and the lake, the heat pumps, the composting toilets and the rainwater storage tanks. Any questions so far?"

"Yes," a young girl asked. "Have you been able to eliminate fossil fuels entirely?"

"That's a good question, and yes, we have, except for plastics. We haven't bought a litre of gas for years, not since we sold the tractor and bought our first two Belgian draft-horses from an Amish farm in Ontario. We've had several since. Today we've got Cavallino and Jerry—that's short for Jerusalem, after the artichokes he loves. They're out doing the haying with my husband Joey. You'll see them shortly. There's a forecast for rain so we need to get it in while it's dry.

"The horses do a lot of our work and they're a lot more fun than a tractor. Our heat comes from the solar heat pumps and our electricity from fifty kilowatts of solar—you'll see it on the lake, over by the forest. We're grid-connected, so there's no problem in winter, but we generate enough to meet our needs. The floating solar was really easy to install, and it's cheaper than roof-mounted."[704]

Floating solar... the same technology Dezzy was telling me about.

"The sixth circle is the farm village. It took a while to get it approved, but it was worth every meeting."

"Can any farm build a village?" someone asked.

"No. You need a minimum thirty hectares and there are strict zoning conditions to ensure that the people who live in the village farm the land."[705]

"Can anyone live in the village?" another person asked.

"As long as they farm the land or earn their income from farm-related activities. If we didn't have those requirements the homes would all be snapped up by commuters.

"The final circle represents our friends and neighbors in the local farming community. They help us, and we return the favor through our monthly feasts, our summer dances and our support for community economic development. We're working with the Fraser Valley Farm Trust to build a local economy that can support everyone who wants to live here. Any questions so far?"

"Do you grow the biofuel to power your tractors and other farm vehicles?"

"No. It made more sense to use horses and to go electric for the farm equipment. As well as saving a tonne of money, the horses protect the soil, while tractors are heavy and compact the soil, making it harder for plants to grow. We have a small electric tractor for runaround chores, and an electric quad bike.

"In any case," Mina continued, "the government has banned growing biofuels on good agricultural land, and since we don't have any marginal land, we couldn't if we wanted to. The horses need six hectares of pastureland; they're our living biofuel machines. Via Campesina, the global movement of peasants and small farmers, sees biofuel production as an enormous threat, since it displaces food production and causes soil erosion."

"Can you tell us about the horses?" someone asked. "How do they compare to tractors in terms of work and efficiency?"

"Let's get going first, so that you can see the farm and not just listen to me."

The summer greenery of the farm was lush and calming. We passed endless rows of vegetables, an enormous compost heap and a cluster of natural beehives, which Mina said gave better protection against the varroa mite.[706] There were strips of meadow that were being protected for their wildflowers, hedgerows that provided habitat for wildlife, an area of woodland, the lake with its solar panels, and pasture for the horses. In one of the fields Mina's husband Joey was at work with the horses pulling a red three-part contraption, gathering rows of dried hay and baling them.[707]

"You asked about the horses," Mina said, as we paused to watch. "We made the switch as part of our commitment to stop using fossil fuels. It took a bit of learning, getting used to the horses and how all the tackle works. We spent a week on a farm in Washington State learning the ropes. It's not just the horses. We had to learn about the plows, harrows and cultivators, the carts, the haying equipment and the cover crop rollers, as well as the harnesses, trace-chains and all the rest of the tack. Then we had to build a forge to fix things when they break down.

"The horses are good for everything a tractor used to do. The Belgians know exactly what to do and as long as nothing spooks them they're wonderful to work with. There was a study that showed that if you grew biofuel for all the tractors in America, you'd need twenty-six percent of America's cropland, whereas with horses you'd probably need eleven percent.[708] About a hundred years ago, when farmers started giving up their horses, someone calculated that it didn't make sense to switch to a tractor unless you had more than seventy-five acres under cultivation. That's about thirty hectares. One horse per twenty-three acres, or nine hectares, that was the formula.[709] On that basis we should have five horses, but with the electric tractor, the quad bike and all the volunteer labor we manage with two. Do you see the haying machine?"

Mina pointed to three bright red contraptions that the horses were pulling. "The first of those machines used to be gas powered, driving the equipment that turns the loose hay into bales. We converted it to electric drive."

"What's it like working with the horses?" a young girl asked.

Mina paused, searching for the best words. "If you really want to know," she said, "it's like falling in love. I had never worked with horses before, and they took some getting used to. But when I discovered how friendly they are, how much they want to please, and how timeless it is when you're working a field together—just you and the horse, the soil beneath you and nature around you—it's really quite meditative. When I'm out in the fields with the horses I experience a peace and wellbeing I didn't know existed. It's like feeling that all's well in the world, but so much stronger than I've known before, and throughout my body. There's no longer a barrier between us. We become a single harmonious whole."[710]

We stood in silence.

"I never got that feeling on a tractor," she continued. "A horse is not a thing. It is part of God's creation. It's a living being. And that makes me feel very humble, and grateful."

The girl flushed and nodded vigorously. I think Mina's response was more than she expected.

"So let's get to the farming," Mina said. "As I said, we practice five-star organic, zero-till, agro-ecological, horse-and-worm cultivated, bio-intensive, zero net energy, community-supported polyculture. Got that?

"Let me unpack it for you. It starts with the soil, which is the foundation for everything that grows outside of the oceans. We never call it dirt. That's a slander the settlers adopted when they first came to North America. It is earth: the planet herself is named after it. Before humans started farming, the soil was always clothed with trees, plants and grasses. It was never naked, and if ever it was it was quickly reclothed with new plants. Nature does not like nudity, even if our visitors from Seattle did."

There was a chuckle. I pictured the nudists weeding.

"In the past when people farmed in the conventional manner they stripped the topsoil off the land, which got washed away by the rain and blown away by the wind until there was little of any value left. So then the farmers would abandon the land and move on."[711]

"How do you look after your soil?" someone asked.

We were standing in a field covered with the stems of a dead crop of some kind, the stalks aligned in the same direction as if the field had been visited by aliens making a crop circle. Mina bent down, pushed her hands between the stems and pulled up a handful of dark black soil.

"Six things," she said. "Worms, zero-till, nitrogen, compost, cover crops and crop-rotation. The worms are the guardians of the soil. Can anyone guess how many worms we have here on our farm?"

"Er, about a million?" someone said.

"Actually, about two hundred million—a million per acre. And a visitor once worked out that we have three billion trillion soil bacteria.[712] I'm glad I don't have to count them all. We do a worm test every year, digging a hole and making an estimate. It's the worms that make the soil, along with the bacteria. They convert the subsoil minerals into a digestible form, carry them up to the top and carry the organic matter down. Over a hundred years, if they're left alone, worms will build anywhere from six inches to two feet of new topsoil.[713] That's why ancient ruins are always buried. So we do everything possible to keep our worms happy, including breeding them to build the soil and to sell as a business.

"Next, there's zero-till," Mina continued. "When I first farmed we used to plough the fields, just as the other farmers did. But when we learned about soil erosion, we moved to zero-till. We no longer plough the land or rototill it. Instead, we plant cover crops—ryes, vetches, millets, barley, soybeans, cowpeas and clover—and just before they ripen we flatten them with a machine called a roller-crimper that's pulled by the horses. We rolled this field five days ago, flattening the rye, and already you can see that it's dying. If you look closely,"— and she bent down to show us—"you can see where the roller has crimped the stems to stop them from growing. This lays down a thick layer of organic matter that replaces the nutrients removed by farming. It also keeps the weeds down—in fact it almost eliminates them, except the ones that blow in, like thistles and milkweed.

Our summers used to be a constant battle with weeds, always hoeing and disking. Now we enjoy a leisurely lunch, and life is more relaxed."[714]

"How do you sow, if you're not plowing and clearing the soil?" someone asked.

"The roller-crimper lays the stems down in the same direction and we use a no-till planter to drill the seeds through the rolled-down ryes and vetches. It works just fine, and the horses don't compact the soil. We also rotate the crops, including the pastureland. That's not enough to return the nutrients we remove, so as well as our own compost we get a monthly delivery from Vancouver's compost centre, and we add the compost from our composting toilets. So make sure you eat a lot tonight, and go before you leave!"

There was a burst of laughter.

"Can anyone remember what the final component is in caring for the soil?" Mina asked. There was silence in the group.

"Nitrogen. It's a critical nutrient. In the past it came from constant manuring. A hundred years ago farmers got it from guano, bird droppings imported all the way from islands off the coast of Peru where there were millions of seabirds. The birds were there because there were so many anchovies and the anchovies were there because the cold water from Antarctica made it great for fish. It's all so interconnected.

"With so many people taking it, however, the guano began to run out shortly before World War One so the scientists developed a way to extract nitrogen directly from the air, using natural gas as the source of hydrogen to make ammonia, fixing the nitrogen and enabling it to be used as a fertilizer. There's an exam afterwards, in case you were wondering." Mina smiled and we all laughed.

"The challenge for organic growers is to get the nitrogen without using chemical fertilizers. The way we do it is through animal and human manure, and cover crops that fix nitrogen through their roots such as peas, beans, lentils, clovers and alfalfa. Follow me...."

We walked over to a field of fava beans. "You see these?" Mina continued. "The plants growing between the rows are hairy vetch and crimson clover. They're great nitrogen fixers."[715]

She bent down and dug under one of the beans, pulling the plant up with its roots. "You see these tiny white nodules? That's where the beans fix nitrogen out of the air. Most plants can't do it, so this makes them special. Later, soil bacteria will break the nitrogen down, turning it into nitrates the plants can absorb. If this didn't happen, we would all starve. There would be no agriculture, no advanced civilizations, no literature, no Verdi or Puccini. Does anyone know the name of these humble bacteria?"

"Rhizobia?" someone said.

"Yes. It's from the Greek words *rhiza* for root and *bios* for life. The nodules form a symbiotic relationship in which the beans allow the rhizobia to infect their roots in exchange for nitrogen from the sugars."

"Do you use biochar to enrich the soil?" Sarah asked. "I saw it being made on a farm in Oregon."

"Yes. We collect woody wastes that are too tough to compost and we charcoal them and bury the charcoal. This locks the carbon away underground where it enriches the soil.[716] One of the benefits of our way of farming is that it increases the soil carbon, drawing it down from the atmosphere by up to a tonne per hectare. If every farmer in the world did as we're doing I'm told we could remove thirty percent of the carbon that's still being added to the atmosphere each year by burning fossil fuels. It's impossible to over-estimate the importance of soil conservation. It's not just about this small patch of earth. The future of our whole civilization depends on it."[717]

I'd never really thought about the soil, or how this tiny slice of nature was the foundation for so much.

Mina continued. "We grow a variety of crops close to each other, which creates habitat for birds, insects and soil organisms. With school groups, we do a species count in the meadow every June, and we're up to seventy-three species of grass and plant. That's good news for the horses, for they never need to go to the vet. When they get sick they find their own medicine. We also raise ducks and chickens, and there's aquaculture happening in the lake with tilapia fish. Does anyone have any questions?"

An older man put up his hand. "Why do you raise so few animals, compared to some other farms?"

"We used to raise cows," Mina replied. "We cut back because of the impact that eating meat has on the planet. The world's farms could feed ten billion vegetarians, but not ten billion meat-eaters. Some of our neighbors still farm pastured beef and dairy. They use biogas digesters to turn the manure into biogas that they feed into the grid, and they return the compost to their fields."[718]

"To who do you sell your food?" It was José, the young Filipino man.

"When we started we sold it to our community farm members. They'd come on a Friday to collect their boxes and we'd enjoy catching up on their news. But as we grew, it became ridiculous with all the driving, so now we do a twice-weekly delivery to the Agora in Vancouver and they supply the stores and restaurants, the hospitals and schools. And all the big superstores take local food now, following the boycott."

"What happened in the boycott?" I asked.

"The superstores thought they could ignore us, but they didn't reckon with how many people were foodies. When the Farmer's Co-op declared a consumer boycott until the superstores agreed to accept local organic food, the store's owners were shocked at how many people stayed away. After a month they gave in and signed a long-term sales contract with the Co-op."

"This Farmer's Co-op. How does it work?" It was José again.

Mina put a hand on her hip. "Now, that's a big question that goes to the heart of the food movement. The Co-op is everything. First, it's a co-operative in which most farmers are members. It has regional branches, but we're all one big family.

"Second, it's a bank, providing us with loans and crop insurance. Since we are all members, we share an annual dividend. It also issues Farm Bonds, which are an important source of capital when we need it.

"Third, it's our means of distribution, getting the food to market. It's become so big that the government has imposed regulations to make sure we don't abuse our monopoly. It would be awful if our leaders became corrupted, which is a danger when you're large. It's modeled on the co-ops in Mondragon in northern Spain, where they have managed to stay true to their roots for over seventy years.[719]

"It's also our source for supplies, whether it's seeds or packaging, and it's our center for training and problem solving. The quality of the on-line material is fantastic. Three weeks ago one of our workers found what we thought was a new insect on a spinach plant. The Co-op's staff identified it as a variant of the green peach aphid and gave us a source for the ladybird larvae we needed to attack it."

"I can vouch for that," Sarah chipped in. "We had potato blight at the farm I worked on last year and the Co-op told us exactly what to do, how we had to burn the leaves along with any contaminated soil."

"Two years ago," Mina continued, "when our beehives got infected with a new virus from California, they were able to provide us with an inoculant from a partner organization in Cuba. The Co-op's also important for forward planning. Their database tells us what every farmer is growing and what the market is thinking. Last year we were considering growing lentils. Through the database we were able to estimate the tonnage and income we'd get per row and sign a contract for distribution before we bought the seeds. It also helps us connect with our neighbors. If a bit of equipment breaks, I can post a message in the morning and I'll usually get what I need by the day's end."

Sarah chimed in again. "They were an enormous help when I needed a grant to finish my farming degree. But switching topics, how does your productivity relate to the number of people you have working on the farm, compared to, say, a hundred years ago?"

"That's a really good question," Mina replied. "I have no idea how many people would have worked here a hundred years ago; that would have been during the Great Depression. We've got eighty hectares, with fifty under cultivation. There's seven of us who do the main farm work, and twenty who are employed on various projects including the educational work, the bedding plant sales, the seed-saving, the jams and preserves, the worm farming, the aquaculture and the honey business."

"How is it for you as business? Can you make good living?" It was José again.

"Our goal is to realize a profit of $25,000 a hectare, or $10,000 an acre. With fifty hectares under cultivation that's just over a million a year. The full-time workers earn around $80,000."[720]

———✦———

"I CAN'T WAIT to get onto the land," Sarah said to me, her eyes sparkling as we resumed our walk around the farm. "I've got one course to complete, then I can join a co-op and start working full-time."

"Does your degree cover the things Mina has been telling us about?"

"For sure! It's a three-year course. The first year is distance education with tutorial groups. In the second year we work on a farm while we continue our studies, and in the third year we do a placement on three different farms, depending on how we want to specialize."

"Where will you farm?" I asked. Back in my time, farmland was so expensive it was completely out of reach for young people unless their parents were millionaires.

"Oh, that's not a problem. The Farmers' Hub lists plenty of opportunities. I'm going to start by renting a place in a farm village."

"How many farm villages have been built?"

"Around fifty, with many more underway," Sarah replied. "It makes all the difference for someone like me. They help the farmers too, since many are getting old and looking for ways to keep the farm going. They're bringing life back to the rural areas, with new businesses, and the children going to local schools. There's a whole rural revival happening because of them."

Our next stop was a large solar greenhouse, filled with tomatoes, peppers, cucumbers, basil and eggplants. Mina explained that the north wall was lined with four-litre milk containers painted black and filled with water to absorb the sun's heat. In summer, the surplus heat was pumped into an insulated underground storage pit filled with sand and gravel. When the temperature dropped in the fall, the stored energy extended the growing season, and they used night curtains on light-sensors to keep the heat in at night.

"It gives us a good start on the year," Mina said. "It lets us get our bedding plants started early and have them in the ground by April, and we can grow salad greens throughout the winter."[721]

"How are you affected by global warming?" It was the young Mozambican woman, Isabella, speaking in translated Portuguese through her device. "In Mozambique, where I live, we have terrible time with flooding," her voice app said slowly after she spoke into it. "Then this year the rains they fail altogether, so now we have terrible drought. My mama, she phones me, and she say, they don't know what they are going to do."[722]

"That's awful," Mina replied. "Last year, we had very unusual summer storms here with hailstones the size of golf balls. They made a terrible mess of our fruit crops. If you look up there," she said, pointing to where the greenhouse roof met the north wall, "you'll see that we've fitted a wire mesh screen we can roll down to protect the glass when a storm threatens. One of the girls is working on a plan to automate it, so that it'll roll down whenever there's a sudden drop in air pressure. There was a hurricane last year that did a lot of damage on Vancouver Island, but it didn't affect us here. They say we may get a thunderstorm tonight; we never had them this early in the year when I was a child.

"Which reminds me—have you all signed the Climate Compassion petition? It's very important if we're going to make an end to troubles like this. And watch out for mosquitoes! We have to be very careful about West Nile virus. It likes the warmer summers. It's only serious in one percent of the people affected, but if you get it it's miserable, because it attacks the brain, causing swelling, paralysis and memory loss. So watch out around dusk. You'll need to be fully covered up if you're outside. Lyme disease is also a problem if you're hiking in the forest, since the ticks like the warmer weather—so no short pants!"[723]

"I've seen the West Nile virus at work," Pelly whispered to me. "It crosses the blood-brain barrier and causes no end of havoc. My colleagues at work have created a replica virus in the hope that they can learn enough to persuade a real virus to carry a nanobot, which could tell us which neurons are being attacked. Once we know that we could work on encouraging the brain to develop new neurons to fight back. The nanobot is amazing. It uses a motor that's only a nanometre wide. That's sixty thousand times thinner than a human hair."

"That's amazing!" I whispered. "Can we talk later? Maybe at tea this afternoon?"

"Will that be with scones, cream and strawberry jam? If so, my mouth is watering."

"You don't prefer pickled rhizobia with bacterial jam?"

"Delicious. I can't wait!"

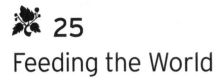 25

Feeding the World

AFTERNOON TEA WAS still a long way off. We hadn't had lunch yet, but first we had work to do.

The work stations had been set up by the greenhouse where our job was to sow the winter vegetables—sprouting broccoli, Brussels sprouts, winter cabbage, chard, cauliflower and collards, spacing the seeds evenly and covering them with soil. They would be planted out in August and harvested next spring. As we worked, Mina continued to answer questions.

"Do you know how many people you are feeding with produce from the farm?" someone asked.

"When we did the boxes," she replied, "we filled five hundred boxes a week, each enough for a family, so about a thousand people. That's on fifty hectares, plus six hectares of pasture for the horses."

A thousand people getting most of their fruits and vegetables off fifty-six hectares—a quick mental calculation told me it came to eighteen people per hectare. Wondering how it would be if the whole world farmed this way, I asked a woman with a device on her belt if she could help. She stepped aside, spoke some questions and tapped in some numbers.

"If all the world's farms achieved this level of productivity," she said, "ten billion people would need 550 million hectares for their fruits and veggies. The world's arable farmland is 1.4 billion hectares, so in theory farmers could grow the fruits and vegetables to feed everyone on thirty-six percent of the farmland."[724]

"Is that really possible with organic farming?" a young man asked. "I read that organic yields were lower, which is why the farmers needed to use chemical pesticides and fertilizers."

"It's not true," Mina replied. "Long-term studies show that organic yields are just as good as yields from conventional farming, and in drought years they are actually better, since organic soil stores more moisture. They found much the same for agro-ecological farming in the developing world."[725]

"But what about the other food people need, like meat and cheese?" a woman chipped in.

"Those numbers are just for the arable land," Mina replied. "There's another 3.5 billion hectares of pastureland and farmland that's used to grow fruits and berries. There's no shortage of land."

"It all depends on what you eat," Sarah said. "In one of my courses we learned that on a conventional North American diet, with hamburgers and hot dogs, you can only feed *three people* per hectare. If you give up most animal products and adopt a vegan or a traditional Chinese diet, you can feed up to twenty people off the same hectare. John Jeavons, who is famous for his bio-intensive methods, has shown that if you farm bio-intensively, you can feed twenty-seven people per hectare. So using his methods you could feed thirty billion people off the world's arable lands."[726]

"But that's not healthy," someone said. "I need my meat and cheese."

"That's fine if you want a higher risk of cancer, heart disease and dementia," Sarah replied. "In the largest study ever done on food and health, called *The China Study*, the research team found that the people who lived the longest with the least disease were those who ate the most fruits and vegetables with the greatest variety and the least processing, while those who ate the most meat got the most diseases and had the shortest lives."[727]

"Is that really true?" someone asked. "The thought of getting dementia scares me stiff. My mother had it. It was horrible the way her life collapsed. We equipped her home with all sorts of sensors to keep her living there. She had an automatic pet feeder for her cat; a lawn moisture sensor linked to her irrigation system; self-closing drapes; and motion sensors that tracked the progress of her dementia. We even installed a webcam and a permanent Skype connection. She used to be so witty and adventurous, but she turned into just a shell of who she used to be."[728]

"My mother was the same," Mina said. "It was horrible to watch. That's one of the reasons why we sold all but two of our cows and switched to fruits and vegetables.[729] It's better for the climate too, because of all the methane the cows burp, and as I expect you know methane is a far more powerful greenhouse gas than carbon dioxide."

"What about grains?" someone asked. "Shouldn't you be growing them too if we're going to feed ourselves locally?"

"We looked into it," Mina replied, "but the prairie farmers can grow grains like wheat and barley so much more efficiently. They can use GPS for accurate harvesting and satellite technology to target their fields for organic amendment where it's needed in a way that would never work here, since we're on a much smaller scale. And now that the railways are being electrified, transportation is less of a climate concern. It's also important to grow soft foods locally, so that they can ripen and be exposed to fungi, producing the phytonutrients we need for our health. Most imported fruits are picked before they are ripe, so they're not attacked by fungi and they don't carry the salvestrols we need to defend us against cancer."

"Salvestrols?" someone asked.

"Yes. They're a phytonutrient some plants develop when they are threatened by fungus. When we eat them, the salvestrols generate an enzyme that attacks any cancer cells. That's one reason why local fruits and veggies are so important, especially the red and black fruits like blackcurrants and grapes, and vegetables like broccoli, cabbages and kale."[730]

I remembered that Aliya had told me about salvestrols, when she was placing her food order through SPUD.

"What about all of British Columbia?" someone asked. "If every farmer grew fruits and vegetables this way, could we feed the whole province?"

"If we used every hectare of protected farmland for fruits and vegetables, grown intensively, BC could feed twelve million people," Mina replied.[731] And that's not counting the food people grow in backyards and community gardens. Detroit is growing close to all of its fruits and vegetables within its city limits.[732] They say Vancouver could meet at least a tenth of its fruit and veggie needs if it really intensified its urban farming."

"If that's so, why do people say there's going to be a terrible famine as the world population grows?" It was an older man, wearing a blue T-shirt.

"I've heard the same dire warnings," Mina replied, "so I appreciate where you're coming from. If everyone ate a traditional North American diet, there would not be enough land, since it takes so much more land to raise beef and dairy and to grow the crops needed to feed them, plus all the water they need. When we switch to a low-meat diet the amount of land and water needed falls dramatically, while our health improves. It just makes sense in a world where we are facing such dramatic water shortages and where children are still going to bed hungry. And because of the climate crisis, things are going to get worse, not better. The droughts are becoming longer and the rain is coming in sudden deluges that wash the soil away, not gently, the way I remember it when I was a child.

"Don't get me wrong," she continued. "I love the taste of slow-roasted lamb, and I used to love crispy smoked bacon for breakfast. But it doesn't seem right to use so much land to raise livestock when, indirectly, it causes others to starve. I have friends in other countries who have difficulty feeding their children, and the less meat we eat the more land is available for people to grow food for themselves."

"So are you saying we should stop eating meat and dairy altogether?" the older man asked.

"That's always a personal choice. There's lots of pastureland that's better suited to raising sheep or cattle, and open grazing is far better than those awful factory farms. Have you seen the live video-streams from the factory farms? Imagine what it was like before they were required to be transparent."

"What about fish?" someone asked.

"Yes, what about them?" Mina replied. "Who would have predicted that humans would fish the oceans so intensively that we'd almost wipe out the fish in their entirety? Not just the big fish, but the small ones too, leaving us with just plankton and jellyfish."

"Is that true?" a young woman asked, looking rather frightened. "Will there be no fish left in the oceans at all?"

"That was the way things used to look. Twenty-five years ago there was a report that said there might not be any fish left in the ocean at all by 2048, since so many of the world's major fisheries were being overexploited. It was because of the high tech fishing vessels, the huge number of trawlers, drift-netters and long-liners, and the failure of fisheries management. The fish never stood a chance.

"I've got friends in Senegal I met through Via Campesina who told me about the poverty and hunger they experienced because of the theft of their fish stocks by the European fishing fleets, and the inability of local fishermen to compete. I gather things are improving now, however, thanks to the *Global Oceans Treaty*. If forty percent of the ocean can become a Marine Protected Area where no fishing is allowed that will allow the breeding stocks to recover, seeding the rest of the ocean with the fish people need. We're only a third of the way there, so there's a long way to go, but you don't need to worry that there won't be any fish. It's ocean acidification that's the bigger concern, because of all the carbon dioxide the ocean is absorbing from fossil fuels."[733]

———✦———

When the seeding was done Mina gave us a break for lunch. Dezzy had packed some sandwiches for me and Sarah and Pelly shared their lunch, so I ate well. Seeing Mina alone, I went over to introduce myself.

"So you've visiting from Sudan?" she asked. "Do you like what you see?"

"Very much. I wanted to ask you about the OMEGA Days. Were you involved? Or were you too busy on the farm?"

"Oh, those were the days!" she replied. "We're always busy, but how could we resist? A Vancouver urban farming group reached out to farmers across the province and urged us to get involved. The G in OMEGA stands for a Green Future, and one of its five solutions was 'to cherish and protect the earth,' which included sustainability for the farmlands, grasslands and forests. So yes, we got involved."

"Do you remember what the other solutions were?"

"They were nested together. They were to cherish and protect nature, the atmosphere, the water, the earth and the humans. I was involved with the earth solutions, but I remember the others.

"The first solution, to cherish and protect nature, called for legislation to give nature a safe, secure home, with legal rights, a lawyer to represent her in court and the restoration of habitat in land-use developments. The global statute that made ecocide a crime against peace followed later. The first case is going to the International Court in The Hague this fall, now that the ten-year amnesty is over. For the CEO of a company who has committed a crime against nature to be on the same criminal footing as a genocidal dictator—that's huge.[734] And on the local level, the various things we're doing here like the woodlands, the hedgerows and the wildflower meadows, the nesting boxes for the bats, the habitat for bees, frogs and snakes—these are being done now on many farms. It's a big step forward."

"And the other solutions?"

"Let me see," she said, shaking her head and loosening her hair. "The second was to cherish and protect the atmosphere. It called for phasing out all fossil fuels and the use of renewable energy instead."

"And the third?"

"To cherish and protect the waters. It called for every river, lake and stream to have the legal right to exist in a pure, unpolluted condition; for strict development codes on aquatic and waterside development; and for a *Global Oceans Treaty*. That's still a really big concern. As I was saying, the oceans are becoming progressively warmer and more acidic, which is why the plankton, shellfish and coral reefs are in such bad shape. The way things are going, by 2050 the ocean will be more acidic than it has been for twenty million years."[735]

There was silence among the group of people who had gathered around to listen to Mina.

"Most people don't realize that there's a forty-year delay between fossil fuels being burned and the rising surface temperature. The impacts we are experiencing today are the result of fossil fuels that were burnt in the 1990s. When the ocean's too acidic, the plankton can't build the limestone shells they need—and it's the plankton that ultimately feed the world's fish and store most of the carbon in the oceans, pulling it out of the atmosphere. Does anyone know how much phytoplankton has been lost so far due to global warming?"[736]

More stunned silence. This was not what we had been expecting in an otherwise upbeat day.

"More than forty percent," Mina continued. "The marine biologists started to notice the decline in the 1950s. It's just one more reason why ending the use of fossil fuels has become so urgent. But where were we, before I got sidetracked?"[737]

"You were going through the five solutions in OMEGA G for a Green Future," I replied. I was also thinking 'how does she know all this stuff?' Maybe public awareness about the climate crisis had increased dramatically; or maybe she was just brilliant.

"Right," she said. "That was the third solution, to cherish and protect the water. The fourth was to cherish and protect the earth, which I've told you about, and the fifth was to cherish and protect all humans. That led to legislation that enshrined the right of every child to have uncontaminated blood, to breathe uncontaminated air, to drink uncontaminated water and to eat uncontaminated food. It was *arrivederci* to food additives, colorants and junk food and *benvenuto* to fresh organic food.

"We're happy, since the changes have brought a reduction in dietary disorders," she continued. "I hope it will also lead to a reduction in cancer. Whatever were we thinking? However did we allow such poisonous chemicals to get into our children's bodies? I've got two lovely daughters and they've always been healthy. I took great care when I was pregnant and we've always fed them wholesome food. They've also been exposed to plenty of germs here on the farm, which is good for their immune systems."[738]

"How were you involved during the OMEGA Days? Did you block the roads, the way they did in Vancouver?" I asked.

"No. That would have been ridiculous here, since everyone's a neighbor. We got a group together at our church and we went around to the other local churches making presentations. Then we invited people to join us in a Pilgrimage of Life in support of the OMEGA movement leaders who were in jail."

I thought of Derek, and the time Lucas had spent in jail. I wondered what he and Aliya were doing on this sunny Saturday afternoon.

"We walked all the way to Vancouver, stopping to pray and talk at churches along the way. It took us three days, staying with supporters at night, and when we finally arrived at the Church of Our Lady of Sorrows in East Vancouver on the Saturday night we were greeted by the entire congregation, including the choir, who took over the street for a spontaneous evening concert. The police were very amenable, and left us alone. The next morning the church was packed to the rafters and Mother Angelina, the newly appointed priest, spoke passionately about the needs of the poor all over the world who were struggling to put food on the table because speculators were forcing up food prices. No wonder there were so many food riots."

Mother Angelina? So the Catholic Church had finally reversed its ban on women becoming ordained priests.[739] I wondered if they were also allowed to marry, but I never found out.

"After the mass," Mina continued, "we proceeded down Hastings Street to the Holy Rosary Cathedral, a thousand strong, where we set up a vigil that people kept up for weeks, long after we'd returned to the farm. I remember it well. Joey looked after the farm and I took Isabella and Bianca with me; they were ten and twelve at the time. They still talk about it. We had an impact, too, since so many people phoned their legislators, urging them to free the OMEGA leaders and support the movement's objectives."

By this time there was a quite a cluster of people sitting around Mina. For the younger people the OMEGA days would have happened when they were in daycare, so this would have been a rare opportunity to learn about them first-hand.

"Do you think the Dhaka Agreement will make a difference now that it's been ratified by so many nations?" someone asked.

"We're all frustrated that it took so long for the agricultural clauses to be ratified," Mina replied. "Britain and America were the big hold-outs, trying to protect their financial interests. Even today, with every ten percent rise in food prices, another ten million people fall below the poverty line and are forced to choose between food and school, food and electric light. Now that the food commodities market has been closed to everyone except those directly involved in agriculture, and the speculation has ended, food prices have fallen somewhat. The new global commitments to agro-ecological farming and to remove the remaining subsidies are so important. So we'll see. Many countries need big land reforms to break up the mega-farms. We need an agricultural renaissance, without interference from agro-business. If only we'd started twenty years ago."

"What made the difference?" I asked. "What could people have done twenty years ago?" This was getting personal.

"I don't know. We tried. Via Campesina was always organizing, but we were never able to grasp all the manipulations that were going on—the power of big money, the games the commodities speculators were playing. We were all about stopping biofuels and fighting Vivendo. First generation biofuels were a problem, but they were not the cause of the rise in food prices. It was the commodities speculators who caused that, and nobody nailed them for it.[740]

"Who knew that the Wall Street lobbyists had worked so hard to get the financial controls on commodities speculation lifted?[741] We were financially illiterate. We didn't understand how banking and finance worked, so it was easier to blame capitalism in general. It wasn't until Via Campesina formed a partnership with Economists Without Borders that things began to change.[742] Then we hired our own economists and joined the global initiative to train ten thousand Chartered Financial Activists. That's when the shift began. With help from our CFAs we were able to do our own precision lobbying, which finally led to the regulations being tightened, the controls on food commodities speculation being re-imposed and all the other changes."[743]

Ten thousand Chartered Financial Activists? That would make a difference, to have that many people trained to understand global finance and be activists for change.

"Did it help that there was a wider movement for social justice and sustainability going on in the OMEGA Movement at the same time?" I asked.

"Absolutely. The OMEGA Days were happening all around the world, not just in the big cities like Vancouver and London, New York and Barcelona. The financial collapse changed the way people thought. The idea that the free market thrived at its best with a minimal burden of regulation was revealed for what it was: a selfish charter for the power-hungry, regardless of the harm they did to humans or to nature.

"The new synthesis has also given us a new philosophical foundation. With socialism gone on the left and neo-liberalism gone on the right there was a vacuum, which the new synthesis is filling. Could it have happened twenty years earlier? I don't know. What we need now is a new OMEGA Days, to shake the world up again and get people mobilized behind the solutions that are still needed, the way we did fifteen years ago. If you know of any way to make it happen, please let me know!

"But look," Mina continued, "I've a lot to do before the day is out, and one of our cows is due to calf over in the barn. I'm going to leave you with Carlo, who will show you the village and put you to work. He's my daughter Bianca's husband, so be nice to him!"

 26

The Farm Village

THE FARM VILLAGE was a clean, well-kept cluster of small square two-storey townhouses built around a maple tree, at a distance from the farmhouse. Carlo told us they'd built the homes from hempcrete blocks that lock together easily. Each house had a composting toilet and rainwater storage, and they shared a collective greywater system.[744]

"Where does your heat come from in the winter?" someone asked.

"The houses are built to the passive house standard, so they need very little heat. One small coil in the heat exchanger is enough. When it's dark and it rains a lot our homes are very cozy."[745] He was a sturdy man with bare feet, black hair and a thick black beard, I'd guess in his thirties.

"How long did it take you to build the village?" someone asked.

"It took us a year. There were fifteen of us, and we had lots of volunteers who enjoyed our community style of living, with all the music and theatre. We used Passipedia, which helped with the building details and things like the zero-energy fridges. Someone had loaded everything there is to know about passive building into a chatbot, a little Japanese girl called Botiko, and when it was only us men we used to flirt and say some pretty outrageous things, to see how she'd respond."[746]

A chatbot, I realized, was a robot that could speak and answer questions.

"You men!" one of the women responded.

"Yeah, I know." Carlo said. "When the women heard about it they contacted the owner and got him to program some blunt new responses. So the women definitely had the last laugh." There was a chuckle from the women in the circle.

"You mentioned theatre. Is that still part of your life?" Pelly asked in a serious voice that made me think she was deliberately changing the subject.

"Absolutely. We use it to release our creativity. Ha!" Carlo suddenly leapt onto a garden table making everyone jump and transformed himself into a wizened old man, staring at us like Gollum in *Lord of the Rings*.

"Be careful, my precious!" he squealed at us. "Seize the ssecrets! Seize the she-wolf! Ssqueeze her! Ssqueeze her tight, tight her ssqueeze! She will BITE!"

Sarah let out a squeal and cowered behind Pelly while everyone else moved back a few steps in alarm. Carlo seized us with his gaze, then smiled and climbed down off the table.

"I learned about it when I was a student of architecture in Barcelona," he said calmly, while the rest of us regained our composure. "As humans, we are deeply programmed for stability and tradition. The satisfaction we get from an orderly existence lays down a path and sometimes it's hard to get off it. The plates live here, the cups live there. It would be exhausting if they were in a different place every day. But it builds biological conservatism, which limits our freedom to design for the future. Drama helps shake it up. We can become new people and embrace new stories. *Ssqueeze her tight!*"

We were sitting in the shade of the maple tree in a scene of rural beauty that could have been five thousand years old. Even the houses looked old and comfortable, as if they had been here forever—except for their shiny solar roofs.

A woman came out carrying a blue glass jug and a tray of glasses. She had flowing blonde hair and she was wearing a long white sleeveless summer dress with a strapped back, a deep cut V-neck and a gold belt around her waist. She looked like a goddess.

"Elderflower champagne, anyone? I hope Carlo isn't boring you. Has he persuaded you to act the part of Napoleon yet?"

"This is Bianca, my wife," Carlo laughed. "She hates Napoleon."

"He was a villain and an impostor!" Bianca replied with a shake of her hip. "King of Italy! Who did he think he was?"

"But without Napoleon, my dear, Europe would have remained a landed aristocracy stuck in the middle ages. Napoleon was the world's greatest liberator!"

"Liberator, my arse. He was a tyrant!"

"This elderflower champagne tastes marvelous," Sarah said, self-consciously. "Did you make it yourselves?"[747]

"Yes," Bianca replied. "It's just one of lots of wild shrubs that are easy to grow. I love its fragrance." She took a drink directly from the jug and when it dribbled down her chin and into her cleavage we men didn't know where to look. I could see that Carlo might have his hands full with Bianca—and that her flirting with us might also be a way of flirting with him.

"What else are you growing?" someone asked.

"What aren't we growing!" Bianca answered. "Everything from figs and apples to peaches and persimmons. We've got plums, damsons, nectarines, greengages and cherries. We've got lemons, which love the hot dry summers we've been having. We've also got kiwis and walnuts and hazelnuts and heartnuts. Hickories, pecans, almonds and butternuts. Black walnuts, quince, blueberries and redcurrants. Blackcurrants, tayberries, strawberries and blackberries. Uh... what have I missed? Mulberries, chokeberries, Saskatoon berries and Goji berries. And a plethora of vegetables."

"What a memory! Isn't she a genius?" Carlo glowed. "Don't forget our maryjane. She's important for the soul."

"You and your soul! You mean your bank balance?" Bianca teased.

"You're growing marijuana for sale?" Pelly asked.

"For sure we are!" Carlo replied. "Best ratio of THC and CBD, $10 a gram

($280 US an ounce), so it's a better business case than the raspberries. We've a new batch in the aquaponics lab growing with the tilapia fish."

"Does that mean you have to pay a lot of taxes?" someone asked.

"Well, that's our private business," Carlo said, flashing a cheeky smile. "But let me say we are happy to do so. Taxes are our tithe to the community. We celebrate them, now that we know they're fair and nobody's cheating."

"All those fruits," Isabella from Mozambique asked slowly, using her language app, "do you own the land that goes with the village? And how does your land relate to this farm as a whole?"

"When the *Organic Farm Act* passed," Bianca replied, speaking slowly so that the app could translate, "it allowed every farm over thirty hectares to sell or lease a hectare for a farm village. Because we designed the village to be so compact, we're only using half a hectare. Each household also has a hectare to farm as long as they uphold the farm's five-star status. That's where most of the fruit comes from. It's amazing what you can grow on a hectare of land when you put your mind to it. Mina's letting us develop a permaculture forest to see how much food we can grow while protecting the forest. But Carlo, why don't you show them the workshop? I've got to look after the boys. Pelly, do you want to come and help me?"[748]

Pelly left with Bianca and the rest of us followed Carlo to the workshop, a circular single-story cob building with undulating walls. Above the entrance there was a quote by Einstein engraved on a piece of wood:

> *Imagination is more important than knowledge.*
> *Imagination embraces the entire world,*
> *stimulating progress, giving birth to evolution.*

Inside, the curved walls created various workstations. There was a fig tree in the middle, under a glass roof surrounded by an amazing mosaic that seemed to capture the whole Universe, from the galaxies to the microbes, the trees, the oceans, and humans and their children.

"This is where we design the future," Carlo said proudly, waving his rough hands. "This mosaic is our inspiration. My wife Bianca did it, with help from many friends. She trained with Lillian Broca, the famous Byzantine revival mosaic artist from Vancouver."[749]

I was stunned. Bianca designed *this*? I examined just one small area and estimated that it must contain a thousand pieces of glass and tile. What beauty! Move over, you Romans and Byzantines!

"Yes, she is very talented," Carlo said. "Her sister is a poet and a playwright."

When we had finished marveling at the mosaic Carlo told us about the workshop and its role in the village.

"We are Makers," he said. "Proud members of the Vancouver Maker Foundation. Regular exhibitors at the Maker Fayres. It's our karma yoga: the way we find union with the Universe. We are born creative. There's nothing wrong with the things other people make, but there's something special about the things you make yourself."[750] Carlo led us to one of the workstations, full of test tubes and metal contraptions.

"This is our peepee project. We take urine from the composting toilets and the cows in the barn to make hydrogen from the ammonia and urea, which goes into a fuel cell to generate electricity. It's not a lot, but we're still learning. So please drink a lot of water. We need your pee![751]

"We're also planning a green burial project using hot composting. We're doing a trial with a deer to see how long it takes for the bacteria to achieve complete breakdown. If it goes well we have a list of several hundred people who want to be composted here when they die, so it could be a thriving little business."[752]

"I'd love to be composted when I die," Sarah said, fervently. "Being eaten by worms and bacteria might be a wonderful final orgasm for the cells in my body. It would be a nice way to say thank you."

Whoa! But clearly no-one else was bothered.

"Hey!" someone said with a laugh. "So when they say 'death is coming,' there might be two ways of looking at it!"

"Oh My God!" someone else said. "Fifty trillion cells all coming at the same time!"

"Woo-hoo!" laughed another. "All together now...." and I watched with astonishment as the group groaned and moaned, making the hilarious sound of orgasms and then bursting into uncontrollable laughter. I noticed that Isabel, the Mozambican woman, was not joining in and looked rather embarrassed, as did José, Serge (the African) and a few other people.

Carlo couldn't help but laugh. "Okay, you young folks," he announced in a loud voice. "Over here we've got our sayak project." He led us to an unusual looking boat that looked like a kayak but was not a kayak. "It swims like a fish when you pedal it. We call it a sayak because it's a cross between a salmon and a kayak. We're hoping it can beat the world kayak speed record. There are two on the lake we're doing trials with. You're free to try them if you want to."

"It swims like a fish?" Pelly asked, having returned from whatever she was doing with Bianca.

"Yes. The pedals drive the tail, which is made from two thousand waterproof scales, giving it the flexibility of a salmon. Each scale is coated with sharkskin microriblets to reduce the friction. Fish have had time to perfect the technology over a thousand million years. We're working with a design studio in Vancouver to optimize the algorithm that governs the scales. They're each bounded by a different set of flow conditions and they need to work together seamlessly.[753]

"We're planning to launch this fall with a race in Vancouver's False Creek," he continued enthusiastically. "Marek Losos is on board—he's our Olympic K-1 kayak gold medalist. Once the kayakers have tried it we hope they'll become our champions, but we're up against the flyaks. They're kayaks with hydrofoils that lift them above the water, removing virtually *all* the friction. They're really fast, but they only elevate at speed, while the sayak is faster at a steady paddle-pedal. We think we can beat them if the paddlers can do thirty kilometres an hour. But that's just the beginning. We're also designing an electric fish that can swim underwater."[754]

"You mean, like a submarine?" someone asked.

"Yes. The same as the sayak but fully enclosed with thirty minutes air supply. We're just not sure anyone will want it, so it's probably an indulgence."

"Do you do this all yourselves?" I asked.

"Yes and no. We believe in collaboration. We don't want to be like the alchemists of old, working alone in their cellars. Most of our projects are open-source, so anyone can chip in to improve the design and share the patent. If it leads to a business, so much the better. If we think we're onto something really good, we keep it to ourselves. Some ideas we share with an open-source company; some we file at half-baked-ideas.org. If we need to manufacture something complicated, we've got our own collection of makerbots.

"During the winter we do a lot of tinkering in the forge, fixing and repairing the farm equipment,[755] and we're learning to make our own equipment using Open Source Ecology. We started out by making a compressed earth block brick press that can make five thousand bricks a day, then we made the electric tractor, which cost us $15,000, instead of the normal $75,000."[756]

"You made your own tractor?" someone asked.

"Yes. It took some trial and error, but we followed the instructions and we put out a call for help when we got stuck."

"What else have you made?"

"The solar water heaters, the solar dryers, a small electric seeding machine and a garlicator."

"A garlicator? What's that?" I asked.

"It's a simple device that allows you to make the holes for twenty-one cloves of garlic at the same time. You place it on the soil and then you do a dance on it. It's definitely fun!"[756B]

"What about capital, when you need it?" someone asked.

"We have a network of friends who invest in a project if it looks promising. They're our Village Angels."

"Does the income from the workshop qualify as farm-based income?" someone else asked.

"Some yes, some no. Only half our income needs to come from farm-based activity. For every ten projects we pursue, two or three will usually earn us some income and one will make it big.

"Take our solar dryer," Carlo said, leading us over to a glass-covered dryer. "We did a lot of work to improve the design, amplifying it with vacuum tubes and a heat pump. A company in China is going to manufacture them, and we'll get 1% from every sale."

He may have bare feet, I thought, but Carlo was clearly very sharp.

"With all these solar systems and electric batteries," a young girl asked, "do you have any concerns about the shortage of rare earth minerals?" She can't have been more than sixteen.

"Good question. By volume, no, but some rare earth mining can be very damaging and involve a lot of exploitation, so that's a concern, and not all mining companies have signed the *Bella Terra Mining Code*. There's also a growing interest in nanomaterials, replacing the indium in solar cells with tiny carbon nanotubes."[757]

"Isn't that interfering with nature?" Sarah asked.

"The way I see it," Carlo replied, "all these rare earth minerals were made in a supernova explosion billions of years ago, just like the gold in my wedding ring and the atoms in my body. We are all stardust from exploding suns. So if the atoms in our bodies were created in the same supernova explosion as the atoms in nanomaterials, who's to say what's natural and what's not?"

That sounded convincing, but not to Sarah.

"You could apply the same argument to genetically modified tomatoes or arsenic tablets. Just because something was made in a supernova explosion doesn't make it okay. The raw ingredients for a nuclear bomb were made in the Big Bang too, but that doesn't make them good."

"Hmmm, I believe you're right," Carlo agreed. "Sometimes I can fool myself with my own arguments. Nanotechnology has to be examined just as carefully as industrial chemicals and genetically modified algae."

"So we may have a problem regarding the rare earths?" the girl persisted.

"Yes, I suppose we may—but speaking of rare earth, there's some weeding needs doing, and various other tasks. Any final questions?"

"Yes," someone said. "How do you find the time to grow all these crops, pick all the fruits *and* reinvent the world?"

"We get up early," Carlo replied. "We have a lot of volunteers, and we're well organized. Everyone's got a responsibility and each week we load the work into the village matrix. Then we sign up for our tasks, list the volunteers we know are coming and the matrix sorts out who gets paid for what. It also tells us where we're going to be short-staffed, so that we can make adjustments.

"We load it up a year in advance for crops where we know the pattern, which helps us choose early or late varieties to spread the work. June is always our busiest month. All the weeding, planting out and harvesting—it all comes at once. We have no time for other projects in the summer. That's why the workshop's empty. Everyone's out planting tomatoes or cooking for tonight's feast—and we've also got band rehearsal. Any other questions, before we get to work?"

"Yes," someone said. "What's the biggest difficulty you've had to deal with?"

"Well, that's not a quick one. If I had to choose, I'd say it's our relationships. And the pollies."

"The pollies?" someone asked.

"Yes, the pollies. The ones who are polyamorous, who are in and out of each other's beds. They say their kind of love is more rich and fulfilling than monogamy, which they see as boring and stultifying. My concern is with how much time they take sorting out their various issues, all their loves and hurts. The pollies are great, and I love them, but having them living so close makes it hard for those of us who are trying to be monogamous. It's unsettling when there's so much sex going on. Sometimes it's too tempting, and people do things they regret. And it takes so much time, sorting out all the jealousies and misunderstandings. Monogamy's so much easier if you can make it work. Maybe monos and pollies shouldn't live together. Maybe they should have a pollyvillage of their own where they can do their polly thing without bothering the rest of us."

"I've been in a polyamorous relationship for almost a year now," Sarah said, lightening up. "I feel so much more free having two main relationships, and knowing we're all okay with it."

I felt a shock of surprise.

"I'm feeling jealous already!" someone said.

"And I'm feeling excited!" someone else said.

"See what I mean?" Carlo laughed. "But enough with all this. It's time to work. Who'd like to pick the garlic scapes? Weed the beans? Help bring in the hay before it rains? Or join the cooking crew for tonight's feast?"

I volunteered for the beans and strawberries, since I figured it would be quiet, and it might give me a chance to have a conversation with someone new. I was well into day three of my journey, and there was still so much that I still wanted to learn.

27

The Demise of the Oil Industry

RYAN WAS A tall, heavily built older man with a silver ponytail and an earring, who had been listening quietly during the tour without saying much. We were hand hoeing, cleaning out the weeds between long rows of snap beans, onions, carrots and strawberries. The summer afternoon air was still, but alive with bees, butterflies and ladybugs.

"So what brings you here, young man?" he asked in a kindly manner.

I explained about my visit, and the things I was hoping to learn. "What about you?" I asked.

"Well, I'm no spring chicken. I was seventy this spring, and my repurposing coach said I might find the farm interesting. I've got too much juice in me to be sitting around all day playing bridge."

"Your repurposing coach: is that like a therapist of some kind?"

"No, not at all. Julia: lovely young woman. Seems to know what she's talking about. I met her on the repurposing course I took when I retired and she's been helping me ever since. Just what I needed. Put me on a new path, freshened up the brain cells. Helped me look at my purpose in life, and what gives me satisfaction. It was she who persuaded me to look into becoming a Personal Learning Coaching Assistant in the local primary school, which I've been doing for three years now."

"Is that a paid job?" I asked.

"No. I'm a volunteer, but I do receive a payment in Time Dollars, which I can spend on anything from getting my hair cut to having someone clean the gutters. I work with the special needs kids who have learning difficulties, helping the full-time learning coaches two days a week. There are quite a few of us in the team. We have one woman, Carolyn. She's over eighty, but she's very sharp. She has a great sense of humor—the kids love her. She's our gardening guru. She shows the kids how to grow food and collect seeds from different plants. When she's in the staff room, away from the kids, she tells the raunchiest jokes."

"What kind of thing do you do with the kids?"

"I help with their reading and writing, and with their music, sports and kinesthetic exercises. Once you understand how they prefer to learn it's incredible the difference you can make. Each kid is so different. You could sit them in rows and try to do it the old way, but it just doesn't work; they'd be bored in five minutes.

Having us work with them one-on-one allows the teachers to get on with the rest of the class without interruption. So much of the learning is self-directed these days, and before they changed the rules to allow us to help, the special needs kids were making life hell for the teachers. It's not their fault. They have so much unexpressed energy and when it's not channeled it comes out sideways."

"Do you know what's been causing all the learning disorders?" I asked. "It sounds like there's a lot of it."

"There are plenty of theories," he replied. "I've done a bit of reading and I'm still not sure what to believe. Some say it's due to the kids spending too much time glued to their screens when they're really young, under the age of two. They say it wires the brain to a speed that's faster than normal, making them think all reality moves at that speed. There's another line of thought that blames environmental neurotoxins in the womb, such as mercury, or some chemical that was present in air pollution before the electric car took over."[758]

"I've noticed how clean the air is. But if that's the cause shouldn't the disorders be declining?"

"Yes indeed, and they are. Some people think they may have been caused by a deficiency of essential minerals and trace elements during pregnancy, and mothers not getting enough fresh organic food. With most food now being organic, that could explain the decline too. Autism spectrum disorder used to affect as many as one child in thirty before the decline. ADHD was even worse—one child in five. That's almost a quarter of the children in a class having a learning or behavioral disorder. How could the teachers begin to cope?"[759]

"What did you do in your previous life, before you discovered the joy of retirement?"

"Ha! That's an amusing turn of phrase. Do you know about the Immortality Clubs, with their blood transfusions, super-foods and mitochondrial jump-drugs?[760] But who wants to live forever? I'll grant you one thing: it's amazing what they can do for you these days. I've a friend, Harry, who'd lost all his hair by the time he was forty and for less than a thousand dollars he's been able to buy himself a new head of hair using hair follicle transplants from his own skin stem cells. I've another friend in her eighties whose balance isn't good, so she's had her house fitted with smart-carpets that can tell if she's had a fall and call for help, or identify an intruder and sound the alarm.[761] And look at me! My wife has me wearing this wristband to track my exercise and sleep. Just in case, she says. Maybe she thinks it'll tell her if I have a little fling on the side. It even beeps at me if I leave the house in the morning without having taken my meds!"[762]

"You seem very young for your age."

"When the spirit's young, they say, the world feels young with you. When my spirit tells me I'm ready to go I'll put my things in order, have one hell of a party and then dial E for Exit. Better that than expect some poor nurse to keep me alive for a few drug-induced months while spending tens of thousands of dollars of the public's hard-earned money to do so.[763] That's my philosophy. It's all in my living will. If what the scientists say is true then everything in the Universe is conscious anyway, so death may just be a change to a different kind of consciousness. It's my life, so I reckon it's my human right to end it when I want."[764]

"That's a very positive way to look at it."

"And why not? I'm so glad they've dropped the term 'assisted suicide' now that it's legal. I prefer to think of it as my final pilgrimage, my Ultima Via. Vianauts—that's what we are. Ultimate Vianauts. Death: the final frontier."

Ryan got up from weeding and stood like a Shakespearian actor with his feet planted firmly in the soil: "This is the voyage of the starship *Ryan*. My final mission: to explore strange new worlds, beyond life, beyond civilization. To boldly die where everyone else has died before!"

"That's hilarious! It almost makes me want to join you. Do you believe there *is* another world beyond this one?"

"That's the big question—don't we all want to know? I'm very persuaded by the evidence from near-death experiences, and things that can't be explained by neuroscience. If everything is ultimately made of consciousness, as they are saying it is, what could be more natural? Once we understand things properly we may find that consciousness follows the same laws as thermodynamics, and can be neither created nor destroyed. So maybe the reality we perceive is only a very small part of reality as a whole, and in that sense, yes, I suppose I do believe in another world."[765]

"What did you do in your previous life, before you began preparing to be an Ultimate Vianaut?"

"I was an oil company executive," he said, with an abrupt change of mood. "I used to commute between Calgary and Houston, searching out the next big play. When the industry slowed down they kept me on as they tried to diversify, but then we had a big blow-up in my family so I finally handed in the keys and walked away. I must say, I'm happy to be finally making some small difference in the world. Here—have another strawberry."

"Oh my god!" I said, savoring its intensely sharp sweetness. "They're so good. No wonder they have so many volunteers. But those are harsh words."

"Now that I'm out of the business," he continued, "I can finally hear what my children were telling me all those years—how we've made such a mess with the Alberta tar sands, how you can see them from space, how we're turning the world's oceans to acid, how the Arctic oil blow-out will pollute the ocean up there for hundreds of years. What a mess! We should never have been drilling there in the first place—and all for a few more months of satisfying our addiction to oil. My kids were literally weeping for the world we were losing and the future their children were going to inherit. That really hurt. That was before the OMEGA Days, and everything that followed. They feel more optimistic now."[766]

"What was it like working for the oil industry when the OMEGA Days were happening?" I asked. We were starting down the next row, heading north toward the mountains. Everything felt great, from the sun on my face to the taste of strawberries in my mouth and the dirt under my fingernails.

"In the beginning we thought we were fighting a worthy war against environmental organizations with deep pockets. Ethical oil, we called it—and we believed it, too. Far better to use Canadian oil, we argued, than oil from some evil dictatorship in the Middle East. But as the protests wore on and the civil disobedience grew, I began to realize that I was on the wrong side of history. The thought of Monday mornings began to make me feel heavy, knowing that I had to drag myself through another week of work I no longer believed in. I was relieved when I finally packed my bags and walked away."

"You walked away...?"

"You're full of questions, aren't you?"

"I'm sorry. I don't mean to intrude. I apologize."

"No matter. I suppose I still find it a bit painful." He stopped weeding and sat down in the row between the strawberries, trying not to damage any of the plants.

"It was when my youngest daughter Samantha started dating her future husband that things began to be difficult. He—Wilfred—he's Cree, from the Athabasca Chipewyan First Nation.[767] He had lived at Fort Chipewyan all his life, downstream from the oil sands, so any pollution from our operations flowed right past his door and through his drinking water. Soon after he and Samantha started dating she phoned me one night and told me about the unusual cancers they were seeing among his people, the deformities she'd seen in the local fish, and how they never touched the water since it had a strange, unpleasant smell. 'Surely, it must come from the oil,' she said. 'You've got to do something, Dad,' she insisted. 'You work there. Surely, you can do something?'[768]

"I tried, but I got stonewalled at every turn. Then a friend said he'd overheard a conversation where the directors were questioning my loyalty, and wondering if they could trust me. From then on, it was only a matter of time. I was getting it from both sides. One side had to go, and it wasn't going to be my family. So I quit. I took quite a cut on my pension, but at least I walked away with a little bit of my pride intact."

"That must have been hard. How did your friends react?"

"Most of our industry friends cut me off cold. I was damaged goods. If I was seen in their company it put them under suspicion too. It was really hard on Renée, my wife. She'd been a part of the wives' circle for almost all our marriage—the cocktail parties, bridge parties and charity events. Suddenly the phone stopped ringing. I know she blamed me, though she said she didn't. It was alright for me; I was the one instigating the change. I had my sense of being ethically justified to comfort me. Renée, she had nothing, and she had suddenly lost her girlfriends too. For a woman like her, who's so social, it was devastating. She continued to meet a few in private, but that only made things worse because they would tell her about the things going on back at camp, as it were, from which she'd been exiled. So we moved. Sold our home in Calgary and moved to

Anmore, just across the river, where we have a lovely passive home surrounded by the forest and the mountains."

"Did she get over it, with the new home?"

"Yes, eventually. That was eight years ago. She's been able to keep in touch with the friends who were closest now that they can see each other privately, and it's actually made us closer as a couple. It's also made us closer to our children and grandchildren, which is what matters. It was a tough time for my son as well. He was struggling with his second marriage while holding down a big mortgage, credit card bills and $30,000 of student debt he'd been unable to repay, even though it was ten years since he'd graduated.

"I needed to earn something to help my son and make up for my loss of income so I got involved with a Canadian mining company. Big mistake. They were doing this barite project down in Mexico; it's used for drilling in the oil and gas sector. There was tension, with local protests about water pollution and poor labor conditions, and at the same time there was a global campaign going on to persuade companies to sign the new *Bella Terra Mining Code*, but my company was dragging its feet. Suddenly, I was having a terrible sense of déjà vu.

"There were mothers picketing our Vancouver offices with their children. I got a call one morning from the government in Ottawa asking if we wanted to sue the Mexicans for breach of the *North American Free Trade Agreement*, and I thought: 'This is all wrong. They're the ones who should be suing us.' So for the second time, I called it quits. Instead I offered my services to the watchdog organization, Mining Watch, to help them pressure Canadian mining companies to clean up their act and stop behaving as if they were some kind of colonial occupying power."[769]

"Did the new *Mining Code* succeed?"

"Yes, but only later, thanks to the global campaign to divest from companies that refused to sign the *Code*.

"Behaving badly is like a habit; it gets under the skin. You feel you're entitled to do it, that earning all that money gives you the right to behave like a social and environmental jackass. I know from conversations with my fellow directors that for them, being made to sign the *Code* felt like being taken before the school principal. They greatly resented that anyone should have the power to do that to them.

"Well, they soon got their come-uppance. A year after I quit there was a shareholders' revolt and the whole board was turfed in favor of directors who specifically *wanted* to sign the *Bella Terra Code*. Two years later they turned themselves into a B Corporation, and last year the magazine Corporate Knights gave them an award for being among the top ten most socially responsible foreign-owned companies in Mexico. That's quite the turnaround. And now we've got the *International Convention on Responsible Mining*, which Canada has signed along with most other countries."[770]

"When you were in Calgary, did you hear any talk about a man called Derek

Brooks, who was assassinated in Vancouver during the OMEGA Days?" I thought I'd seize the moment.

"Yes, I do remember, vaguely. Wasn't he one of their leaders? I know they said they didn't have leaders, but no-one gets to be as successful as they were without having leaders. Why do you ask?"

"They've never solved his murder. I'm staying with his sister, and she's been telling me about it. Do you have any theories about who might have been behind it?"

"Follow the money, is the best advice I can give. Follow the money."

"How did people in the oil patch react to the talk of closing the tax havens, and publishing the names of people who were using the havens to avoid paying taxes?"

"We thought it was preposterous. Most of the offshore accounts were totally legal. We used a very prominent firm of tax accountants who advised us on ways to stay competitive and reduce our taxes. It was all above board, and within the rules. It was the rules that were corrupt, that needed changing, so in that regard I supported what the folks in the OMEGA movement wanted. I took care not to say so in public though, and definitely not where I worked. My company had a huge amount of money in the Cayman Islands and other tax havens, and if we'd been blacklisted it would have hit us very badly, affecting our contracts with the European Union."

"So they would have been happy if something was done to stop it?"

"Oh for sure. It was like Little Texas in the oil patch. People liked to wear their boots, if you know what I mean."

"But they wouldn't go so far as to assassinate someone...."

"Hell no. This is Canada, after all. But I remember a meeting with three American oilmen and them telling us we had to stop pussy footing around if we wanted to stay in the business. Our American partners were watching us closely, they said, and they expected us to do something. 'If you let this thing drift, next they'll be seizing our money,' they said. 'We either have to hang together, or we'll all hang individually.' Talk to Syd, they said. Syd Brockle. He was apparently a man who knew how to get things done."

"Do you know who he was?"

"I only met him once. Aussie billionaire, big-headed bastard if ever there was. Like a character from a TV series. Total racist, misogynist, libertarian. Wouldn't trust him with a single honest dollar."

"Do you think he'd do a thing like that? Go so far as to get someone assassinated if he thought his money was in peril?"

"Probably not for himself, but he had these grandiose ideas about power, and who should hold power in the world. Some of those Aussies, you never knew what had gotten into them, apart from the whisky. Strictly old school. Super-entitled. Always arguing for lower wages for the workers while giving bigger bonuses to themselves and their friends. If he thought the oil industry was in danger, who knows what he might have done."

That gave me plenty to think about. Tangled web indeed.

"Can you tell me about the OMEGA Days protests? I only know what I read as a teenager." Another white lie, but that was my alibi.

"Oh, we had fasts, sit-ins, every kind of protest you can imagine, right outside our offices in Calgary. 'Ethical cancer,' the posters said. 'Ethical war,' 'ethical climate change,' mocking us for our claim that the oil sands were ethical. The oil sands themselves were in northern Alberta so it was relatively easy to police them, but we were much more vulnerable in Calgary, and the police were reluctant to be heavy-handed against protestors who were so polite. They were young themselves—the police, I mean—and some of them probably sympathized with the protestors. They kept rolling out these grey-haired professors and innocent-looking children who would stand there in silence as we came and went, offering us flowers. That really got to us. Very few of our female staff could take it. They'd rather stay bottled up inside all day than go out for lunch and face the children.

"The schools had this project called Caribou Love that tracked the caribou in the boreal forest using microcams. Their habitat was directly threatened by the oil sands and they had hundreds of schools following the herds over the Internet, learning to identify them by name, based on their features. The cameras are tiny and they're recharged by the caribou's motion, so they transmit live. The footage during the rutting season is incredible, especially if you get to see it wearing one of those virtual reality helmets. So as you can imagine, the children didn't think much of my company's work in the oil sands."[771]

The very same caribou rutting season that I had just experienced first-hand. Or kind of first-hand.

"They were also going after our investors," Ryan continued. "I remember feeling like I'd been kicked in the stomach when my son told me his university was disposing of its investments in coal, oil and gas because of the harm fossil fuels were doing to the planet. It was like the campaign to end slavery, he said. It was immoral, and it was wrong.

"And they had these trained activists—Chartered Financial Activists, they called them, who bought our shares and knew how to use a Bloomberg terminal, tracking down our investors and approaching them as fellow shareholders. Every year they organized shareholders' motions at our AGM, accusing us of exposing them to unnecessary financial risk by not accounting for the carbon listed in our reserves that was unburnable without frying the planet. They even persuaded celebrities like kd lang to join them. After the Global Carbon Cap came down, of course, everything began to change. My company was obliged to disclose the value of its reserves that could not be burnt without putting the planet in the danger zone, and we had to bid in the annual carbon permits auction to win the right to produce under the cap. You could read the writing on the wall when that happened—especially after Canada's Supreme Court ruled that the government was obliged to act—and rapidly."[772]

"Was it true that the environmental organizations were well funded?"

"Not in the beginning. From what I gather they were pretty much hand-to-mouth, especially compared to the millions the oil companies were pouring into climate denial and marketing. But as the protests grew they did get some big donations. It was Michael Bloomberg, who was Mayor of New York at the time, who set the trend when he gave $50 million to the Sierra Club to help them close down America's coal-fired power plants.[773] Coal, oil and gas: we were the three ugly sisters, along with our evil cousin the nuclear industry. The gas industry tried to claim they were different, they were clean and natural, but it was so far from the truth. When you include the fugitive methane that escapes during fracking and distribution, natural gas is sometimes worse than coal because of all the heat the methane traps in the atmosphere, and that's on top of the water contamination the fracking fluids cause, and the small earthquakes they sometimes trigger."[774]

"What was your company's line on renewable energy, and the solutions to climate change?"

"We had all bought into the belief that it wasn't our fault that people wanted so much energy to support all the stuff they owned, and that renewables could never provide the energy the world needed. What was solar in those days, one percent? Something like that. Things have changed so much since then. These days, if you don't have solar on your roof, the neighbors think you're some kind of dinosaur."

"So what made the difference?"

"I think it was the accumulating evidence that renewables could do it. There was a website where you could click on policy changes for any country and watch the emissions fall. That had a big impact. I had bought into the belief that you'd have to cover half of America with solar panels and install wind turbines in every neighborhood, and I didn't realize how rapidly renewable energy was progressing, or how fast the prices were falling. We scoffed at the idea that renewable energy was economic, even when we were investing four dollars in capital expenditure for every dollar our projects were earning, and when solar investors were earning four percent a year and our investors were losing four percent. We were strictly short-termers, stuck on the quarterly results and the annual budget. No historical perspective. And no capital discipline, either. We were like gamblers, depending on junk debt to finance the projects we believed would save our bacon by earning us billions. When the first electric vehicles came on the market that had decent range and cost the same as a regular vehicle, while costing far less to run—that sealed it. The market saw where things were heading and oil got stuck at $20 a barrel, condemning every project under development to closure. Globally, I believe something like a trillion dollars ended up being written off, to the shock of investors everywhere. But you know what? It was just as well, for things seem to be working out far better with renewables than I'd ever thought possible. Maybe these things have a way of working out for the best in the end. But we were blindsided. Very few of us saw it coming."

"Is that what they mean when they say it's hard to understand something when your salary depends on not understanding it?"

"Something like that. Whenever the price of oil rose we would laugh, thinking about our bonuses, and when it fell we would shrug, trusting that it would soon flip back. We never grasped the fact that people *wanted* change; they *wanted* to drive their electric cars and e-bikes and to crowd into buses instead of burning more oil. None of us made a serious attempt to ask whether the world could operate without fossil fuels. Not Shell, not Exxon, not BP, not Statoil. And look at us now! Look at all the cycling, car-sharing and solar arrays, entire communities heated by stored solar energy in the cold Alberta winter.[775] I suppose it's all normal to you. When I was born the environmental movement hardly existed and NASA was just putting the first solar panels on its satellites. We've come a long way since then. All those years, drilling for the remains of ancient fossil fuels. It's scary when you think how shortsighted we were. It still scares me, now that I'm out of the bubble and I can read the science objectively. The rising sea-level, the melting Arctic, the ocean acidification, the droughts, the superstorms. Shit, what a mess. It could be centuries before things return to normal, if they ever do."

We worked for a while in silence, turning the corner at the end of the field and heading down the next row towards the farmhouse and the lake, with the forest to our left. What a stunningly beautiful afternoon, with the velvety blue sky, the smell of strawberries and the fresh black earth filling my senses.

It suddenly struck me that if Ryan had worked in the US maybe I could find out what was happening there, hopefully without coming across as a total ignoramus.

"You must have spent a lot of time in the States," I said nervously. "When all the changes started to happen, how did Americans react to them?"

"Well, I've not been down for a while, but I spent a lot of time in Texas. Great people. Most Texans were so far inside the bubble, they couldn't see out. The protests and goings on—they were far away, and there had always been protests against something or other. And Texas was producing scads of wind power, so it was doing its bit for a greener world."

"Is Texas still producing oil?"

"A bit, but it's mostly history now. Houston is remaking itself as a Green Gulf Resort, growing organic cotton and restoring its seafood industry. It's got some of North America's best bikeways, and the place is full of farmers' markets."[776]

"How has this played out politically?" I asked, hoping to get to the heart of the matter.

"Oh, I get it. You want to know if America is going to become a gun-toting mobocracy. Is that it?"

"Well no… and yes, actually. Canada's such a different country. It doesn't carry as many burdens as America does."

"You've certainly got that right. Canada doesn't have America's burden of class and race antagonism, and the fear and distrust that fuels so much public anger. Slavery left a huge mark on American society, pitting whites against blacks and unleashing all sorts of fears, especially among the white working class. They often felt they were being ignored, or getting the rough end of the changes. It's almost as if America has never been able to address its two major challenges at the same time: the need to end racism against the black and colored community and the need to end poverty among *all* working people, whatever their color. It's like quantum theory. If you took a firm position on race you lost your momentum on poverty; if you took a firm position on poverty you lost your momentum on race."[777]

"Is that still true today?"

"Well, I've got many friends in America and from what I gather the movement to build local cooperative economies is beginning to make a difference. It's enabling people to come together, and it's succeeding in creating jobs. It's got a lot to do with public banking—and local banking. There's quite a big movement with smaller communities cracking down on the chain stores and working overtime to help local businesses. If they can solve the economic tensions, the racial and class tensions will hopefully be reduced. So I'm hopeful that America's going to pull through and discover a more sustainable version of the American dream, without the terrible divide between rich and poor, and the two different versions of justice, with violence, humiliation and jail for the blacks and Hispanics and a gentle wrist-slap and a fine for the corporate executives who steal billions and get away with it."[778]

"Were you in America during the second financial crash?"

"Yes, we had a home in Houston for when I was there on business. It gave me a front-row seat for the revolution. What an incredible time that was. I would never have predicted it, not in a thousand years."

Revolution? Front row seat?

"What was it like?"

"It was amazing. I'd been coming and going for years and I thought I had a good grasp of what was happening. I knew America had a flawed constitution that granted corporations far too much power over the democratic process, that big money ran the show and the electoral system was badly corrupted. Here in Canada we take it for granted that our electoral governance is non-partisan, but that was not the case in America. I also saw how many Americans were disenchanted with their political representatives, and how many people were disturbingly ignorant, and easily influenced by the right-wing shock-jocks. They didn't understand how the world worked, and they indulged in far more extreme and magical thinking than I was used to—blaming floods and droughts on personal misbehavior, for instance, the way people did in the Middle Ages. I sometimes wonder whether the Enlightenment ever arrived in parts of America. So I was really surprised when things turned out the way they did. Really surprised."[779]

"What happened?" I asked, itching to know.

"When the crash happened none of us were very worried in the oil industry. We had seen it coming and we'd built up huge cash reserves in a complex network of offshore accounts to keep us safe from damage. We knew periodic corrections happened, so it was best to be prepared. But we never anticipated the size or scale of the correction, or how so many Americans would react. I don't think there was a single commentator or policy analyst who got that one right.

"Our home was in Jersey Village, in the northwest suburbs of Houston. Very pleasant place: lots of trees, backyard pools, quiet cul-de-sacs. I knew one or two of my neighbors, but nothing more. It was mainly a base for when I was away from home. Renée preferred Calgary, where her friends lived. Anyway, a lot of people in Houston were having a tough time following the crash. Unemployment, bankruptcy, dispossession, suicide—they were all on the rise. The food banks were struggling to feed the line-ups.

"The banks were taking it hard too: they had armed guards to try to prevent the fire-bombings and drive-by shootings that were happening elsewhere in America. People were that pissed.

"But anyway, several months into the crash I came home one evening and I felt a different energy on the street, with more coming and going than usual. I knocked on a neighbor's door to ask what was happening and he told me there was an eviction scheduled at a home down the street the next morning and some of the neighbors were organizing to try to stop it. I went down to the house and introduced myself. Very pleasant couple, two children, family dog, well-kept yard, but they had both lost their jobs and they had huge medical bills because the wife had cancer and their insurance only covered a portion of the cost.

"Several of their neighbors were really determined. Three were also in arrears, and they knew they could be next, and while I was there they decided to go door-to-door telling everyone what was happening, asking for their support. They had no more plans in their head than I did: they were just acting instinctively, doing the best they could.

"By the following morning they had erected a barrier around the house with garden stakes, lawn furniture, whatever they could gather, and they were set on manning it twenty-four hours a day. I called into the office to say I'd be working from home and I joined the guard, hoping there'd not be any photographers. When the eviction crew arrived they tried several times to push their way through, but they were easily outnumbered, so they took off to try their luck elsewhere."

I continued to weed the strawberries while Ryan was simply holding his hoe, deep in his story.

"It was a Friday, and over the weekend there was a non-stop discussion about what to do. Two of the neighbors went door-to-door again and they found ten more families living in fear of eviction with debts they couldn't pay, some with scarcely enough food to get by on. Someone cooked up a big pot of soup, and someone else took the kids out on a neighborhood bike ride to let the adults talk things through.

"But here's the fascinating thing: these were not people you would ever have thought would organize together. They were Republicans and Democrats; blacks, whites and Hispanics; Occupy sympathizers and Tea Party supporters. Some were committed Christians; one was a committed atheist. But they shared two things in common: they were neighbors on the same street and they knew that whatever the politicians were doing in Washington, it wasn't working. Some of these people, they would probably have been screaming at each other if they had discussed politics or abortion, but that was all secondary now. Even though most had not known each other a week ago, they were neighbors, and they faced losing their homes. As one woman put it, 'We've had it up to here with Washington. We pay our taxes, but they do nothing for us. It's time we did something for ourselves.'

"They all knew America was still recovering from the last financial meltdown eight years before, and whatever their politics, they knew something was badly wrong when it was the banks and finance houses that got the bailouts, not the ordinary working people who were being evicted. And they knew that unless something happened to make it different they would probably lose their homes. For some who were renting, the landlords who were pressing for eviction were actually Wall Street hedge funds, if you can believe it.[780]

"When the eviction crew showed up at six o'clock on the Monday morning they weren't expecting to find an all-night guard sounding air-horns to alert the neighbors. Long story short, my neighbors kept it up for months, and during that time they changed their views about each other, changed their thoughts about politics, and changed their beliefs about what ordinary people could do. They spread the watch to the whole street to prevent other evictions they knew were coming; they taught each other how to grow food; they organized teach-ins to try to fathom out what was really happening; they shared tools and equipment; they helped each other file wrongful foreclosure lawsuits against the banks; they looked after each other's children; and they gave each other advice on how to cope with their debts and reduce their energy bills. And after the word got out and the local paper did a story on them, they inspired people on other streets to do the same.

"By December, with Christmas approaching, they were still holding their ground. The city had a real crisis on its hands, same as cities all across America. Neighbors all across the country were banding together to protect their homes, saying no to the system that gave tax-breaks and bailouts to the rich and evictions and jail sentences to the poor. Even in Houston, one of America's more prosperous cities, five per cent of households were facing foreclosure and eviction. That was more than twenty thousand families, and people everywhere were asking the same questions. Where were they supposed to live? In their cars? On the streets? Without a home or a regular address, how were they to find work? Without jobs, how were they supposed to feed themselves? How were their children supposed to go to school?

"By this time I'd persuaded Renée to join me, I was having so much fun. Not 'laugh and make merry' fun, but fun in that I looked forward to going home every day, engaging with my neighbors and joining some kind of activity, whether it was a potluck or a protest. Many of the parents were stressed out about Christmas since they didn't have the money to buy their kids the normal gifts. When they talked about it, someone suggested they organize a late night toy swap in someone's garage. They asked their children to part with the toys they didn't play with any more and they traded them, so everyone had something they could wrap.

"And then the snow came—a big winter blizzard that covered the city with six inches just before Christmas. An Arctic oddball, they called it, something to do with the jet stream being pushed off-course by the warming Arctic. By now the kids all knew each other and they were building barricades for snowball fights, having a wonderful time. We made a special effort to decorate every house, and at night the street looked wonderful. I remember being out with Renée late one night sharing a hot apple cider and a tot of rum, and one of the women, Paulina, who was still under threat of eviction, said it was the best Christmas she'd ever known. 'This is what the American dream is supposed to be,' she said to us. 'What went wrong, that we forgot how to be kind and help each other? This is how it was meant to be, how it should always be.'"

"That's amazing," I said, pausing in my strawberry-weeding. "How did it end? Did they get to keep their homes?"

"When the city got a report on the cost to local charities and social services if all the evictions went ahead, and the damage to marriages, to children's lives and to the community as a whole, they imposed a hefty weekly tax surcharge on any house that was unoccupied for more than a month. That meant the banks or whoever was trying to repossess the properties would have to lease them out almost immediately, or lose more money than they were owed in unpaid mortgages. But lease to whom? Only to the same people who had just been evicted. Then the Galveston Bay Foundation, a local Land Trust, put together a deal to buy most of the threatened houses off the banks for a fraction of their value and lease them back to their original owners.

"I can tell you, when that deal was signed, and the homeowners got their letters in the mail offering them a lifetime lease, with up to five years of lease payment deferral in cases of hardship, you could hear the cheers go up right across the city. There was such a party that weekend, with music and conga dancing through the night. If this was revolution, I thought, it was pretty good!"

"I wish I could have been there," I said. "What about America as a whole—has there been progress with the larger problems you talked about?"

"Yes—tons of progress. The new Congress rolled in some huge changes, justifying the term 'The Second American Revolution.' The most important change was the constitutional amendment redefining a 'person' as a living human being, ending the corporate 'personhood' that gave corporations unlimited political influence through their claim that they were 'persons' and that money was a form of free speech. It also ended the ability of the Supreme Court to overthrow enacted legislation, and it empowered federal, state and local governments to regulate and control campaign contributions. Getting the money out of politics has been the most important change, handing power back to the people's elected representatives."[781]

"Did the Democrats and Republicans actually agree?"

"It was the women in Congress who made it happen. They worked across party lines. It was they who achieved the unbelievable, the two-thirds majority needed for the 28th constitutional amendment. They were also very prominent in the campaigns to get the amendment passed by three-fourths of the states."[782]

"That's a big very change for America."

"Yes. It has really contributed to changing the overall matrix of power."

"What about the inequality and the deep-rooted hostility to the Hispanics, the blacks and the poor? Is that getting any better?"

"Maybe, slowly. The women in the new Congress really had balls, if you'll pardon the expression. As well as supporting the global treaty to close the tax havens, the global cap on carbon and the financial transactions tax, they closed all the domestic tax loopholes, imposed an absolute annual income ceiling of $20 million and taxed everyone in the super-rich bracket at 80%.[783] And there have been a slew of changes to weaken monopolies, regulate the banks, permit public banking and strengthen local cooperative economies.

"Now that I'm retired I can look at these things more objectively," he continued. "Do you know the fairy tale about the princess who was held captive in a tower for a hundred years, who had to be kissed by a prince to wake her up? That's how I feel about America. For some reason, there's a weakness in American culture that allows an evil monster to gain control every so often, locking the princess away. The monster is the elites who crave the power that money brings; the princess is the American dream in which people live in peace and liberty, and who work together to protect the dream from being stolen and locked up in the tower. That's what happened in the period leading up to the Crash of 1929, and again leading up to the recent financial crash. People have this way of forgetting, so the bad things happen all over again. The elites seize power and they use it to lock up the princess. It is only when the princess is released that Americans get back to achieving amazing things and being a power for good in the world. When the princess is locked up, Smaug rules, and America's influence becomes dark and devious, supporting all kinds of wars and nasty dictatorships, while oppressing its own people."[784]

"So now the princess has been freed?"

"Yes! It'll take her a while to find her feet, but yes, she is free. Many of

America's actual prisoners are being freed, too, under wide-ranging amnesties. It was awful how many people America held in its prisons, often under intolerably cruel conditions. Far more than in Russia and China, proportionally. There were more than two and a half million people in jail before the changes.[785]

"Will the princess remain free?" Ryan continued. "I don't know. It depends how willing she is to defend herself, and to resist the temptations of greed. She needs to build a stable economy in which everyone gets to benefit, including nature, rather than the jewel-encrusted version in which Smaug once again climbs to the top and locks the princess away."

"You said you thought there was a fatal weakness in American culture," I asked. "What do you think it is?"

"That's a really big question. My personal theory is that it's got something to do with America's historical origins, and the deeply embedded sense many Americans had that they had been chosen by God to lead the free world, and that they deserved to have it all. I reckon it came from conquering and taking possession of a new land. All they had to do was kill the troublesome Indians and push the survivors onto reservations and they could help themselves to everything they wanted. I think the very act of conquest gave individual glory a special place in the American soul. Maybe it does that to conquerors in any culture: it corrupts the soul and distorts our humanity, making us a worse version of who we would have been if we had not become conquerors.

"We all have a little bit of greed within us, but in America it became a virtue, instead of something to be ashamed of. It's the only reason I can think of why poor Americans so often voted against their own interests, supporting cuts to education and tax breaks for the rich. It's because they believed that one day, they too could be rich."

I pondered Ryan's thoughts, and his image of Smaug and the princess in the tower.

"Do America's First Nations—the Indians—have a role to play in this?"

"Now that's another interesting question. From listening to my son-in-law and his friends talking, I've come to see them as the memory-keepers. This goes for Canada as well as the States. They inhabited North America for fifteen thousand years before the white man arrived, so they knew the land far better than we will ever know it. Maybe longer than that, judging by the human remains they keep finding in South America.[786] But as long as Americans live in denial about the killing, and the plunder and theft of land from the Indians, they will continue to dream of the wide open prairie and unlimited greed, supported by Hollywood stereotypes. If people really acknowledged that satisfying this greed came at the price of cultural genocide and the deaths of millions of people, maybe they would understand that this kind of greed can never be satisfied without paying a horrible price. Maybe it will help them heal their souls and put their greed back in the box, where it can be managed and controlled."[787]

Wow. There was enough here to keep me thinking for months.

"Well, this has been a great trip down memory lane!" Ryan said, standing up and looking around the field. "I'm glad you kept working, because it doesn't look like I've done much!"

"What's next for you?" I asked. "Are you going to continue coaching, or does your being here mean you're thinking of something else?"

"I don't know. My grandkids keep telling me what a wonderful time they have on their school trips to the farm, so I thought I'd check it out. Maybe I'll do one of those storytelling courses they teach at the village. I love the way kids get so enraptured by live storytelling. 'This is so much better than TV,' little Dominic said to me a while back when I was reading him a story. He's one of my grandsons. Maybe I'll talk to Carlo, see if I can do some of his theatre stuff… see what might be hiding inside me. Looks like it might be fun. Look—you missed a dandelion."

I turned around to catch the weed, and was struck once again by the beauty of the afternoon and the peaceful farmland around me.

"That sounds like a great idea," I said. "You have a very easy presence, and kind eyes."

"Are you saying I'm a big soft suck?"

"Well, your words, not mine. But big and soft? Yes, I reckon you'd qualify."

"You know," he replied, "I have often noticed that people become a lot more human once they've been through a personal crisis of some kind, and come out the other side. It teaches you a bit of compassion, gives you empathy for what other people have to go through. Bad times are sure hell when it happens, but they can be good for you in the long run. I never knew what a jerk I was being, chasing all those deals so aggressively. My kids were always trying to tell me, but who were they to tell me how to live? And who am I to be telling anyone else? It's up to your generation now to steer us to a better future."

We stood at the end of our freshly weeded rows, looking east over the wide-open fields and woodlands. My hands were filthy and my back was aching, but my heart was singing.

"Isn't this wonderful?" Ryan said, as if he had tuned into my feelings. "I'm so grateful to be alive in a world that is so beautiful. You've been good company. What say we go and find a glass of that delicious elderflower wine? I could do with a drink. Maybe La Bella Bianca will serve us personally. We might even find her without that man of hers!"

I wanted to remain in that moment forever, feeling Ryan's kindness and his hopes for the future in the warm afternoon sun. It was surely true that anything was possible. Just think how far we had evolved, from cave people who would gaze up at the Moon to space people who could walk on it. And here we were, two people who were total strangers just over an hour ago, now good friends.

"You're crying," he said to me. "You can't be. Here, have a hug. I'm good at them."

I have seldom been hugged by an older man with such warmth and stillness. This wasn't a quick obligatory squeeze. It was a real, heart-to-heart soul-hug.

And so we stood, me aged twenty-five and he aged seventy, embracing under the afternoon sun in the middle of paradise, the blue heavens above us and the freshly weeded soil below. I'm sure Pelly could have told me what was happening in my brain, but as I experienced it my rational senses were overwhelmed by my feelings, and tears of gratitude rolled down my cheeks.

28
The Secret Garden

AT TEA THAT afternoon I felt particularly mellow. Maybe it was the work in the sun, or my conversation with Ryan. I saw a man with two young kids so I went over to see what they were up to. Just my luck! He turned out to be a school-teacher. I was conscious that my time was running out, and I had been wondering what schools were like in this future.

His name was Serge Mtoko. He had come to Canada with his parents as refugees from the Democratic Republic of Congo, in Africa, when he was just seven years old, and he was here with his sons, Tempe and Tusayan, who went to the same school where he taught in a lower-income area of East Vancouver. All three had rich, dark chocolate skin and bright shining eyes.

"I love the chickens," the younger son Tusayan said when I asked what he liked most about his school. He looked to be around seven years old. "I feed them every morning and I collect their eggs. My favorite is Bessie, the buff one. She lets me hold her."

"What about you, Tempe?" Serge asked. The older son seemed to be around eleven.

"I like the physics. We're learning about electricity and I'm making a solar car—a small one, but one day I'll make a real one, a big one. I've designed a way to give it a solar shower each time it goes through the dirt and gets all messed up. It's really foul. Did you know that the Sun's light takes only eight minutes to reach us? My teacher says gravity is the Earth's way of being in love with the Sun and that's why the Earth circles the Sun. But that's just her thoughts. She says we should find our own ways to think about it."

"That sounds like good advice," I said.

"Yeah, and when you put the Sun's light through a piece of glass it breaks up into all the colors of the rainbow, and like, they scatter all over the room. It's pretty cool. I never knew colors could hide inside of glass."

"They don't hide inside the glass, Tempe. They're in the light itself," Serge said. "The glass breaks up the light and lets us see the full spectrum of color within the light."

"Oh," said Tempe.

"Why don't you two go and play on the lake? Maybe the sayaks are free. We'll come and watch while I talk to Patrick here."

"We can take them out? That's totally foul. C'mon, Tusa. Let's go!"

"The sayaks?" I asked. "The boats they're making in the village?"

"Yes. They've got a pair on the lake. You should try one!"

We walked over to the lake and sat down on a bench while the kids put life jackets on and got into the sayaks, trying not to fall in.

"What made you decide to become a teacher?" I asked.

"I had planned on being a doctor, but then something changed, and I realized that what I wanted more than anything was to be a teacher. I was studying medicine, and had a chance to return to the Congo, to the village I came from. They've had a really hard time with the civil war, and the continuing conflict. That's why we left. There were a lot of people who remembered my parents, and aunts and uncles keen to make a fuss of me and find me a wife. They showed me the house where I'd been born, the place where the rebels attacked, the place where my brother was buried. Some of them had had their hands hacked off. There's a lot of really hard history there.

"But don't get me wrong—it's a lovely village. When I told them I was training to be a doctor I had a line-up of people wanting attention, but I didn't have any supplies. But when I talked with them I discovered that what they wanted more than medicine was education. Congo has so much promise, but the government was so corrupt and they didn't see much future apart from the coltan mines and the army. That's when I realized I had it all wrong. I didn't want to be a doctor—I wanted to be a teacher."

"How did you come to be here in Vancouver?"

"When I graduated from Montreal I got a job here as a substitute teacher and I fell in love with the city. Then I met the beautiful woman who would become my wife, Adeola. She's from Nigeria."

We chatted for a while about my life in Sudan, and my reasons for visiting Vancouver, and then I asked Serge about his school and what it was like to be a teacher.

"It's nothing like the teaching I trained for," he answered. "It's so much better. Our school is nothing special, just a square building surrounded by housing, but when you get to know it, it's like another world."

"What do you mean?"

"We're like a big family—or a garden, where the children are the plants and we are the gardeners. They're each unique, and they each have their own way of growing. We have children from so many different ethnic backgrounds and kinds of family, including single parent families, gay couples and foster parents. Our job is to provide the best conditions and to help them discover who they are. Do you remember being a child? Do you remember how you maybe had a secret garden, a place filled with magic where your imagination could flourish?"

"Yes," I replied. "I would go up on the roof in Khartoum. I remember being there at sunset watching the Moon rise over the city, thinking how magical it was."

"We want our whole school to feel like a secret garden," Serge said. "We want the children to feel safe to explore their world. We don't do any standardized tests, we minimize homework, and we have nooks and crannies where they can feel at home and work on their various projects, and where those who are autistic can hide away for a while when they need it. When I was at high school in Montreal it had *less* magic than the outside world; it actually felt flat and two-dimensional. If we can achieve our secret garden vision, maybe they'll realize that the whole Universe is wondrous. It's all in how you see things, if you are able to hang onto your inner child and not bury it away."

"What do you mean by a school filled with magic?"

"Our philosophy is that when someone is fully alive, and not carrying a burden of dysfunction, defeat, or a deficit of love, not carrying negative attitudes that warp the brain, and not carrying a burden of poverty or prejudice—or the sense of self-importance and entitlement that comes from having too much wealth—that when someone is fully alive it is normal to experience the Universe as wondrous and magical. When I ask myself where humanity is going, and what our true potential is, the feeling I get is one of secrets waiting to be discovered. That's why the children need their secret gardens, to keep the feeling alive when the adult world tells them to sit down and stop asking endless questions."

"This must seem very far from your village in the Congo."

"Not really. There were times when it was terrible, when my father had to hide us, but there were lots of times when it was great. It didn't matter that we were poor, because we were all poor. I was full of wonder, and I wanted to explore the world."

While we were talking, Tempe and Tusayan were racing around the lake, steering their sayaks around the solar array.

"Is this approach being used in all the Vancouver schools?" I asked.

"All the way from Kindergarten to Grade 12."

"So what caused the change?"

"At the time of the OMEGA Days there were many protests and demonstrations by school-children as well as by everyone else, which focused people's attention on the kind of schooling they were getting. I was at high school in Montreal at the time and we had all sorts of things going on, generally making a stir. So after the OMEGA Days there was a lot of rethinking. Here in British Columbia the Ministry of Education and the teachers' union stopped arguing and started cooperating over how to build a better system of education. They decided to gather the best of what was happening around the world, and their first stop was Finland, since many of us had read *Finnish Lessons* by Pasi Sahlberg, and we had heard him speak.[788] The Finns changed their education system a long time ago, organizing it around the child's natural instinct to learn, abandoning tests and trusting their teachers to teach. Their students do extremely well in international tests that compare 15-year-olds for literacy in reading, math and science. Canada does well too, but Finland is regularly among the top countries. And here's what's so amazing—they do no formal classroom work until the age of seven, they have fewer hours of teaching, they do less homework, they do no standardized testing until the teen years, and they have no drop-outs."[789]

"That's pretty incredible."

"It's all about cultivating the imagination, and the natural desire to learn. Another place they learned from was the Italian region of Emilia-Romagna, between Venice and Florence. Coming out of the war, when Mussolini smashed the hell out of them because they were anti-fascist, they asked a crucial question: what kind of education would ensure that their children grew up to become strong, healthy individuals who thought for themselves, and never became fascists, obsessed with power and hating joy?

"It was in answering that question that they developed the Reggio Emilia approach to early childhood education. It places a big emphasis on respect, responsibility and self-directed learning. One of their founders was a teacher called Loris Malaguzzi who believed in children discovering their own ways of self-expression, and not having someone else's view of reality imposed on them. He used to say that a child has a hundred languages, but traditional schooling steals ninety-nine of them away. It shows them a world that is flat and dull, not one full of wonder.[790]

"The model we have adopted engages the teachers and parents in a constant process of learning, encouraging us to become more effective. We have a professional culture of peer review and collaboration, including 'observe and help' sessions when we go into each other's classrooms and give constructive feedback."[791]

"What's your school like, physically?"

"Well, the first thing you'll see on entering the school is a poster of the Afghani woman, Malala Yousafzai, who was attacked and nearly killed for daring to stand up to the Taliban and their efforts to stop girls receiving an education, who is now her country's ambassador to the United Nations.[792] It's one of the *Change the World* series of posters that you might have seen around Vancouver."

"Yes, I have—I find them very inspiring."

"It carries these words of hers, which we hope will inspire our students:

> *One child, one teacher,*
> *one book, and one pen,*
> *can change the world.*

"Our school is very homelike, with natural light and warm colors. We're close to the Downtown Eastside, where there's not much green space, so we shrunk the sports field and created a garden where we grow food and keep chickens. We've built a solar greenhouse that doubles as a classroom, and a nature trail with trees, ponds, climbing frames and a wild play area. Kids need an element of danger, so it's important to provide it."

"How did the parents respond?"

"Some were hesitant. The insurance rules were getting so ridiculous that we had even been told to take down the climbing ropes in the gym in case a kid got injured. Luckily, the Supreme Court ruled that children need to be exposed to a degree of risk, which made it possible to restore them."[793]

"What about organized sports?" I felt as if I was an interviewer, asking all these questions, but it was proving effective.

"We had a big debate about that. When we asked the children, most said they preferred cooperative games, with no winners and losers. So we do both. Personally, I think it's important to promote competitive sports—giving young people the experience of team effort, speed, agility and success, so that they have that memory throughout their lives, a reminder to keep working to stay fit as they grow older. When you look at evolution, the evidence shows us that competition and cooperation are equally important, so we need to nurture both. Personal fitness is also really important. Every student gets a workout every other day, and we have a five-kilometre running route through the neighborhood that we all have to do once a week, staff and students alike. Here, let me show you some photos."[794]

Serge picked up his screen and showed me pictures of the school grounds before the changes: a large empty sports field, a large expanse of concrete, some play equipment and a few trees dotted around the edge. And then after: the square edges had disappeared and the grounds felt more like a garden, with a greenhouse, wandering paths, vegetable gardens, fruit trees, and a small amphitheater for outdoor classes and concerts. There was a cob climbing wall, an organic climbing frame and a wooden palisaded castle.[795]

"That's amazing."

"It took a lot of fundraising and weekend work bees, and we had help from the City, the Neighborhood Association and the local Watershed Society. The creek has now been restored all the way from Robson Park to the ocean, with instream geothermal cooling. We were involved with the salmon spawning project, and now the fish return every year, after an absence of a hundred and fifty years. Here's the street before the changes, and after."[796]

The first photo showed a humdrum city street with worn-out paving and rough parking on either side. In the second photo it had been transformed with a pond and a creek that flowed down the street, and a single lane for cars with trees, shrubs and picnic tables.[797]

"We have several outdoor learning spaces," Serge continued. "For some reason the kids learn much better when they're outside. Maybe it goes back to our past, when we lived outdoors all the time. That's certainly how I grew up. We also encourage the older kids to help the younger ones. They enjoy doing it, and you get better retention when you learn something and then use it to teach someone else."[798]

"How did all these changes come about?"

"From what I've been told, after the research phase there was a big gathering at a retreat center up on the coast where teachers, school trustees and administrators from across the province met with staff from the Ministry of Education for a week. They did some really deep listening, which someone said was the first time in living memory, and they came up with the *Hollyhock Manifesto*, which became the basis for the new schooling. It included all teachers needing a Masters Degree, as they do in Finland; allowing us to teach without tests and assessments; placing a limit on class size; encouraging us to collaborate to improve our skills; encouraging children to learn in their own way guided by their curiosity; allowing schools the freedom to experiment; encouraging more arts and outdoor education; and increasing the pay of the teachers' assistants and giving them help from trained volunteers. They also called for reducing the voting age to sixteen, and measures to tackle child poverty and hunger."

"Is hunger still a problem among your students?"

"Not really any more. I used to be able to tell if a child had come to school without breakfast because they would be listless and unable to concentrate. Someone did a survey before the OMEGA Days and fifty-four percent of the students had either had no breakfast at all, or perhaps just a donut or a bag of chips. Even in my village in the Congo we always had food for breakfast. So the teachers launched an appeal and started serving a regular school breakfast."[799]

"Do you still serve it?"

"Yes, but far fewer kids need it."

"So how do all these ideals translate into the kind of teaching you do?"

"We don't think of ourselves as teachers any more. We see ourselves as learning coaches. We place a big emphasis on the arts, because they seem to reach parts of a child that other kinds of education can't. When you take the arts out of education, you take the soul out of learning. We need our artists to speak passionately to the souls of our children, parents and executives alike, saying, 'Yes, there is more.' We also have a great school choir, and an orchestra, which I conduct myself."[800]

"You didn't tell me you were musical."

"We're all musical in the Congo," he replied with a laugh. "We have a saying that you can take Africa out of the music, but you can never take the music out of Africa. Do you know about Vancouver's SongFest? Our choir has been participating for several years, along with thousands of singers from all over the province. It was started by a couple of people from Estonia who were inspired by their country's Singing Revolution. It's wonderful how it inspires the children.[801] Some of us have also adopted the Ron Clark style of teaching, using singing and rap to learn the basics of math and some science."

"How does that work?" It was not something I'd heard of.

"He's a teacher from the US who developed a very successful way of teaching in which the students sing and dance, repeating formulae and other important things the way they do with advertising jingles. It's really effective, and the students love it, but I only do it when a coaching assistant is able to take the autistic kids away to a quiet spot, since it overloads their sensory systems."[802]

"It sounds like a lot of fun."

"It is! We also use on-line courses from the Khan Academy, Coursera and the like. The quality is excellent, and it allows them to progress at their own speed. We do team projects using real world math, like weighing the carrots they grow, calculating how many hectares you'd need to feed the whole of Vancouver, and then going out in groups to interview people, asking if they grow their own food.

"The time they spend here at the farm helps them understand how food grows, and where it comes from. They come here for a week at a time, staying in the dormitory, and as well as helping on the farm they do all their learning outdoors. For some it's the first time they've been out of the city. We use microscopes to look at the creatures that live in the soil and to see what lives in the lake. A lot of them have never handled a worm before, or a snake, but by the end of the week they are going around with pet worms in their hands telling me about all the birds they've seen.[803] It's one of our goals that every child should develop a good understanding of nature, including things like how the carbon cycle works, and what Earth's ecological carrying capacity is. It's that kind of ignorance that got us into such a mess in the first place. You can't get into university these days unless you have completed Earth 101 at high school and have a basic grounding in planetary and local ecology, climate change and ecosystems restoration."

"You can't get in at all? How did *that* come about?"

"After the OMEGA Days several of Canada's top universities got together and made it a requirement, because of the climate crisis. The schools had to scramble, but now they all do it."[804]

"So it works, mixing the computer-based learning with outdoor learning?"

"Yes, thanks to the coaching assistants and their helpers. We have a Code Club where they learn how to make computer games, animations and websites, and many of the kids use Minecraft to build 3-D palaces, ruins and spaceships on Xbox.[805] You can't get more self-directed than that.[806] The whole school is wired for cable, so we've no need for Wi-Fi or Li-Fi.[807] To balance the cyber-learning we do traditional courses in reading, handwriting and poetry recital. Last year one of my students came top in Vancouver's poetry-reading competition—I was really proud."

"How do the coaching assistants and the volunteers fit in? I just met one, an older man called Ryan. We were weeding the strawberries together."

"Oh, Ryan!" Serge replied. "The kids love him. He helps the ones who have ADHD, fetal alcohol syndrome or parental loss syndrome. It took the OMEGA Days to make the government realize how difficult things had become, and pay for more coaching assistants. They help the special needs students develop new neural pathways, bypassing those that have been harmed, perhaps by a family dysfunction or a toxic brain incident during pregnancy. We're always on the lookout for ways to overcome the negative messaging that results from neglect or abuse. It's so destructive to a child's self-esteem."

"Do many of your kids come from single parent families?"

"Quite a few. But they do seem to be fewer of late. Maybe it's the relationship support groups, or the love-drugs that couples are using to increase their bonding. It could also be the street changes, because when neighbors know each other there's a much stronger support network."

"What do you mean by love-drugs?"

"You don't know? The oxytocin nasal sprays. They give you the same feeling that you get when you cuddle, or are intimate. They work wonders in my marriage!"[808]

Well, that was different.

"Spiritual and emotional development is important for children as well as adults," Serge continued. "It's one of our goals that our children grow up able to enjoy happy, healthy relationships, and to participate in the global extension of empathy that's so critical for the planet as a whole."

"Does this mean you're bringing religion back into the classroom?"

"No, there'd be an uproar if we tried to do that. Religion's best left to the parents. We use two approaches that allow them to explore their inner worlds without religion. They are meditation, and the Virtues Project. We use a program called MindUp that was founded by one of your Hollywood movie stars, Goldie Hawn. At the start of each day we have five minutes of silence when the students listen to their breathing, and the sounds around them. I might set a coin spinning, or release a drop of water into a bowl. It teaches them to be aware of their senses, and helps them calm down. Then we discuss what happens in the brain if they get into a fight, or if they're feeling angry or distressed. Using MindUp prepared the ground for full meditation, and now we have a regular weekly class. There's good evidence that meditation lowers stress and improves attendance, so the School Board has allowed it in spite of complaints from some parents.[809]

"We also use the Virtues Project—it enables us to discuss virtues such as bravery, persistence and kindness. We encourage every child to keep a Gratitude Journal, so that they become familiar with the idea that they can control their inner space and choose whether to fill it with irritation and resentment or with gratitude. It helps them understand that they can change their brain chemistry, and choose to feel good, instead of getting caught up in anger and stress.[810]

"What about bullying?"

"It has been our experience that the more the students are able to talk about their feelings and develop empathy for each other, the less bullying there is. We are committed to being a bully-free zone, and also to the children not bullying when they are away from school, but it still happens. Some of it comes from children feeling neglected, and trying to compensate by picking on children they think are weaker. And sometimes it's a form of displaced anger against a dominant or an abusive parent, though the child usually has no idea where it comes from. That's why talking about feelings is so important, to make it normal to have conversations about anger, jealousy, feeling neglected and feeling love.[811]

"We also have a secret weapon that we learned from a math teacher in America," he continued. "Every so often, on a Friday at the end of the day, we ask our students to write down the names of the children they'd like to sit with the following week, and to nominate a student who has been an exceptional classroom citizen. Over the weekend we analyze the answers, and we look for who's being neglected by the other students, who has no-one they want to sit next to, and who had lots of friends last week but none this week. It gives us an insight into what's really going on, and it enables us to do subtle interventions."[812]

"That's really smart."

"Let me tell you a personal story. A few days ago Tusa got upset about something over breakfast and he went off to school in an angry mood. Normally, the bike ride and the fresh air would clear it up, but on this occasion it didn't. He was being aggressive to the other kids in his class and his teacher could see that something was up. So she asked everyone to make a paper hat, color it the way they were feeling and write a word on it that said what they had drawn. Tusa colored his hat in big red splashes and wrote ANGRY on it. His teacher told me he was actually quite proud of it. Then after a while he just took it off, and it was gone. Emotional intelligence starts with the recognition that you're feeling something, and naming it. Then you can talk about it, and it no longer rules you."[813]

"You spoke about widening their sense of empathy. How do you do that?"

"We celebrate our differences, and we talk about the ways people oppress each other because of racial, national or gender differences, and through bullying. We use an activity called Roots of Empathy, in which a volunteer parent brings a baby into the class on a regular basis and the students learn to read the baby's emotions. It's really different, and it teaches them a lot. We also use an empathy game called Feeleez—it's a set of cards that show pictures of children with different expressions. The children discuss the cards and share what they think the kids are feeling. It helps them get used to discussing their feelings, including hurt, fear and anxiety."[814]

"Do you do any wider social activism?"

"Yes, quite a lot. Here in Vancouver we participate in We Day, which was founded by Craig and Marc Keilburger, the Canadian activist brothers," Serge replied. "Every year the entire school attends the We Day gathering along with 50,000 other students, where they experience the passion of young people who are doing amazing things to make the world a better place, mixed with great music and dance.[815] It has inspired many of our social change projects. Our main one is Roots and Shoots, based on the work of Jane Goodall, the famous primatologist. We have adopted a troop of chimpanzees in my home country, the Democratic Republic of the Congo. There's a webcam in the forest that enables the students to follow the daily activities of the chimps; they even know their names. The chimps used to be so threatened. They were being hunted for bush meat, and the forest they depend on was being cut down by the farmers, loggers and mining companies. But now that the fighting has ended, and the Democratic Republic of the Congo has joined the E-70 group of nations, I'm a bit more hopeful. One of our top Congolese singers has been campaigning to end the trade in bush meat, and there's a new chimp sanctuary in the northeast. It's such an enormous country—it's two and a half times larger than British Columbia, three times larger than France."[816]

"How on earth do you find time in the school day to do all this?"

"We don't waste time on exams and tests, and when the children are motivated everything seems to happen more easily."

"What about the high schools—do they follow a similar approach?"

"Yes. Many are basing their education around the students' natural interests, and spending more time on the arts. It turns out that when you take time away from the sciences and shift it to art and music the students do just as well in math, science and software skills. It's all to do with motivation, awakening the desire to learn, and touching the spiritual core. The competition from countries like China and India is fierce, so we need young people who know how to work hard and retain their edge. If we don't, Canada could go into a slow decline, with a falling standard of living."

"You think so?"

"For sure. It's a tough world out there, and our students are going to need all the initiative they can muster. There's a great program many schools run called The Leader in Me, which teaches the habits of really successful people.[817] Most high schools also run a cooperative enterprise program, where the students form small businesses, raise the share capital, make or provide whatever they're offering and then go into voluntary liquidation at the end of the school year, sharing out the profits. The needs of our children and of Canada's future economy are really very similar: they both involve initiative, innovation, motivation and teamwork.[818]

"And did I mention our politics program?" he continued.

"You teach politics to primary school children?"

"Not the way you're thinking. We invite them to discuss and then make small decisions together, and we hold opinion polls so that they can see how different people have different ideas. And we encourage deep listening, so that they get used to the idea that when you really pay attention there's a deeper form of communication that can happen, which takes things to a different level. It's important for their personal lives, as well as for the future of democracy.[819] In Grade Six we hold elections for various positions of leadership, such as looking after the chickens, and we make an annual visit to the local high school where they observe a proper political debate. The idea is that when they get to high school they'll be ready to vote when they're sixteen, and they'll understand how the process works. If Canada is to keep a healthy democracy it's absolutely critical that young people become engaged in politics, and don't sit on the side feeling cynical or left out. Ryan was telling me how poorly informed many people in the States are about their history and geography. He thinks it contributed in part to the decline of their democracy before the Second American Revolution. He likes to talk about the connection between education and democracy—it's something he's very passionate about."

"Hey, Dad! I beat him! Hey hey, hey hey!" Tusa came running up from the lake, followed by Tempe.

"I wasn't trying—and who cares who's fastest? That's a stupid winning thing." Tempe threw the words back at his brother, tossing his hair to one side to show that being older, he knew better.

"Like I said," Serge said, "it's not always easy. They're always looking for an edge, something to push off. Sometimes it's not even personal. Who knows what challenges they may be processing, in preparation for the future? When I look at the world, there are days when I feel like shitting my pants, if you'll pardon the expression. The kids know it too."

"Is it that bad?" I asked.

"Don't *you* think so? With the knowledge that the Amazon rainforest is burning up, and could be gone by the end of the century? With last year's failure of the monsoon in India, and half a billion people facing starvation? With the oceans becoming ever more acidic? With the East Antarctic ice-sheet on what scientists say is an unstoppable meltdown? With all the forest fires and droughts? I keep asking myself what more we could be doing, but I don't have any answers. I've two boys to look after and a school full of kids I've dedicated my life to. But I worry. I really do."[820]

"Something will come up," I replied. "Maybe some of your students will go on to do something amazing that will make a global difference, and all because you cared, because you wanted them to be their best. People won't know it was you who gave them the foundation, but I'll know, and so will the forest, and the salmon. They'll know, and they'll thank you."

"Hey—I didn't know I was talking to a poet. Thank you. That makes me feel better."

"Come on, Dad!" Tempe said, dragging on his father's sleeve. "Let's go play some soccer. I'll get the ball."

"Looks like I'm needed," Serge said. "It's been good talking with you. Are you staying for the feast and the dance tonight?"

"Wouldn't miss them for anything!"

But before I did any feasting, I had an appointment for tea with Pelly.

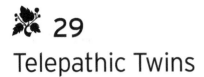 29

Telepathic Twins

IT WAS GETTING on for six o'clock, and I found Pelly and Sarah in the barn with the horses. A young man with long black hair and a dark complexion was showing them how to brush the horses, their necks caked with sweat from their afternoon's work on the farm. First a circular rubber brush to loosen the dirt, then a hard brush and finally a soft brush, while the horses chomped happily on their oats.

I leaned on the barn wall watching, and Pelly let out a pleasurable sigh when she finished her task. The twins' faces were effectively identical, and they both had reddish hair. Pelly was wearing blue jeans and a light green blouse, while Sarah wore brown dungarees over a blue denim shirt. They were both wearing bright yellow socks.

"Time for tea?" Pelly suggested.

We walked back to the farmhouse patio, which served as a café for visitors and volunteers. Sarah made a pot of tea while Pelly and I started talking.

"You were saying how you could make a virus carry a nanobot to fight the West Nile virus," I said, jumping in where we had left off earlier in the afternoon. "But what would happen if one of the nanobots got out of control? Might that not be very dangerous?"

"You sound like a science fiction writer from twenty years ago," she laughed. "That kind of nanobot would have to be self-reproducing, and there's a code of conduct that prohibits the development of self-replicating nanobots."

"But what about an unscrupulous company? Would you trust them not to develop a self-replicating nanobot if it could eliminate the need to use pesticides on genetically modified crops?"

"Now you sound like Sarah! She thinks I'm going to morph into the evil queen of the nanobiotic empire and infect the world with genetically modified Martians. But meanwhile, the West Nile virus is breeding away, infecting more people every year. It's serious. We've got to do something. It might get Sarah one year. She's my twin sister, so I've got to be there to protect her. Whatever she feels, I feel."

"What do you mean?"

"She means we're teletubbies," Sarah said, returning with a pot of tea made from herbs grown on the farm—rose hips, anise hyssop, yarrow and bergamot.

"Telepathic identical tubby twins. That's what we are. Teletubbies. If Pelly cuts herself really badly I feel it immediately, wherever she is."

"And if Sarah falls in a composting toilet," Pelly responded, "I'm the first to share the experience. We're telepathic, telemorphic, telesensational *and* teletubby."

"*Very* telesensational, if I fell into a composting toilet," Sarah joked back. "Tele-get-the-fuck-out-of-here!"

"Do you remember when you asked your friend Martha to pour a pack of ice-cubes down your back, to see how I'd react?"

"That was a scream. How about the time when I stuck a needle in my bum when you were sitting on the toilet to see how you'd respond?" They burst out laughing.

I was getting mightily confused. It was hard enough to keep up with who was speaking, without the telepathic double-talk.

"I've got two questions, if I may interrupt," I said, looking at Sarah.

"Sure, fire away," she replied, still laughing. "We had a lot of fun with it when we were kids."

"First, what do you mean when you say you're teletubby?" The twins were well padded, but not obese.

"It's two things," Pelly said. "When our mother was pregnant, she had difficulty carrying us, so the doctors told her not to exercise. That caused her to put on a lot of weight, which got passed onto us in the womb, and set us on the path to obesity.[821] And secondly, it's the chemicals she absorbed when she was pregnant. Obesogens, they were called. Bisphenol A was the worst, but there were lots around. She worked in a restaurant for the first part of her pregnancy, and every receipt was coated with it. She absorbed it, so we absorbed it, and because it's an obesogen it increases the fat cells in our bodies. So now if we eat too much or get the tiniest bit lazy we pack on the fat like pigs in a cake shop. Bisphenol A has now been banned, along with all the obesogens, but we have to live with the consequences. It's only because our parents learned what was happening and made a special effort to feed us healthy food and make us take all sorts of sports that we're not total tubs of lard."[822]

I was shocked. Obesity was a huge problem back in my time, but most people blamed it on too much junk food and not enough exercise. I had no idea there might be other factors at play.

"Yeah, it's tough to live with," Sarah said. "But there we go."

"What about the telepathy?" I asked. "I thought that was a myth that had never been proved."

"Not at all," Pelly replied. "It's a known fact that a third of all identical twins are reliably telepathic, demonstrating mental transference of sudden pain or threat and a variety of casual behaviors. It doesn't happen between regular twins, or the other two thirds; only between monozygotic identical twins who shared the same amnion, chorion and placenta, who spent the longest time conjoined in the womb before they separated."[823]

"Do you experience it on a regular basis?"

"Yeah. Twins like us have always known it. Until about fifteen years ago, however, when I was studying neuroscience, it had been pushed into the broom-closet because it didn't fit with the laws of science, or any known theorem. In science, we say exceptional claims require exceptional evidence, and that was used to dismiss the accumulating evidence that telepathy was real.

"The evidence has actually been strong for years. I'd even say super-strong, but none of it registered because it was based on studies that revealed telepathic contact through probabilities and statistics. When you look closely at the studies, the odds against the results that demonstrated telepathy having happened by chance are a billion or even a trillion-to-one, but the evidence was still ignored."[824]

"What happened to change things?"

"About ten years ago the Institute of Noetic Sciences organized a series of scans of telepathic identical twins. They took fifteen sets of twins and placed them in separate fMRI chambers, continents apart, and they monitored the flow of blood to different areas of the brain. Then they subjected one of each pair to a sudden extremely nauseous smell and observed what happened to the other twin. And guess what? The transmissions consistently showed up in a specific area of the brain. If we had functional Neurosynaptic Pattern Imaging we could get even closer, but it was enough to break the dam. It forced the scientific community to take a fresh look at the nature of thought and consciousness. The physicists who were working on the Theory of Everything were thrown for a loop. What value was a Theory of Everything if it ignored an entire dimension? When similar results were obtained using animals it threw the cat among the pigeons, so to speak. Do you know about Schrödinger's famous cat?"[825]

"No. Was it telepathic too?"

"No. Well, yes. Schrödinger was a famous Austrian physicist who did some of the fundamental work on quantum theory in the 1930s. One of the paradoxes of quantum theory is that a particle can be both a particle and a wave, in what's known as a superimposition of states. As soon as a human observes it, however, it immediately becomes one or the other, collapsing into a definite state. It used to be thought that the act of observation caused the change of reality from one state to another. That was considered pretty weird by many scientists, so much so that Richard Feynman, the famous American physicist, said that he could safely say that nobody understood quantum mechanics.

"Schrödinger wanted to show his fellow-scientists that they couldn't ignore the problem, so he devised a thought experiment in which you put an imaginary cat in a box, along with a glass flask of poisonous gas and a Geiger counter containing a tiny amount of a radioactive substance. If one of the radioactive atoms decays the Geiger-counter discharges, releasing a hammer that smashes the flask, releasing the gas and killing the cat. If no atom decays, nothing happens, and the cat lives. The atoms exist in a superimposition of states until they are observed, being simultaneously particles and waves, which means that the cat must be simultaneously alive and dead."

"Did Schrödinger consult the SPCA before he created his thought experiment?"

"Ha! In theory, you could put SPCA volunteers in the box and they would also be simultaneously dead and alive. But since it is now accepted that telepathy is real, this suggests that consciousness may be omnipresent, so the cat is conscious too, with the ability to do the observing that collapses the quantum state into the classical state we're used to. So there's no need for an external human observer."

I wasn't sure I understood this very well, but one aspect jumped out at me.

"Are animals telepathic too?"

"The evidence indicates yes," Pelly answered, "but it goes a lot further than that. The theory that's creating waves is that consciousness is a fifth dimension, and that it's as omnipresent as space-time. If this is so, then every particle is its own observer, and the philosophical difficulty associated with quantum theory, that nothing exists objectively until it is observed, vanishes. In reality, nothing is ever unobserved, since every particle or wave is its own observer and its own agent of change. Maybe space-time itself is an observer, but that's beyond my understanding."

"Mine too! I have a hard enough time understanding that space and time are part of a single entity called space-time. Are you saying that we should be thinking instead of a single entity called space-time-consciousness?"

"That's one of the theories being floated. If the new hypothesis about consciousness is correct, then observation, intention and agency are as omnipresent as gravity and space-time. In our department at the university we have a team of physicists who are using the Amazonian mixture of plants and roots known as *ayahuasca* to observe atomic activity while in a heightened state of consciousness. Their observations indicate an inseparability between particle, wave and space-time-consciousness, and a generic uncertainty that may pervade the entire Universe, since wherever there's consciousness there's uncertainty, since consciousness brings the potential for agency and choice wherever it goes."[826]

My brain felt really stretched, but if it had been demonstrated that sensations and thought could be transferred over distance in a replicable manner it would be the biggest breakthrough in science since Newton's discovery of the laws of gravity in the 17th century.

"How are scientists interpreting the telepathy findings?"

"As a neuroscientist, I interpret the telepathy that Sarah and I experience as raw consciousness field melding. What I find really exciting is the emerging evidence that telepathy also occurs between cells, even between atoms. For some reason it remains unconscious in most humans, which led us to believe that it didn't exist. So the question changes. Instead of asking why do some humans experience telepathy, we need to assume that *all* humans are potentially telepathic, and ask why so few people experience it consciously. One theory goes back to the development of language, when the evolution of the neo-cortex pushed the telepathic whisperings of the older brain down into the unconscious. Language was far more effective than telepathy, so our experience of telepathy atrophied and now only rarely emerges into consciousness.

"But why *does* it rise into consciousness for a few?" she continued. "That's the tantalizing question. One hypothesis is that it's because of our special bonding as identical twins. We're chummy-love, right down to our atoms. It can't be zygotic, or it would happen to all identical twins. So why do two-thirds of identical twins *not* experience it? We know that it's sometimes experienced where there's a very close emotional bond, such as between some mothers and their children, some owners and their dogs, and some breast-feeding mothers who know when their babies are hungry even when they're in a different building. So now there's research underway to see if there's a correlation between the experience of telepathy and the levels of oxytocin and vasopressin that promote bonding."

"How do you interpret this in terms of fundamental physics?"

"The latest thinking is that telepathy is linked to the dense coded entanglement of sub-atomic particles sharing information at ten thousand times the speed of light. God doesn't need to play dice with the Universe after all, since every atom knows what every other entangled atom is up to. One possibility is that when Sarah and I were in our earliest zygotic state, sharing one cell, our entanglement would have been complete, and when we divided into separate embryos our zygotic memories retained their original linkage. But that doesn't explain why it doesn't happen with other identical twins: the difference is probably connected to the time the twins were conjoined in the womb before they separated. What's particularly interesting is the way we become telepathic when there's a physical emergency that could be life-or-death, or when one of us experiences sudden unexpected pain."

"Or sudden compost immersion experience," Sarah chipped in.

Pelly rolled her eyes and continued.

"Nor does the zygotic linkage theory explain why we should be telepathic over bizarre and even silly things like the books we read and the color of the T-shirts we wear. Why did we both buy these bright yellow socks last Wednesday without any knowledge that the other was doing the same? Why did we both show up wearing them today? It's too weird. Some of my colleagues think there may be a continuing ovogenetic field of consciousness that triggers the same concept cells in each of us, causing us to make the same choices. Others think that our thoughts don't reside in specific concept-cells at all, but in a field of consciousness that's not limited to space-time, which expresses itself locally in our brains' concept-cells."

"What do you think?" I asked.

"I think that when one of us, say Sarah, makes the decision to buy yellow socks in a totally normal way, her brain creates a 'yellow socks' concept cell. As her telepathic twin I enjoy a random and seemingly serendipitous sharing of the contents of her consciousness: we share a waveband in the wider field of consciousness. So the decision by Sarah to buy yellow socks, accompanied by the pleasure she experiences at making the decision, would be experienced by me as an unconscious prompt, resulting in what would appear to me to be an independent decision to buy my own yellow socks."

"Does that fit with what you think, Sarah?"

"I'm a farmer, not a neuroscientist; I leave those things to Pelly. As far as I'm concerned, I experience buying yellow socks no differently from buying milk or condoms."

"Sarah, do you *have* to?" Pelly exclaimed. "So as I was saying, some scientists are trying to pin telepathy down using functional neurosynaptic pattern imaging to track molecular change in synaptically connected groups of neurons; but it's very elusive. The neuroscience community has been thrown into a strange kind of ecstatic frustration by the discovery that cellular, atomic and particle-level telepathy appears to be real. The trouble is, we're not getting consistent results in any specific groups of neurons. That points to the theory that it's happening in a non-material field, maybe similar to whatever connects entangled particles."

"You talk about the field of consciousness as if it were an accepted thing."

"Yes, it's true," Pelly replied. "There are various theories. Some think it's all-pervasive and omnipresent, going right back to the Big Bang—that it's a fifth dimension. The religious types have jumped all over that. There are books being published every month linking the fifth dimension to God and spirituality. Others think it's an emerging property that results from advanced material complexity. It's an ongoing debate. What's no longer in dispute is the reality of telepathic communication in certain specific circumstances. It has moved out of the sphere of the anecdotal and into the realm of observable science.

"A rainbow is also a specific circumstance," she continued. "It might seem like an anecdote if you've never seen one, yet it was by observing and analyzing rainbows that Aristotle, Ibn al-Haytham and Newton were able to explore the nature of light and color. This may be the same. In science, it's the exceptions that force us to take a closer look. I think we'll see some amazing breakthroughs over the next few years."

Ibn al-Haytham: the hotshot Islamic scientist, mathematician and physicist from Basra in Iraq who was known as 'the second Ptolemy' in the 10th century AD. I only knew about him because my grandfather had worked as a contractor with the British forces in Iraq, where he acquired an appreciation of Islamic history and science that he loved to impart to anyone willing to listen.

"One of the big questions is whether telepathy requires the presence of a field," Pelly continued, "and if so, whether it meshes with the gravitational or electromagnetic fields or whether it is an independent fifth dimensional field. That's what I love about science. I like the fifth dimensional theory myself. It makes more sense to assume that everything has five dimensions. The animal studies prove that telepathy precedes human existence in evolutionary terms. Some of my colleagues are studying telepathy in flatworms, while others are looking for it at the cellular level. If it's present there, who's to say it doesn't go all the way down to the particulate level, and beyond? And if it's there, it's reasonable to conclude that it was present at the moment of creation."

"Are you saying consciousness was present at the time of the Big Bang?"

"Maybe even before, if such a thing is possible. Either raw consciousness was created at the same time that energy, matter and space-time were created, or it pre-exists it. That would fit with what quantum theory says about entanglement. So instead of consciousness residing only in the human observer, which never made sense from an evolutionary perspective, it resides in all reality. So it's impossible to talk about energy or matter without also talking about consciousness. It's certainly got everyone talking. There's a big demand for degrees in consciousness studies these days. Our physics department runs a regular course in it."

"I can tell you one thing," Sarah chipped in. "None of this stuff is available to me telepathically. I often haven't a clue what Pelly's talking about. What use does telepathy have if it tells me what color socks to wear, but not where she is when I want to meet up with her?"

Pelly, however, was in full stream.

"The exciting thing about all this is the incredible complexity of the brain. We're talking as many as two hundred billion neurons and a hundred trillion synapses, a thousand times more synaptic information pathways than there are stars in the galaxy. The computers we're using for our part in the human neurome project are getting close to having more processing capacity than the human brain, so we may soon have the potential to map the brain's entire connectivity in real-time 3-D, using diffusing imagery down to the nanometre—that's the size of an atom or thereabouts. It will take another twenty years, but everyone thinks we'll get there. Now picture the same research happening in the context of the ubiquity of raw consciousness, and its impact on the measurement and observational problems that underlie the uncertainty principle. The new field of psi-quantum research is exploring the hypothesis that all reality is imbued with causal agency and choice, right down to the quantum level."[827]

"But doesn't choice imply an active, conscious participant who can exercise what they believe to be free will?" I asked.

"Exactly. This line of research suggests that there is not only an omnipresence of consciousness, but also an omnipresence of agency and choice. Consciousness is the matrix for experience, agency, intention, purpose and free will. The belief that only humans have consciousness is just another of our delusions, like the belief that the Earth was the center of the Universe and that only humans could feel pain, not animals and fish."[828]

"What are you working on yourself?"

"My research team is working on the multi-level integration of syntropic organizational algorithms, or SOAs, as we call them. Every particle, atom, molecule and amino acid has its own organizational algorithm. Whenever they self-organize to a higher level, their SOAs have to integrate. If our purpose-built nanotech atoms have consciousness, like other atoms, this might mean that they have their own autogenic SOAs, with their own inner willpower and choice. The implications are kind of interesting."

"You mean… conscious robots?"

"Yes. Conscious robots. But only if they self-construct in a way that allows the syntropic self-organizing impulse to express itself. Not if they're assembled by humans from a kit of parts."

"Nonetheless, isn't that a bit scary? What do you think, Sarah?"

"It boggles my mind," she replied. "But if a tiny seed can grow into a tomato plant or an oak tree, who knows what else is possible?"

"But conscious robots?"

"We wouldn't see them as robots," Pelly replied, "any more than we see test-tube babies as anything less than human. One of my best friends was a test-tube baby. My personal work is focused on the SOAs of the olfactory center, which governs smell. I'm working with mice, which can sniff out the tiniest change in molecular presence. That's why they're used in the early detection of cancer, Alzheimer's and Parkinson's disease. I keep expecting them to self-organize in the syntropic field and start demanding delicatessen cheeses."

We both laughed.

"What I find amazing about all this," Pelly went on, "is that we're not outside the box. The human brain that we're researching is inside each of us, or rather, we are inside of it. If there is raw consciousness at the nano-level, and raw consciousness is the lived experience of what we used to call quantum uncertainty by atoms at the nano-level, then our experience of free will may be the syntropically organized accumulation of nano-experience at a vastly more integrated level."

If you could have seen inside my brain as I listened to Pelly, I'm pretty sure you'd have seen my neurons scrambling. It felt simultaneously wonderful and weird to be sitting outside on this sunny summer evening, with the swallows flitting around, the trees so calm and stately and the chickens scratching around in a patch of dirt, discussing these ideas about the deepest level of reality.

"Don't worry if you feel confused," Sarah said. "She does it to us all. It all goes back to a biology project we were doing at school. We had to dissect the brain of a mouse, and while Pelly cut calmly away I threw up all over the teacher. That's why she's a neuroscientist and I'm a farmer. And I *never* trap mice, by the way. I encourage the cats to do it for me."

Out of my fascination with what she was saying, came an idea.

"These syntropic organizational algorithms. Do they have an opposite, like an entropic organizational algorithm?"

"Yes, they do. How did you know?"

"Oh, just my natural genius," I joked.

"I've a colleague at the Brain Institute who's working on positive and negative mindsets in humans," Pelly said. "There's a particular organizational algorithm that seems to be associated with negative attitudes, which we see being activated in an area of the frontal cortex. When we compare it to people with positive attitudes, we can clearly see the difference. People with positive attitudes show activity in more neuronal centers than people with negative attitudes, whose brains seem to be sending entropic messages to the brain as a whole, switching off the areas associated with creativity and decision-making. The entropic algorithm discourages the use of free will, while the syntropic algorithm encourages it."

"Do you think this could apply to a culture, or to a civilization as a whole?"

"I've never really thought about it. But given that our brains are acculturated by the attitudes we absorb from our parents and from our culture as a whole, I would say it's a reasonable hypothesis."

"And the positive brainwork people are doing. Could that change the cultural algorithms?"

"I see where you're going. And yes, that's exactly why some people are doing it, though they don't call it that. I had a boyfriend a couple of years ago who suffered from depression, and his therapist had him use a brain-app to stimulate the growth of new neurons. Every time he became conscious of a negative automatic thought he used the app to counter it with a positive thought. I certainly saw a difference in his attitude, but it could also have been his change of diet, the long-distance running he took up, or the gratitude journal he started keeping, in which at the end of every day he would write down five things that happened during the day that he was grateful for.[829]

"So to speak to your idea, it would certainly be true that if most people in a civilization were to get into the habit of positive future-imaging, it would generate new neuron pathways. You might even say that they were laying down new neural pathways in the global brain, or the noosphere."

"Okay, that's enough from you two," Sarah said. "We're here for a farm visit, not a doctoral presentation on neuro-algorithmic syntropic quantum theory. Time's up!"

"Okay, Sis. It's hard to stop once I get going, and Patrick keeps egging me on."

Pelly leapt up suddenly with a scream, making her chair fall over. "Look! There's a mouse! Quick! On your sleeve!"

Sarah leapt up, shaking her arms frantically.

Pelly looked at Sarah and burst out laughing.

"Charley Brown, missed the ball again!" she said, and joined in the laughter.

When their laughter petered out, I asked, "Who wants to come to the lake to try out those boats they're making—the sayaks?" I needed some exercise, and a change of pace.

"I do," Sarah said. "I think I've had enough of my dear sister for the time being!"

 30

Mid-Summer's Eve

SARAH WAS MUCH less intense than Pelly, and we chatted casually as we walked toward the lake.

"Thanks for rescuing me," she said. "I know we're twins, but sometimes she's too much for me. That's why I want to be a farmer. It's easier on the brain, talking to ducks and horses."

The evening grew quieter as we moved away from the farmhouse, where people were preparing for the evening's events. The sunlight coming in from the west was striking the trees, and the field where the cows were grazing glowed with warmth.

"Have you been here before?" I asked.

"No. I live on a Gulf Island and I rarely get away. This place is gorgeous."

"How does it compare to other farms you've been to?"

"It's larger and better organized; they've really got their act together. It's like its own little kingdom, with its village and all. It feels like paradise."

"Do you ever wish that everyone could live like this?"

"All the time. But so many of my friends are stuck on the city. They think shopping and going to a restaurant or a nightclub is what life is about. It takes more than a day on the farm to wean them off that frame of mind. Those things seem so insignificant once you've fallen in love with the land."

"Is that how you are—in love with the land?"

"Totally. I love nothing more than to get up early, feed the animals before dawn when the world is quiet and then rest up in a hammock in the middle of the day. And at night I sleep like a baby!"

"Maybe it's because we have lived close to nature for most of our human existence, not hemmed in by streets and buildings," I suggested.

"It's all going to come around again," she replied. "In the future, I reckon people will look back at the cities, compare them to the tranquility of rural existence, and ask themselves whatever were they thinking? That's probably why there's so much support for all the initiatives to green the city and restore some of the wildlife that was there before we paved it over."

"What about the larger crisis? Does that trouble you?"

"Yeah, I worry. I would probably worry more if I wasn't so busy farming, and finding a place where I can realize my dreams. It's amazing what they've achieved here. Compared to how it was, I feel so much more confident about the future. It's so good to know that people *can* change the world when they are organized and persistent enough.

"When I was twelve," Sarah continued, "our teacher gave us a series of classes on the global environmental crisis. I was so frightened I told myself I would never have children. No wonder we felt depressed and hopeless. It was all about the problems, and how terrible everything was, not about how people could make a difference. I don't feel that way anymore."

"So now you're going to have a big family to help you on the farm?"

"Just two. But maybe I'll adopt another dozen. But first I've got to find a man who wants to share them with me. I don't suppose you're free?" She threw back her head and laughed.

"Oh, you mean a quick fling in the hay, and then twelve, no, fourteen children?"

Sarah laughed. "We could always have a short honeymoon before we started making babies."

We were almost at the lake. "Look, the sayaks are free. Do you want to try them out?" I asked.

We put on the life jackets, pulled the boats into the water and I placed my feet in the pedals that flipped the tail, driving the boat forward. The front cowling was higher than on a kayak to make room for the knees, but apart from that it was much the same, and wow, was it fast! The sayak swayed to and fro as I pedaled, but I was soon zipping through the water as I got the hang of it. If the lake had been any larger I could have switched into top gear, but in no time we had rounded the solar panels and were completing the circuit.

"Watch out! Here they come!" Sarah said, as we circled the lake for the third time.

"What?" I asked, peering around.

"The mozzies! We'd best get ashore and into the farmhouse."

West Nile fever—I had forgotten. Misery and death at the sting of a tiny insect. We scrambled to the shore and ran back to the farmhouse, laughing at our adventure while slapping at anything that felt like a mosquito.

"They're not really that bad," Sarah said, gasping for breath once we were safe inside. "You just need to wear long sleeves and trousers when you're outside in the evening, and make sure the screen doors are shut."

The barn was a bustle of activity in preparation for the evening feast. I made myself busy helping lay out the tables and carrying in a keg of beer from the farmhouse. People were arriving, and when everyone was seated there must have been two hundred of us. I sat with Sarah and Pelly, since I was mentally exhausted and happy not to meet any new people. The barn was filled with happy chatter as friends greeted each other and caught up with their news.

Then Mina welcomed everyone. It was her pleasure, she said, to host this monthly feast for their neighbors, and for all the farmhands and volunteers who had worked here during the week. This evening we had some special guests, including Finn Donnelly, Canada's Minister of Fisheries and Oceans; Andrea Reimer, the Mayor of Vancouver; and Isabel Makonde from Mozambique. Then I heard my name—"and Patrick Wu, who is visiting us from Sudan," and I had to stand and receive the applause.[830]

We lined up at the buffet, and before long I was enjoying buttered new potatoes, freshly picked fava beans, roasted sunchokes and New Zealand yams, a beetroot salad and a quinoa walnut casserole, followed by rhubarb pie, much of it cooked on an amazing series of SunStove solar cookers, the likes of which I'd never seen before.[831] There was also tilapia fish, local venison and braised Canada goose marinated in mead for those who wanted it. When I asked, one of the servers told me they came from the fall cull, when farmers controlled the populations of deer and Canada geese due to the absence of wolves and cougars.[832]

I was finishing my pie when someone's hands covered my eyes.

"Guess who?" a familiar Scottish lilt whispered into my ear.

"Laura!" I turned rapidly around and gave her a hug. My heart was in my mouth, and I had a sudden pleasant feeling that all was well in the world.

"I can see I'm not going to be getting my twelve babies from you," Sarah sighed with a smile.

"It was fourteen, actually, but I might have settled for twelve," I replied with a laugh.

"Where did you come from?" I asked Laura. "Have you been here all day?"

"No, I just arrived. I got a ride with the Mayor. She wanted to hear about our green business certification program, and how it's going. She says they're considering giving a tax holiday to the first ten businesses that achieve five-star status before the end of the year."

We were interrupted by a hand-bell.

"Listen up!" Mina called out in a loud voice." Due to all the rain we've had the slugs are having a feast, so while the band's setting up there's a Slug Picking Contest. Mayor Reimer and Minister Donnelly have promised that they'll open the dancing with whoever picks the most slugs. Be sure to wear long sleeves, pants and face-nets against the mosquitoes, and enjoy the contest! Tomorrow morning we'll take them up to the river and throw them in."

"Are you coming?" Laura asked.

For the next half-hour we scoured the fields for slugs, using trowels to collect them in buckets. Then we climbed into an old oak tree and watched as the sun slowly slipped below the horizon, casting a rosy red glow across the land as the stars began to appear in the east. And then we lifted our face-nets, and kissed.

I had no sense of time, just the awesomeness of the kiss and the stillness of the midsummer evening—and a growing discomfort from a branch that was sticking into my backside.

When we finally moved apart, Laura looked at me and swept the hair from my eyes. "Well, you're quite the surprise, aren't you! I come all this way across the ocean to get away from the Celts and what do I find! A Chinese, Sudanese

Celt! Is there anywhere the Irish haven't been? So, are ye ready to go dancing, or shall we do some more kissing?"

Needless to say, it was a redundant question. And the girl from Jerusalem—where was she? Conveniently gone. Not a peep from that direction.

Some while later, we climbed down from the tree and ran over to the farmhouse, carrying our buckets full of slugs. A large tent had been erected, and people were hanging around waiting for the music to begin. So this was the Syntrodance, whatever it was. Mina announced the slug-picking winner, a young boy of about ten, and everyone smiled as Mayor Reimer opened the dancing by waltzing him clumsily around the room. The evening then began with a traditional Celtic tune, and people took their partners for an old-fashioned barn dance with a caller who guided us through the steps.

"Is this it?" I called out to Laura, before being whisked off by another woman.

"Wait and see!" she called back amid the whirl of dancing couples. I was quite at home with a barn dance since I'd been to plenty when visiting my grandmother back in Ireland in County Kerry.

The music then changed to something very Elizabethan, played on recorders, cello and keyboard, and the dancers adopted a formal elegant step. This was followed by a series of waltzes—to be with a hundred other couples all waltzing together was amazing. As it ended, the music changed to an upbeat ragtime and I stood to one side as Laura continued to dance. But when it evolved into a Charleston I couldn't keep my feet still. I was grabbed by an older woman who was full of enthusiasm as she kicked up her feet. In no time at all couples were taking center stage, exploding into body-shaking shim-sham and then the syncopated back-beats and rhythms of swing. As the music moved into the wartime era couples began to show off their stuff, flipping and throwing each other as the rest of us watched—all kinds of couples, including men with men and women with women. Laura had returned to me by now, happy to catch her breath. This was crazily good music, and I was astonished at how the musicians switched genres so quickly.

As the swing music ended a drumroll grew in crescendo and the band burst into 1950s rock and roll, causing everyone to burst back onto the floor. I had the general idea by now—I just didn't know where it would end. What surprised me was how well people knew their moves. The same dancers who had been elegant in the Renaissance dances were wildly proficient with their tosses, flips and whirls. The drum rolled again and we were into the Sixties, with the Twist. Then it was Motown, the Supremes and disco dancing. People started separating and doing their own individual thing. There was some amusing exhibitionism and the music was great, but the dancing was boring compared to before. The 1980s and '90s were no better, in spite of some Michael Jackson moonwalks and breakdancing that two middle-aged men dared to contribute. The dance turned the corner into the 21st century, and electronic trance music. By now everyone was dancing alone, and when the tent roof lit up with flashing strobe lights I felt as if I was back in 2012.

This went on for a while, then the music changed to a slower rhythm and people around the edge of the tent start circling, much as we had the night before in the Song of the Universe. Was it only last night? It felt like a month ago.

The edge-circle broke in two, and the new circle moved in the opposite direction. The trance-dancers in the middle linked hands to form their own circle, and the outer circles began to weave in and out. From then on the dance became more complicated. Dancers self-organized, reaching out and breaking off into new dance-cells. Another cell would seek entry, and if accepted, the cells would merge and self-organize. The music was a combination of trance, baroque classical and romantic, blended harmoniously, even though that sounds impossible. Sometimes it had direction and flow and the dancers would move in an organized way; sometimes it went all misty and confused, as if Debussy or Ravel had taken over, washing everything in an ocean of impressionistic sound. Then the groups became amorphous and individuals broke away, adopting personal moves and seeking new cells and new harmony.

Finally the music simplified down to a solo flute and the dancers formed one large circle, moving calmly one way then the other. Five or six dancers took center stage and began to entwine with each other very physically as the music became more billowy and harmonious. Other people joined them, touching, becoming close, and slowly we folded ourselves down on the ground, everyone touching, becoming a single heap of being. All movement ended, and the music ceased.

And then, such stillness. It was as if we shared a single heart. Nobody moved, except the occasional shift to settle in and become more comfortable.

How long we remained that way, I don't know. Time stood still, and my senses were no longer my own. We remained in silence with scarcely a finger twitching, so close I could feel people's breathing. Finally there was a chime and people started to hum, creating a quiet tapestry of sound, and slowly our organism pulled apart and formed itself back into a circle, holding hands. We raised our hands in final greeting and then people went down on their knees, heads to the ground, expressing gratitude, returning to their individuality. Normality returned, but at a far deeper level. And then it was over, and people were hugging each other.

"That's it," Laura said, many hugs later. "Do you want to go for a walk?"

"For sure," I said with a mixture of reluctance and delight. If only I could rewind time and relive the experience, from the opening moments of the dance.

Outside, the night was dark and the Moon was almost full, low in the sky to the south. I was tempted simply to enjoy the magic of the evening, but I was carrying urgent information I had to share.

I told Laura what Ryan had told me when we discussed Derek's murder, and she immediately pulled out her device and searched for information about the character Syd Brockle. You'll recall that Laura's father was the Scottish detective who had helped Vancouver Police Chief Lui Cheng with the investigation, and he was currently in Vancouver sharing a holiday with his old friend.

"Holy shit! There's a link here that says he was a supporter of the Alphas during the OMEGA Days."

"What else does it say? Are there any other links about him?"

"Bloody hell. His bio says he's a Companion of the Order of Australia. Director of a big Australian oil company. And get this. He was also a director of Jusco, a Calgary oil company. He's got more than 25,000 references. Shall we limit it down?"

"Do a search on Syd Brockle and the OMEGA Days."

"Well, whadda ya know? He's got an op-ed in the *Hobart Mercury* where he's suggesting that the OMEGA leaders' call for full financial transparency was treasonous and communist, and the movement's leaders in Australia should be sent to a penal colony for undermining confidence in the Australian economy. It's pretty extreme stuff."

"Try Syd Brockle and Alphas, see what comes up."

"Alphas. Hey, look at this! There's a report here that lists all the known sources of financing for the Alphas during the period when they were active. And guess what? It says Jusco $30,000, Brockle Petroleum $30,000. He was right into them."

"But that proves nothing. It's one thing to give money; quite another to commission an assassination."

"I don't care. I'll call my dad in the morning. He's in Calgary with Lui Cheng; maybe they can dig something up. It says here that Brockle has a getaway somewhere in the Kananaskis Country outside Calgary, so there's gotta be people who know him."

"Skinny-dipping!" someone shouted, and we ran to the lake. Soon everyone had their clothes off and we were floating quietly, watching the moon's rippled reflections on the water. An owl took off into the night. The horses snuffled and whinnied in the darkness, and time stood still as we floated in silence in the night, only the occasional splash as someone dunked a limb to shake off a mosquito. Laura lay with the top of her head touching mine, our arms outstretched holding each other's hands.

And then, as if this was not enough, Schubert's 'Ave Maria' drifted across the water, a simple soprano with piano accompaniment. Silently the senses abandon their defenses. Someone must have brought a player and chosen this particular piece of music. It was one of the most sublime moments of my life.[833]

All good things must come to an end, but as long as I was in Laura's company no end was in sight. After we had dried off and put our clothes on, Laura asked me quite simply if I would like to spend the night with her in her tent. Would I? Do sun-roasted strawberries picked right off the vine taste good?

Afterwards—that simple, love-filled ocean of a word—afterwards, we slept together, entwined like two twins in a telepathic pod, and when we were woken by a thunderstorm and a torrential downpour we just snuggled closer, warm and cozy under the cover of love.

I knew it would end, and tomorrow I would be gone, but I pushed the thought away. Maybe Laura was a time-traveler too, and our meeting was a projection of an actual future we'd be blessed to discover back in our own time. Maybe she came from three hundred years in the future. Who was I to care? The brain doesn't work well when it's flooded with the hormones of love.

As dawn broke, we awoke to a chorus of birdsong. "Do you remember how many species of birds Mina said they had on the farm?" Laura asked, sleepily.

"Five thousand?"

"Well, here's five kisses before you leave. I know you're moving on. I'll hold onto the rest in case we meet again one day."

Five was good. A hundred would have been better, but five was good.

 31

Into the Forest

THE MUSIC CAME drifting across the meadow, a deep gentle sound.

"It's the Taizé service!" Laura exclaimed. "I'd completely forgotten. Come on—we've got to go. Just one last kiss."

One last kiss....

The big tent was none the worse for the night's thunderstorm, and the grass was deliciously wet. Three people were using long-handled wooden scythes to cut the grass, one harmonious swish at a time, while the music floated over the morning air.[834]

Inside the tent the walls had been darkened with purple and mauve cloth and illuminated with electronic candles. People were sitting on cushions facing large red and yellow flowing banners, singing a soft rhythmic chorus accompanied by guitar and clarinet. We found a place at the back, from where I could see Mina and an older man who I took to be her husband Joey. Carlo and Bianca were there, but apart from them it was a sea of new faces. Where had they all come from?

The singing continued for fifteen minutes, expressing a variety of Christian themes, mostly in Latin, and there was a Muslim prayer sung in English. This was followed by a period of silence when some people prayed, some bowed down and others meditated, the canvas flapping gently in the breeze.

Mina greeted me as we emerged into the morning sunshine. "It's good to see you, Patrick. And you too, Laura," she said with a smile and a knowing expression.

Laura laughed. "Ah, but Patrick's abandoning me; he's taking off after breakfast. Men are so fickle. Can I stay here for the day to mend my broken heart? I'm sure I could make myself helpful."

"For sure. I have a group of men being Monks for a Day: they'll be weeding the vegetables, working in silence. Or do you feel like picking strawberries? That's always good for a broken heart. There's a team starting at ten if you want to join them. What about you, Patrick, you evil heart-breaker?" she asked with a twinkle in her eye.

"What about *my* broken heart?" I replied. "We men have feelings, too. I have to head back to Vancouver, but I'm open to a circuitous route to see where fate might take me."

"In that case, there's a man leaving after breakfast who might give you a ride.

I think you'll find him interesting. He says he's going hiking in the forest north of Maple Ridge."

And so it was that after a lingering farewell with Laura I found myself speeding west along the open road with Govinder Prajit Singh, a good-looking young man in his twenties with slicked-back hair, driving a shiny silver-green electric car.

I had invited Laura to join me, but she declined. My feelings were real, but even if she had come I would have had to leave her tonight. That was my strange reality. From that perspective, the girl in Jerusalem, who might not even be alive, was more real. It still hurt, however, to have known a slice of heaven in Laura's arms and to know that I would never experience it again.

⁂

As soon as we were on the road Govinder opened up the roof and I felt the wind in my hair, promising fresh adventure. This was my last day in the future, and I wanted to make the most of it.

"So what brings you to Canada?" Govinder asked.

I told him about my mission, and he told me that he was a soil scientist who worked on antibiotic-producing intermicrobial signaling agents. He was married with two kids and lived in a large house in Surrey with his wife's extended family.

"How do you like the car?" he asked. "It's the Tesla Model Z.[835] Her graphene battery has a 25-year guarantee, 500-kilometre range and a 5-minute booster charge that makes her good for another hundred, so I can take her anywhere.[836] And with all this space she's good for the whole family, including Jasmeet's parents. She's got anti-crash surround-sensors,[837] cyclist and pedestrian detection sensors,[838] a night-vision windscreen,[839] web-mirrors,[840] and external airbags triggered by photo-sensors in case there's a crash."[841]

"She's an impressive piece of work," I said, enjoying the ride.

"She's got dual motor all-wheel drive, great road-holding, and a fabulous turning circle. Zero to a hundred in two and a half seconds. Here, sit tight...."

The road was empty, so Govinder slowed to a crawl and then floored the pedal and—*holy crap!*—my body was pushed back really hard into my seat.[842]

"She's got a top speed of a hundred and seventy, but I could be dinged a hundred bucks by a traffic-drone if I pass a hundred. I hate those things! Luckily they're not usually around on a Sunday morning. My son Ranvir cost me five hundred dollars in a single week when he started driving before I had the speed control installed, and the police installed a tracking tag in the license plate. The GPS is hard-wired, so I have no choice about the drones, but combined with the V2V surround-sensors they have certainly driven the insurance down. There's far fewer collisions these days. The insurance is distance-based, so the further I drive the more I pay. It seems fair. Look...."[843]

Govinder did something with his fingers and the car's screen showed the distance driven since the start of the year, and the cost of insurance."[844]

"What was that thing you just did with your fingers?"

"This?" He showed me his left hand, and a small plastic device he wore around his thumb. "It's my fin-control. I use it to activate my devices. It's very nifty—some kids in India developed it."[845]

"And what's V2V?"

"Vehicle-to-Vehicle. It senses the presence of other vehicles nearby and it overrides the controls if there's a risk of collision."[846]

"Have you ever been in a self-driving car?"

"Yes, but they're nowhere near as much fun as driving yourself."[847]

The car was spacious, with seating for seven. The screen in the instrument panel showed where we were, the remaining range, and the location of nearby EV charging stations. It flashed the speed limit when we exceeded it, gave directions on an on-screen map, and had a range of smart-phone functions that were disabled when the car was moving.[848] It also had a mandatory alcohol and marijuana detection system, Govinder explained, with transdermal sensors on the steering wheel that prevented the car from starting if it detected a certain level of alcohol or marijuana vapor molecules, either in the air or on the driver's hands.[849] That must cut down on drunk driving, I thought to myself. And on driving while stoned, too!

"Is that solar coating on the hood?"

"Not just the hood—the entire car. It's good for a 12% trickle charge on a sunny day. I've a solar angel in the back that can push it up to 50% when I roll it out."

"A solar angel?"

"Yes—it's a pair of solar wings that fold out on the ground and give me extra charge when it's sunny. We use them on camping trips, when we're far from a charging post."[850]

"What about theft—does that worry you?"

"Nah. They only go after the older cars where they can disable the GPS. A car like this requires a unique key and driver eyeball ID. If a thief did find a way to steal her I'd get an alert on my device that I could switch to the police, and a drone would lock on within seconds."

"So they're not all bad, the drones?"

"I suppose not. But who knows what else they're collecting? They may be spying through my bedroom window for all I know. I don't feel private in my own home any more. When I think what they do at the borders, with their full body scans and genetic ID scans, it makes my blood boil. Whatever happened to privacy? I pay extra to keep my data at P-5, the highest level compatible with my work, but I can't keep CSIS and the NSA out. It's like the OMEGA Days never happened when you get inside those establishments."[851]

I let that pass, and asked, "Can your car's battery send power to the grid if you want it to?"

"Yes, but I never bother. My neighbor Ranjit, on the other hand, he's a penny-pincher. He's set his battery to feed electricity to the grid on a price-optimized basis. He earns a steady trickle from it."

"What about car-sharing? Do you share your car to earn extra income when you're not using it?"

"No way. I'm not letting anyone else get their hands on her. What am I, a pimp? Not while she's so new, anyway," he laughed. "I expect Jasmeet will make me do it in the end. She says I'm a selfish prince, but she's happy to spend the money I bring in. And what's so wrong with being a prince? We don't all have to be monks. She looks after that side of things for me."

Govinder was a charmer, and I could see how he might get away with things. But I wanted to find out what kind of worries he had.

"What's your biggest concern in life?" I asked, as we crossed the Fraser River over a new bridge at Fort Langley and the screen said we'd just been dinged five dollars at a toll point.

"Making sure my children get a good education and get ahead in life. That's the most important thing these days."

"And globally? What is your biggest concern for the world as a whole?"

Govinder paused to think.

"Personally, I worry about Internet monitoring, and the ease with which a democratically elected government can turn into a police state. I only had to visit the Khalistan website a few times and I had a call from security wanting to know what I was looking for. Khalistan is just about independence for the Sikh homeland, the same as Quebec and Vermont. What's wrong with that? I don't even know if I support them, but now they've probably got me listed as a potential terrorist just because I browsed their website. If it goes on like this they'll be investigating me every time I visit nudes-are-wonderful, or whatever.[852]

"And then there's those I-buttons people wear on their collars so that their friends with high reality glasses can catch their personal uploads in real time. It's all too much. We've already got the police demanding that anyone with a criminal record be forced to wear one. It's all too *1984* for my liking. My son Ranvir's got the glasses, and he let me try them. Temperature, weather, time of day, street directions: I don't mind that. Even real-time messages from close friends—they're kind of cool. And the on-board videocam is fun."

I could tell that I was becoming accustomed to this future, because this kind of thing no longer shocked me. I found it strange that no-one else had spoken about privacy, however. Maybe it was just Govinder who worried—or maybe I simply hadn't asked the question before.

"But I don't like what my son Ranvir was telling me about the high reality poachers who read what's going on without your knowing," he continued. "I don't think I'm being paranoid. When I was young the verb 'to google' meant simply to search for something. Now it's far more sinister, with the police and the government deep-googling your behavior. *Doogling*, Ranvir says he and his friends call it. But this is all just personal…" he said, his mood becoming more serious.[853]

"If you want to know what really worries me," he continued, "and my colleagues too, it's the climate crisis. We see lots of data from forests and farms around the world, and it's really unsettling. We're members of the Global Association for Soil Protection, so we receive real-time satellite data on relative soil moistures, collated into regional time-sequences. They're called TerraCharts,[854] but we call them terror-charts because they're so alarming. It's not just soils drying up. It's soils becoming waterlogged, and being lost through erosion due to the loss of forest cover, giving up their carbon instead of storing it.[855] It's like the melting Greenland icecap and the methane that's pouring out of the permafrost—the curves are all heading in the wrong direction.[856] When we screen for land that has been farmed organically for five years or more the decline disappears. So spreading organic farming to the planet as a whole has to be one of our highest priorities, ending the chemical farming that has undermined the soil's carbon base.

"I've a cousin who's a marine biologist, Gagan, and the things she tells me are *really* worrying. The starfish—they've still not returned to the coast after the wipeout that happened twenty years ago, triggered by the warming ocean. Many species of fish are still on the endangered list, and there are jellyfish everywhere because the fishing fleets have taken almost all the predator fish, upsetting the marine food chain. And I don't know about Sudan, but the fish being caught here are really small—yet the fishermen seem to think it's normal. Shifting Baseline Syndrome—that's what we call it. Each generation of humans accepts what they inherit as normal and loses sight of what's been lost. The marlin the fishermen used to pull out of the ocean a hundred years ago were *huge*—sixteen, eighteen feet long. Today, what do they get? Six, eight feet? Max.[857] It's the same with the birds. When I was young I read an incredible book by Farley Mowat, the Canadian writer, called *Sea of Slaughter*. He went through the historical records and found evidence of how things used to be before humans began the invasion and the conquest of nature in North America, five hundred years ago. Beaches in the Magdalene Islands packed with walruses, hundreds of thousands of them. Skies black with migrating birds for days on end. And today—all gone, or almost all gone. The new normal—that's what it is. The new baseline. Each time I see her, Gagan reminds me that the oceans are becoming steadily warmer, more acidic and less oxygenated, and without enough oxygen many species will die. It's awful how fundamentally we have messed with the very foundation of our planet's ecology.[858]

"You did ask, right? Well, that's what I think. So we've got to do everything we can to drive the global carbon emissions down to zero and drive the carbon capture up in forests, farms and pasturelands. That includes putting a heavy tax on beef and lamb, so that they become a luxury, not something people eat every day. If we can live well and prosper without meat in our family, why can't everyone?

Why do people insist on their right to eat meat, as if they were immune to the global crisis? What do they think they are—cyborgs, without any protoplasm or carbohydrates? It makes me furious when I think what they're doing to my children's dreams, and the Earth they're going to inherit."

Beneath his playboy surface, Govinder had a passion for the Earth that was equal to anyone's, either here or back in my time.

"What more should we be doing?"

"We need the OMEGA Days all over again," he said without hesitation. "They were just the warm-up. They were like a trial run, here in Vancouver and a few other cities. We need a huge global OMEGA Days with people in every city and every village rising up to demand change. We need to learn from the OMEGA Days, and apply the lessons globally. We've got the means to make it happen. We're one big global village; everyone is just a click away. We're each of us a particle in the unfolding wave of the future and we've got to come together as a wave to sweep away the blindfolded brutalizers of the Earth, the people who are destroying our children's future, the egomaniacs who believe that ego is more important than eco and that personal status is more important than planetary survival."

Holy shit, I thought. Govinder was giving me the shivers.

"Will it happen?" I asked, as we drove into the forest north of Maple Ridge. "Or rather, what do you think it will take to make it happen?"

"People like you and me. It's got to happen—it's simply got to. I just don't know what will trigger the tipping point, the moment that announces 'This is the Day.' What was it that started the OMEGA Days, back when I was a teenager? I heard that it was a group of seventeen-year-olds who'd had enough when they were thrown out of a coffee shop because they didn't have the ID cards the shop manager demanded they show. They did a spontaneous sit-down, texting and tweeting their friends to join them, and then one of them stood on a table outside the coffee shop and made this incredible speech, calling for change not just there but everywhere, and it quickly went viral. Within ten minutes they were fifty and within an hour they were five hundred. For every ten the police arrested another hundred arrived, and before the year was out the revolution had begun."

I was taken totally by surprise. Govinder was certainly a charmer, but he was also so much more.

"Here's what it means for me," he continued. "I have two homelands: one in Vancouver and one in India, on the beautiful plains of the Punjab. I see all the things that are happening here but I also see the things that are happening in India and Pakistan. We have such a long way to go. The whole world needs its OMEGA Days, both practically and spiritually. Do you know the other meaning of the word 'Omega,' the one used by the French Jesuit scientist and priest Teilhard de Chardin in the last century? He must have been a Sikh in his heart. He had a vision of evolution unfolding towards greater consciousness as well as towards greater biological complexity. At the end of the journey, he believed, we will merge our individual consciousness with God, and know true fullness of being throughout the Universe. There was a great Indian mystic, Sri Aurobindo, who came to a similar conclusion. They lived at the same time, though they never knew each other.[859]

"Teilhard called the end-point the Omega Point," Govinder continued. "It's similar to syntropy, the new theory that the Universe contains a self-organizing matrix that causes all existence to seek greater wholeness and order. We need a global OMEGA Days so that we can continue to evolve towards the Omega Point. The whole experiment with life and civilization might collapse if we put ourselves on the side of entropy, instead of syntropy. Maybe the OMEGA Days were the last determined expression of hope in a world that's spinning into chaos, not the first wave of a new harmonious future—but not if I have any say in it. No bloody way."

Wow, and ouch! It was only rational to keep the possibility of both futures in mind. I could see that too much optimism might blind people to an approaching apocalypse.

By now we had reached the parking lot at the edge of the forest. The sign said we were entering the Malcolm Knapp Research Forest, a project of the University of British Columbia, and in front of us there was a beautiful timber-framed natural history shelter, built from large beams, like a temple to the forest.[860]

"I'm going on a three-hour hike. Would you care to join me?"[861]

"If I come with you for a while and then return, is there a bus that could get me to Vancouver?"

Govinder took out his device and clicked a few buttons. Then he did something I had not seen before. He found a dark part of the forest, out of the sun, and projected his screen into the air in front of me, creating a holographic image. And there was all the data.

"There's a Sunday bus that leaves from Silver Valley at twelve. That's about a kilometre down the road. Or there's a bike-share station at the bus stop if you'd prefer that. There's also a hitching point that might get you there sooner. Have you got your device?"

"No, I left it at home for the weekend," I lied.

"No matter. If you had, you could use it to see if there were any drivers offering a ride and text them where you were."

Hitching? That might be an adventure.

"This is Jasmeet and the kids. Aren't they beautiful?" Govinder projected a real-time holographic video of his family into the air, showing three people goofing around in the garden, having fun with a floating sphere of some kind.

Then "Hey—what's she doing? I told her never to play in the pool unless there's someone around." There, in the quiet of the forest, the hologram clearly showed his daughter splashing around with no-one in sight. Govinder was furious. He changed the angle on the webcam to see if anyone was around and then he called home and had a heated conversation in Punjabi with his wife, berating her for letting their daughter play in the pool unsupervised. Then his voice changed and he said something like "teekay," repeatedly.

"She says I should relax, everything's okay, she was watching from the kitchen window. I don't think she appreciated me spying on them like that."

I smiled. "It's ironic, isn't it, that you hate the government using the technology on you, yet you find it useful to keep an eye on your own family?"

"Ha! I suppose you're right. But what if she got into trouble when there was no-one watching?"

"You're just being a good dad, that's all."

"That's what I say! But we should be getting going," Govinder said, plugging the car into a charging point.

"How do you pay for this?"

"The first hour is free, then it's on my card. In the city there's a charging post wired to every lamp-post and parking meter, which makes it very easy. Are you ready?"

Alongside the trail there was a carpet of undergrowth; above us the Douglas firs and western red cedars grew tall and vertical, partially covered with moss. The silence of the forest was only disturbed by the sound of our feet on the path, and the croak of the occasional raven.

Govinder told me that the trees we were seeing were regrowth after a fire that had happened a hundred years ago, set off by a spark from some logging equipment.

"What does your work tell you about the forest and the soil here?" I asked.

"The older the forest," he replied, "the more carbon it stores, the better it is for the intermicrobial signaling agents and the closer the tree-fibers are, producing stronger quality timber."

"Do they still ship raw logs to China?" Back in my time, the skimpiest of trees were being clearcut and packed onto boats, leaving ruin in their wake and few jobs for local people.

"No. That stopped soon after the OMEGA Days, with the new sustainability policies. Almost all public forests have now been certified by the Forest Stewardship Council, and the forest companies are required to channel an increasing share of their timber to local mills to support the value-added industry. There was also a softening of the American attitude and an end to the Softwood Lumber Agreement, which had made things really tough for the Canadian forest companies. It was timber from this forest that was used in the Agora building in Vancouver's Urban Farm District. Have you seen it?"[862]

"Yes, it's great! How do you log a forest commercially while protecting its ecological wealth?"

"Very gently, tree-by-tree, and with as little equipment as possible. The heavier the equipment, the more it damages the soil. It's like me with my car. If my car can do a hundred and fifty, I feel that she wants me to take her there. If a feller-buncher—a machine that cuts and bunches a tree—can take out three hundred trees an hour, it's like it colonizes the mind of the operator, the way soil fungi colonize the biochemistry of the forest floor. So here in this forest they've gone back to horses, and hand-cutting with a chainsaw."

"Horses? The same as on the farm?"

"Yes. Good heavy draft horses, Belgians and Percherons. An old horse-logger told me once that when you become a good horseman or horsewoman you achieve a kind of mind-meld. It's all about intention, along with body language, and having a clear image of what you want. Pictures, not words. If you can master those, he said, you and your horse can share one mind. You can become a horse-whisperer of the forest."

"And no need for gasoline for all the heavy equipment."

"Precisely. When you work in the forest you learn to think in a four-hundred-year cycle. With everything we do, we ask how it affects the soil, the trees, the canopy, the biodiversity and the hydro-ecological cycle over a four-hundred-year period. There has been a huge shift in attitude. It's no longer taboo for professional foresters to share the reverence they feel for the forests, and to speak of them as highly complex communities of living beings."

"What about timber yields? Do they fall, with the new way of logging?"

"It's the reverse of what people used to think—over a hundred years the forest actually yields more timber than it would if it was clearcut. For sure, if you clearcut a forest and put the money in the bank at a high rate of interest it'll seem as if you're doing well. But what about the soil, and the fungi and bacteria that made it possible for the forest to grow in the first place? What about the rain, and the way the forest protects its moisture? What about the salmon and the bears, which carry nutrients into the soil? A forest can only take two or three rounds of sixty-year clearcutting and then the soil is gone. That's why they made it illegal so long ago in parts of Europe. When you cut for a four-hundred-year cycle you take only the annual growth-rate, and you choose your trees carefully to maximize the growth opportunity for those that remain."[863]

How Lucas would have loved this. I wondered what he and Aliya were doing on this quiet Sunday morning. Were they lazing in bed, or were they out on horseback enjoying the trail through the Urban Farm District, and on to Burnaby Lake Park?

"There was a temporary loss of income when the rules changed and the export of raw logs slowed down, but then the sawmills and the woodworking companies recovered and we began to build a reputation for wonderful architectural designs using wood. These days our wood designers are some of the best in the world, making everything from cathedrals and convention centers to forty-storey towers. There's a workshop located right here in the forest, further up the trail."[864]

"Speaking of which, I should be turning around soon," I said. "But before I go, you said you also worked on farms?"

"Yes. That's why I was at Joey's Farm. I left the dancing early, since I'm not a good dancer. When farmers used chemicals they polluted the soil infrastructure, killing the worms and destroying the complex ecology of soil species. Did you know there are as many microorganisms in a gram of soil as there are humans on the entire planet? And a million different species of soil organism?[865] The more we learn about the way plants assist each other and compete to claim soil space, the crazier it is that we allowed synthetic chemicals and pesticides to be used in the soil. I didn't know how little people knew about it when I decided to become a soil scientist. That makes it exciting, because there's so much to be discovered. We're doing an annual DNA soil genome profile of a sample section of the land on Joey's Farm. The computer power we've got is incredible, but it still takes a month to analyze a single microgram of soil with its millions of bacteria. We hope to be able to do it faster as computers approach the capacity of the human brain."

"Have you found anything special?"

"Yes. Mina and Joey are lucky, because their soil comes from millennia of ice ages and melting glaciers, and it's filled with nutrients that have been washed down from the mountains. We're studying how the worms influence their associated bacterial communities, and how they help with the mineralization of organic matter. That's important, because there are certain critical mineral trace elements that are linked to brain health, things like iron, zinc and selenium, and it's the worms that convert them into a form plants can absorb, enabling us to absorb them. So if your brain's not working properly, you might want to ask if there were enough worms in the fields where your food was grown. Joey's Farm has a higher worm density than a neighboring farm where they use an electric tractor instead of horses, since the tractor compacts the soil. A single gram of earthworm castings can contain as many as fifty million bacteria, from just one worm. And there are seven thousand species of earthworm globally. But am I boring you?"[866]

"Not at all."

"You and I have close to thirty thousand genes. An earthworm has twenty thousand. What functions do they all serve? We don't know. Can you imagine what kind of complexity an earthworm must have to need that many genes?"[867]

"I heard someone say that Joey's Farm has two hundred million earthworms in its fields. Is that possible?"

"A hundred and fifty years ago, Darwin estimated that farmland held up to fifty-three thousand worms per acre. On Joey's Farm we do an annual count that tells us it's closer to a million, so yes, with two hundred acres, to a depth of a foot, they'd have two hundred million. Worms love horse manure, which is another reason why horses are so good for the land. I've never heard of worms loving tractor exhaust.[868]

"Getting back to climate change," Govinder continued, "we know that organic farms store more carbon than non-organic farms, and that it's the earthworms that break down the organic material, locking the carbon away. So the importance of the earthworm for carbon sequestration is enormous. More than anything, it's the worms that build new soil. Over a century, they can build anywhere from fifteen to six hundred centimetres of new soil (six inches to two feet). So long live the earthworms! It's the same with the ants in tropical countries. They weather olivine rocks up to three hundred times faster than would happen naturally, and turn them into limestone in their nests, sucking carbon out of the atmosphere. There's an initiative underway to replicate the way they do it in the hope that we can speed the process of carbon sequestration. But look, hadn't you best be turning round?"[869]

"Yes, I should. But I'm wondering if you can explain one last thing."

"I'll give it a try."

"Mina said that the new farming represented the fifth agricultural revolution. What were the other four?"

"Well," he replied, "the first revolution was when we learned how to domesticate crops and sow seeds, using oxen or buffalo to pull a scratch plow back at the beginning of the Neolithic era, ten thousand years ago.

"The second revolution started in the 6th century AD, when the monks of Europe developed the moldboard plow to turn over the deeper soils, doubling the yields."

"And the third?"

"The third revolution came in the 18th century, at the beginning of the industrial revolution, with new farm machinery, steam tractors and seed drills, the hybridization of seeds and all sorts of other improvements.[870]

"The fourth revolution followed in the 20th century, with oil, tractors, chemical fertilizers, pesticides and genetically modified seeds. So the New Farming represents the fifth agricultural revolution. It will finally put global agriculture onto a long-term sustainable footing, able to provide nutritious food for the ten billion people we'll soon need to feed—provided they don't all want to eat beef or lamb all the time, that is."

"What will it take for the New Farming to spread across the world, the way the other revolutions did?"

"That's the big question. We need farmer education, and we need consistent supplies, from seeds and worms to horses and equipment. We need new methods of financing. We need reliable methods of storage, distribution and marketing. In many countries they need serious land reform. We need national legislation and full-cost accounting, with incentives for sustainable farm practices. And we need a popular movement in every country to support the organic farmers and buy their produce. Think you can manage it?"

"No problem!" I replied.

"The good news is that the revolution is already underway in most countries. As soon as a government makes farmers pay the full environmental cost of conventional farming, as they did here in Canada, most are happy to make the switch. The market alone can't do it, since there are so many vested interests, and the agro-industrial corporations profit off the fact that they've *not* had to pay for all the hidden costs they cause. But once a government gets behind it, things can change quickly. That's why there's a big push to ratify the *Global New Farming*

Treaty, which binds the signatory nations to embrace the change. But you'd best be getting on your way, or you'll miss your bus."

It was true. I liked Govinder and was sad to part, but I thanked him for the time we had spent together and set off back down the forest trail.

I had so much to think about. Back in my time only a fraction of the world's farmland was organic, which meant the worms were still under attack, the soil was still losing carbon and chemical fertilizers and pesticides were still polluting the soil, undermining our food, our health and our climate.

No wonder we were in trouble.

 32

Hitchhiking

THE SUNLIGHT DAPPLED through the forest canopy as I returned down the path. All things are possible, I thought to myself—Govinder's car, the holographic webcams, last night's Syntrodance, the closure of the offshore tax havens. Everything in our human world has been invented by humans at some point in time, from the simplest salt shaker to the entire edifice of law and tradition, so everything can be reinvented, and a million new things invented.

But what did it mean that even though the people in this future were building a world that was so much more sustainable and harmonious, they still worried about the future, and the same issues that concerned us back in my time? Was it simply that my journey had been to the greenest city, not the greenest world? Maybe I would need another journey to a more distant future.

I was stopped in my tracks by an owl sitting low on a branch to my right, its head turned round to watch me. What did it know? Just an inscrutable stare. I walked quietly on trying not to disturb it, but it spread its wings and disappeared into the forest. The beauty of the Earth in a flutter of wings.

How many other creatures inhabit this forest, I asked myself, from the worms and beetles to the birds, deer, cougars and black bears whose ancestors have lived here since the end of the last ice age? How can we escape our delusional bubble and realize our connectedness to all beings? Is it Earth-empathy that we need? Is that what has been holding us back, the illusion that our species is somehow more worthy of empathy than the others? All beings have consciousness, and it's not personal survival that matters now—it's species survival, planetary survival.[871]

Shortly before I found myself on this journey I had been reading *The Better Angels of Our Nature—Why Violence has Declined* by the Canadian/American psychologist Stephen Pinker, in which he demonstrates convincingly—in contrast to what many people believe—that there has been a steady reduction in warfare and violence over the millennia due to a variety of factors associated with civilization and democracy. Not just warfare, but public burnings, floggings, the cruelest tortures. It was the spread of literacy at the end of the 18th century, he argues, that gave us greater empathy, enabling people to imagine their way into the hearts of others and understand their hopes and sufferings—the reality of the poorhouse, the debtor's prison, the life of a coalminer.[872] If he's right, maybe what we need now is more stories about the lives of bears and elephants, owls and worms, cows and chickens, so that we can leave our human bubbles, enter their world, and understand the role they play in nature and the reality of their feelings and sensations.[873]

I continued the rest of my walk trying not to think, just being present in the forest.

At the edge of the forest the road led downhill through a rural area, with houses set back from the road. A few children playing, the occasional car—it seemed like a peaceful contented place. It took me half an hour to reach the bus stop, where I touched the display and discovered that the bus had left twenty minutes ago. There was a hitching post half a kilometre down the road, as Govinder had said, so I carried on walking.

The hitching post was a simple wooden sign. You could either wait, or use your phone to tell drivers you were there, and where you wanted to go. Only I didn't have a phone, and I wanted serendipity to be my guide. So I stood by the shelter and stuck out my thumb whenever a vehicle passed.

After five minutes a motorbike pulled up and within a minute I was riding down the road with Juan Sanchez, heading west across open farmlands on an electric BMW wearing a helmet with good intercom, a rear-view display and a GPS-based map of the area. The bike was almost silent, which felt strange but made it easy to talk.[874]

He was a carpenter who had been living in Canada for six years, he told me. He had made his way across the border at night and managed to get his residency papers thanks to the support of a local church, in spite of being an illegal. As soon as he had his papers he had brought his wife and children to Canada, which was a blessing, since his life in violent, drought-stricken Los Angeles had been so difficult. That was part of the reason he had been granted status, he said, because of the threats to his family.

"Canada has been good to us," he told me. "Now I can work and my children can go to school without being afraid of the gangs."

He dropped me at a hitching post close to the Lougheed Highway and within minutes I was picked up by Larry, who was taking his daughters to their weekly soccer game in an old Chevy S-10 electric pick-up truck. It was a bit of a squeeze, but one of the girls sat on the other's lap and they seemed to think it normal that their father would pick up a stranger and start chatting.

He had converted the pick-up ten years ago, Larry told me, when the price on carbon had made him seek ways to reduce his footprint. He had a minimum wage job at Walmart at the time, and he had joined a course at Douglas College in Coquitlam where he had learned how to convert a car over a six-day period.[875] Afterwards, the Vancouver Electric Vehicle Association (VEVA) helped him and five of his classmates form a small conversion co-op. They rented a space and in less than a year, working one evening a week, they converted six vehicles. The government gave the same incentive they did for new electric vehicles, and

VEVA organized bulk purchasing for the parts. The S-10 had a range of a hundred kilometres, and was still running on the original lithium batteries he had installed.

"He calls it Benjamin," one of his daughters giggled.

"My other cars were all female," Larry said. "I gave it a boy's name so as not to make their mother jealous."

"She'd never be jealous of a car, silly," the other daughter laughed. "She's only jealous when Auntie Lacey comes to stay and you go all soft and mushy."

"I never!" Larry replied, making a major effort to appear affronted. "It's not my fault that your aunt's so pretty. She's not half as pretty as your mother."

The girls looked at him quizzically, as if to say 'yeah, whatever,' but they were too young to roll their eyes and actually say it.

"So was it a good deal, with all the time it took?"

"I had no idea a car could be so complicated. They said it would take one person two hundred hours to do a car, but we did all six in just under a hundred, working together, so we felt pretty good about that. Call it a hundred hours at $25 an hour, plus the cost of the car and all the parts and the battery, less $5,000 from the government incentive, so for around $5,000 I got myself a car that costs just $3 a week in electricity."

"Were many people doing what you did?"

"Yes. There was a big waiting list at the college. But then some people set up a co-op, doing it like an assembly line. They had it down to fifty hours and they were selling the converted vehicles for $10,000, so that pretty much ended the demand for do-it-yourself. No-one's doing conversions now that the Chinese and Indian electric vehicles are so cheap."

"Do you still work at Walmart?" I asked. He was obviously more skilled than to be working for minimum wage.

"So where did you grow up?" He looked at me quizzically, and I replied that I had actually grown up in Sudan.

"We haven't had a Walmart around here for ten years now," he said. "Not since we tried to unionize. When the government told them they had to allow the union they simply shut up shop and walked away. It's a shame in some ways, since they were doing great stuff to make their products more sustainable, and to get the whole company certified as green. But they hated unions, they had dozens of offshore tax accounts to avoid paying taxes, and they cast such a blight on the local economy, causing so many locally owned stores to shut down. I could never get over knowing how wealthy the Walton family was, thanks to the workers they hired on poverty wages. The last time I looked, which was some time ago, the six Walmart heirs had a personal wealth equivalent to forty-two percent of the entire American population, and they were using every loophole in the book to avoid paying taxes. Just six people. There was something really wrong with that.[876]

"At least the local council has done something good with the old Walmart store," he continued. "They've turned it into an incredible library with a farmers market, public meeting spaces, study rooms, computer labs, a café, a used book store and an auditorium. It's a great example of how to deal with all the abandoned shopping malls that are showing up these days."[877]

"So you lost your job?"

"We all did. But I was ready to quit anyway. It was soon after the OMEGA Days. Everything was changing fast, and it was easy to feel optimistic. I used to think the future would be like something out of a Margaret Atwood novel, all dark and bleak, but it has not turned out that way at all. I was one of the first in the new community welfare program, and the purpose counseling helped me stop and think, and ask what I really wanted to do with my life. It helped me understand my skills and realize that there was more to me than I thought. Ain't that right, girls?"

The girls squirmed, not knowing what to say. They seemed around ten years old—too old to be sweet, too young to be cheeky.

"So what came of it? What do you do now?"

"I did an electrical mechanics training program and now I'm the mechanic and general handyman on Mason's Farm, bottom of 150th. Electric tractors, farm machinery, milking equipment, you name it. Always something to do. They've given us a tenth of a hectare where we can grow our own food. But look, I've got to turn off here to get the girls to their soccer game. Can I drop you on the corner?"

I said goodbye to Larry and the girls and watched as the pick-up moved silently down the road. But where was I? All I knew was that we had crossed Pitt River and must be somewhere in Port Coquitlam.

It was around one o'clock and I was feeling hungry, so when I saw a café offering homemade soup and sandwiches I went right in. It was quiet, with circular wooden tables, bright blue chairs and a retro feeling. And on the wall of the washroom, carefully displayed in a glass frame, there was another *Change The World* poster. This one had a photograph of Jonas Salk, the American medical researcher who discovered and developed the vaccine against polio, which used to afflict hundreds of thousands of people around the world, causing paralysis of the limbs.[878] And his words:

There is hope in dreams, imagination,
and in the courage of those
who wish to make those dreams a reality.

Hope. Dreams. Imagination. Wasn't this exactly what my journey was about?

When the waitress asked for my order, I chose the broccoli soup and she asked where I came from. On hearing that I was visiting from Sudan, a young man came over and invited me to join him. He introduced himself as Frank, from the Sto:lo First Nation, the People of the River. He had a smiling round face framed with short black hair.[879]

"So you're chasing after what makes us tick, what makes us want to build a better future?" he asked, after I explained the reason for my trip.

"Yes. I want to know what motivates people to live more sustainably, in harmony with nature."

"Well, let me tell you a story. When I was young, my father used to take me fishing every year to our traditional fishing grounds on the Fraser. Our people have lived here for ten thousand years, fishing the Fraser, so it's very deep with us. Some years were good, but in others there'd be a complete collapse of one of the salmon runs. One year it would be the coho; the next the sockeye. My dad was always working to help protect the salmon, going to meetings, negotiating with the non-native fishermen who thought we were taking more than we should, restoring creeks that had been silted up by bad logging practices, making it hard for the fry to survive. So I became very motivated. It didn't take much to make the connection between the collapse of the salmon stocks, climate change, the way capitalism worked, and the way civilization encourages human greed at the expense of nature. Can you imagine what it would be like for your people if the land was barren and no crops could grow? That's what it's like for us when the salmon runs fail."

"I'm from Ireland on my mother's side," I replied. "Our people survived the potato famine in the 19th century. It killed a million, and drove another million from the land. We only hung on by our fingernails thanks to an ancestor who married again after he lost his wife and all his children to the famine."

"That was really bad," Frank replied. "I'm doing a course on the impact of colonialism for my history degree at UBC—University of British Columbia. What motivates us to want to build a better world? For us, it's either that, or give up and die. We've spent most of the last century trying to get our treaty rights recognized, and helping our people overcome the impact of being sent to residential schools, where they were not allowed to see their parents, speak their own language or engage in First Nations cultural rituals. They also had to heal themselves from the terrible things they endured, including sexual assault and brutality."

"I know what you're talking about," I replied. "Some of those things happened to my ancestors in Ireland too—being forced to stop speaking the Gaelic language, and being sent to residential schools. Many Irish children had to endure beatings and sexual humiliations, too. I know what it does to people. Same British imperial origins, too."[880]

"I'm working to learn my language," Frank said. "It's a big thing with us, rediscovering our heritage. I've got a translation app that can translate my words into Halq'emeylem."

Frank proceeded to say some words into it, and the device responded in Halq'emeylem.

"Here, you say something."

"It's a lovely afternoon, I'm happy to meet you," I said.

His device translated my words into Halq'emeylem. Frank pressed a button and the device translated them back into English. *"I'm happy to pay for your lunch, and have you seen my big dick?"*

"What!! I never said that!"

"No, but it's sure good to see your expression!" Frank laughed. "I'm just messing with your mind—it's one of our sacred cultural traditions."

I laughed. "I can see that I'm not going to have an easy time getting answers from you!"

"About the lunch? I'm sure you can pay with your device."

I laughed again. "No. Answers about this planet and how we're going to save it, and why people here are so motivated."

"Oh, that. Okay, I can see I'd best put my serious hat on. It's the weekend, so it may be buried...." Frank dug around in his pack to see how I'd respond.

"Here, borrow mine," I said, pulling a cap from top of my backpack.

"Oh, I can see you've got me trapped," he said. "You're quite the determined character."

Too right, I thought. I hadn't come all this way just to fool around—though I was enjoying it.

"So what's your take?" I asked. "You're studying history; you must have thought about it."

"Yes, I have. I'm just delaying, since it's such a nice day. Say, which way are you going? I could give you a ride. We could talk along the way."

"I'm easy. I need to be back in Vancouver by seven."

"Oh, that gives you lots of time. I'm heading across the Fraser on the Port Mann Bridge and picking up a friend in Langley before heading east to the Seabird Island First Nation near Agassiz, on the way to Hope. There's a drumming circle there this evening I want to make. Can I give you a ride?"

And so it was, after he had unplugged from the café's parking spot, that I found myself travelling south with Frank in a black and white open-topped electric sports car—at least it looked like a sports car to me, with its space-age streamlining and large rear tires.

"Great little buy. Fifteen years old. Picked it up from a friend for four grand. Only costs me two bucks for a hundred kilometres. How cool is that?"[881]

"Do you use it to go to UBC?" The university was a good forty kilometres to the west, out by the ocean.

"I could, but I don't need to. I'm in the COT stream—Community Online Teamwork. It's much cheaper than the traditional courses, but still top notch. We do lots of MOOCS—massive open on-line courses. Last month I attended a series of lectures by Felipe Fernández-Armesto, who just happens to be the most famous historian of global civilization in the world."[882]

"He was in Vancouver?"

"No. He was in Oxford, England. He's over eighty. There was half a million of us hanging on his every word. I can't begin to imagine how rich he must be with that many students paying a dollar each. He sees civilization as an attempt by people to cope with the environments they find themselves in by asking questions and seeking solutions. He's quite the Darwinian, in that sense. He doesn't believe there's such a thing as progress. He's incredibly knowledgeable about the First

Nations people who inhabited the Americas before Columbus, and the cruelties, betrayals and sufferings our people went through after the colonialists arrived. It makes me feel quite sick about the whole concept of civilization, if that's what it took to earn the British their manor houses and their well-kept lawns, with tea and croquet in the garden."

"So I take it you're not a fan of syntropy," I said, hoping I'd picked up enough to be able to ask the question. "Do *you* believe there's such a thing as progress?"

"Yeah, I do. It's progress when you build a strong local economy, the way the Seabird Island First Nation has done. It's progress when you get yourself off drugs and start a business or get a steady job. There's no contradiction with syntropy. Seabird Island is up the Fraser Valley, just north of Agassiz. They've been self-organizing towards a more harmonious existence for years now, with sustainable housing, solar energy, local food production and a strong community economy. It certainly feels like syntropy to me. Maybe progress is when the increase in syntropy outweighs the increase in entropy."[883]

Interesting thought. Interesting definition.

"So why have people here done so much to build a just, sustainable world?"

"I dunno. Maybe because it's so friggin' beautiful—or because this is the west coast, where new ideas are like spring flowers that people welcome, not anything scary. When you're surrounded by such beauty, and educated enough to know that it's so threatened, when you believe in your ability to make a difference—maybe these are some of the things that encourage powerful vision and powerful action. Fernández-Armesto would argue that all species seek comfort and stability, and evolution is an adaptive change that occurs in response to threat; rapid evolution occurs in response to massive threat. So maybe the real question is why are people on Canada's west coast more attuned to the threat from climate change? Maybe the answer lies in nature's beauty: the contrast is so strong between what we see and love and what we fear could happen."

Thinking back about his comment I find it really insightful, but at the time I was pre-occupied with the practical aspects of his course, so I didn't pursue it further.

"When you're studying online, what do you do for community with your fellow-students?"

"We meet with our learning coach every week in a two-hour online tutorial. She sets us a question, such as 'What were the conditions that led to the French Revolution?' or 'What does history tell us about the psycho-social impact of defeat?' We have to research the question and come back with a cogent set of arguments, backed by sound historical evidence. One week we work on our own; the next we collaborate in teams using our screens. The interactive textbooks are amazing. They enable me to chase down anything I don't understand, and tap into my team's knowledge. There are twelve of us in my team: three in Vancouver and the rest in various parts of western Canada. Once a year there's a three-week residency at UBC where we meet for intensive study and get to know each other.

We can't all live on campus. The online learning makes it easier to care for my five-year-old daughter, keep my business running and keep my place in the drumming circle."[884]

"What does it cost compared to doing it on campus?"

"About half as much—but I won't need to pay a thing until I graduate."

"You mean your student debt has been deferred?"

"No. It's no longer classified as debt. I'll pay through a deduction on my future income for twenty-five years. It's how all college education is funded in Canada these days. I'm doing a full graduate degree course, so my deduction will be three percent; it's less for shorter courses."

"Your business: what does it involve?"

"It's called Sto:lo Soul Adventures. We offer experiential holidays where you get to live with a First Nations band, learn about their history, practice their crafts, make a drum, experience the wilderness and share your personal spiritual journey. We get a lot of young people, and it's really popular among the Germans for some reason. They cross the Atlantic on the Solar Sailor and take the high-speed train across Canada. I have a personal coach, Brent Hammond, who calls me every week and keeps me on my business targets. Trailblazer, his company's called. Great outfit."[885]

"You said drumming was really important...."

"Yes. It's how I listen to my heart. History's important, but when you study too much you can get lost in your head, like you're on a railway and you go off down a side-track. You're going along quite happily and then one day you wake up and realize that the world you're passing through is arid and dry. No juice, no feeling.

"Drumming opens up another track. It takes me to my spiritual home. I can't explain it if you don't know what I mean. Most white people, they're just visiting, moving from one place to another. My people have been here for ten thousand years. They know every bush, every creek—at least, they used to. And my ancestors—they're still here. They're not always happy, but they're very patient. I'm learning how to listen to them so that I can pass on their wisdom. It starts with respect, and gratitude for their existence. I'll join them one day, and if we keep to our traditions maybe in a thousand years the young people will seek my wisdom. That would be a laugh.

"We're brought here and given a job to do, but I need something to remind me who I am and why I'm here. The drumming circle does that for me, as well as some other ceremonies I can't tell you about. Can you imagine if it was possible to understand the history of different civilizations from the perspective of the heart, and the billions of realities people have lived through, each with their own cultures, their own heart-spaces, their own stories of hope and love, passion and pain? That's what I believe history should be. Get behind the stories of trade and conquest, find the common heart deep in the culture. My people share the same past with your people. We all come from Africa. We're all descended from the same primates, the same early mammals, the same first forms of life. How cool can that be? Here, give me your hand."

Frank took one hand off the wheel and held it against mine, his fingers outstretched.

"Your hand descended from ancestors who travelled out of Africa and into Asia and Europe. Mine came from ancestors who travelled up to Siberia, across the Bering Strait and down the coast in canoes to populate the Americas. For thousands of years we were separate. But look at our hands. Do you see any difference? Do you feel any difference? We need to learn to love each other as one family, not as different tribes and nations who are trying to get the better of each other. That's the flow of history I want to see: the long rainbow arc that shows the unfolding heart of all history."

"And they teach you that at UBC?"

"What? Sorry, I was miles away.... Yes, they do, but it's not history as most people know it. I'm doing a course with an incredible woman called Nimah Kaytum. She's Mayan, from the Yucatan. She traces the heart-lines of different cultures around the world to show their inner relatedness. She shows history through a more female perspective, since it has been mostly women who have been the guardians of the heart-lines. She shows the relationship between the strength of the heart-lines, the stability of the family, the connection to the land, and the hearths where people gathered and told their stories. It's fascinating. But look, we're in Langley; this is where I'm picking up my friend. Can I drop you somewhere?"

"Right here will do," I said, trusting to fate. "Or maybe over there, next to that park. It's been great meeting you—I'm going to remember our talk. Good luck with your studies!"

"And good luck with your travels!" Frank got out of the car and shook my hand. Then I strolled across the road and sat down on a bench next to the park to gather my thoughts.

I could feel how satisfying his life was. It made me envious. We still had the church, but hardly anyone I knew went to church anymore. I recalled Friday evening's Song of the Universe, and this morning's Taizé service at the Farm. They had made me feel connected. Maybe that was it—we needed new ways to celebrate our connectedness as well as our differences. But the Song of the Universe had been just one evening. I wanted a whole life that would do it.

33
Sustainability in the Suburbs

THE PARK WAS quiet, but I could hear children's voices through the trees so I went to explore.

Just beyond the park there was a large adventure playground filled with tree houses, wooden castles, rope ladders and climbing nets. Parents were watching from an overlooking slope and there were adults in the playground too, working out on various exercise machines.

"You new around here?" a woman asked, a touch defensively. She was small, with short red hair, blue jeans and a red blouse, very country western, and rather mother bearish in the way she approached me.

"Yes," I replied. "I'm visiting from Sudan to learn how Vancouver became one of the greenest cities in the world. My name's Patrick. I'm on my way back to Vancouver and was dropped off here. This is quite the playground."

After I explained about my trip, she softened towards me. "I'm Jenny—pleased to meet you. That's my daughter over there in the pirate's castle, and that's my husband on the pull-up bars. You've come to the right place if you want to learn about urban greenery. Have you been to Rose Crescent yet?"

"Rose Crescent? No. What is it?"

"It's where we live, about ten minutes from here. It was one of the first suburban retrofits. We've turned a slice of suburbia into a village, and it's going rather well—or so we like to think."

"That's fascinating. Is it the kind of place I could visit?"

"I'll ask my husband; we'll be done here shortly. If he doesn't have other plans you could come back with us and I could show you around. I'm an urban planner, so I live and breathe this stuff. Do you have any children yourself?"

"No, just me."

"I never thought I'd have kids when I was younger. My mother was always complaining about how much time she had to spend ferrying us around between my gym classes and music and my brother's hockey, all while holding down a full-time job. TV was her solution whenever we were too wild and noisy, so I grew up feeling that I was just another thing she had to look after, along with the dog, the goldfish and the husband."

"Is that what made you become an urban planner, to change things so that it would be easier for other parents?"

"Quite possibly. I never thought I'd end up living in the suburbs, of all places, but when I got my job in Vancouver it made sense. So we thought we'd try to make the most of it. Here's Joel."

Joel was large and clean-shaven, wearing a yellow tank top and shorts.

"Hi mate! Pleased to meet you. You been chatting up my lovely wife?" he said in a strong Australian accent.

Jenny introduced me, and so it was that I found myself accompanying them back to their home, with five-year-old Zara riding in Jenny's cargo-bike and me on a rental bike-share on safe, comfortable bike lanes separated from the traffic by rows of young trees.

The streets we passed through were suburban, and apart from the bike lanes and there being more trees, and the solar panels, of course, it was much like a typical suburb from my time. But when we turned the final corner there was a speed bump and a large sign that said *Rose Crescent*, decorated with hand-painted flowers. The street that used to be three lanes had been narrowed down to one winding lane, with passing and parking places. The freed-up space had been filled with apple and cherry trees and raised beds filled with tomato plants and runner beans, and three chickens ran across the road in front of us as we arrived. Two people were working on one of the beds while children played nearby.

"Welcome to Rose Crescent!" Joel called out as we pulled into their home. "We've got to get those chickens under control, Jenny—they'll create havoc with the vegetables."

"Agreed—I'll speak to Pam," Jenny replied. Then to me—"Let me take you on a tour. Then we can have a beer."

We parked our bikes and Jenny walked me down the road. Several of her neighbors were out enjoying the Sunday afternoon, and she greeted them as we passed. It was much like Dezzy's street, but in suburbia instead of the city, with bigger lots and more space.

As the crescent turned I was surprised to see a complete village center with three-story townhouses on either side of a courtyard, with a store, a café and a small canopied stage.

"Whoa—how did that get there?" I asked.

"A lot of hard work!" Jenny replied. "I'll fill you in on the history later."

The road curved around the bandstand to the south and the area to the north was pedestrianized. There was a community notice board, and Jenny pressed a few buttons to show me a photo of the street as it used to be: normal suburbia, same as anywhere.

"Do you know what the joke was when we moved here? 'How do you know which house you live in, when you live in the suburbs?'"

I paused, waiting for the answer.

"You drive down the street clicking your garage door opener until you find the one that works."[886]

I laughed, and Jenny said, "So now look." She clicked to the next image, which showed a group of people standing around a map. "That's us discussing the plans, deciding how we wanted things to be." The next images showed houses being cut in half and towed away and the new village center being built, culminating with a street party with music and dancing.

"What fun! How many new homes do you have in place of the ones you took away?"

"We removed six single family homes and replaced them with twenty-four townhouses, including the shop and café. Twelve on each side. We own four ourselves."

"You own them?"

"Yup, including the shop and café, which have rental units above them. Property of the Rose Crescent Community Development Corporation, which is an arm of the Community Association."

The design was simple, four groups of six connected townhouses surrounding the circular courtyard with its open stage.

"We built them to the passive house standard, as all new homes are required to be these days. They've got carbon-absorbing cement in the south-exposed floors, eight-inch studs, solar-coated super-windows, rainwater capture with storage tanks under the houses, the latest composting toilets, and a shared greywater treatment system. They only use a small amount of power, thanks to the super-efficient lights and appliances, and the solar covers most of it averaged over the year as a whole."[887]

"How much water does the rainwater harvesting gather?"

"Enough for about half their needs. There's a rule concerning the use of water in new buildings called 'no net increase.' Developers are required to demonstrate that they have saved as much water in existing buildings as they'll need for a new building."

"How does that work?"

"There are companies that specialize in saving water, and each time they do a water-saving retrofit on an older house they earn water-credits, which they sell to developers who need them to get a building permit. We had to buy water credits for the townhouses, which gave us a reason to make them as water-efficient as possible. It's a neat system: it pays people to make their homes more water-efficient while saving the municipality millions on new water supply and sewage disposal. We paid for more than a hundred houses to become more water efficient to earn the water credits for the twenty-four new townhouses. Each house gets half its water from rainwater and half from the saved water from four other homes, so there's no additional burden on the city's water-supply."

Water-credits. What a simple idea.[888]

On the north side of the courtyard the ground floors of two units had been combined to create a small store and a coffee shop. "The store's really handy," Jenny said. "There's no parking, so people only come on foot or by bike, and there are bicycle carts you can borrow to get your shopping home. Before it opened we had to drive five kilometres to get to the nearest store.

"The café can seat fifty when we move the tables back. It gives us a place for community meetings, and people can drop in for a coffee anytime, knowing they'll meet someone they know. Francesco, our neighbor across the road, tells me he's written half a novel here because he likes the atmosphere so much. In summer there's a Saturday morning market in the courtyard where people exchange or sell their surplus food. But speaking of refreshments, how about we go back to our place for that beer?"

Later, sitting in Jenny's back garden with a glass of home-brewed ale, surrounded by roses, sweet williams, nasturtiums and rows of fresh young lettuces and beans, I learned how when they moved here it had been a typical suburban subdivision where most people didn't speak to each other much.

"Suicide by silence," Joel said, in his colorful Australian accent. "I couldn't believe people wasn't even talking to each other, just driving into their little boxes and doing whatever they was doing, all on their ownsome. I made Jenny promise that if we ended up living here we'd shake things up a bit, have a bit of a party. Where I grew up, tiny place called Yarragundry, west of Wagga Wagga, it would have been completely unthinkable to ignore your neighbors like that. You'd have been run out of town if you kept it up for more than a week. Right drongo, you'd have been."

They started by organizing a block party, Jenny explained, which led to the formation of the Rose Crescent Community Association, and then they formed a Transition Street Group that became the hub for new ideas. The raised beds and the street narrowing were inspired by what was happening in Vancouver. A cross between a pocket neighborhood and a Better Block: that's what people said they wanted.[889]

"But how did you go from a few raised beds to redesigning the entire street?" I asked. "That's a huge leap."

"We needed a place where we could meet together," Jenny said, "like a village center. But none of us knew how to do it, not in the middle of suburbia. Then one day, Jordi, who was an architecture student at the time, son of the Petersens over on Camas Way—he arrived with a set of drawings he'd been working on with some local children. They were not short of ideas, he said. They'd removed two of the houses and replaced them with a playground and a barbecue pit. Needless to say, it didn't go down very well with the people whose homes they wanted to remove.

"Then one of the owners said that if someone was willing to pay him for his house, he wouldn't mind the sacrifice. So that got us thinking and we asked Jordi to come back with some more ideas. To most people it was ridiculous to think of moving six houses and replacing them with twenty-four townhouses. When Jordi took the idea to a developer he was told it would be a ten million dollar project.

How could a bunch of neighbors take on something that big? Here—have another cookie. Joel made them this morning.

"The global Transition Town movement was pretty advanced by then, and every year they published a book of success stories. In one, a group of neighbors in Milan, in northern Italy, bought two houses and turned them into three by joining them together, and they used the profit to create a workshop where young people could develop their skills, working alongside retired people. So we took the idea to Vancity, our credit union, who are well known for supporting community endeavors, and the woman there suggested that we form a not-for-profit community development corporation and hire a project manager.[890]

"We were a bit overwhelmed by the idea, but at the same time we knew that the only way we could persuade six people to sell their homes was if there was something in it for them. So Vancity gave us a grant to hire a consultant, who crunched the numbers and showed that if we offered the owners ten percent above market value and re-invested the money in twenty-four townhouse units we'd be able to clear a profit."

"So you became the developers? What persuaded the owners of the houses to sell?"

"The money, basically. Ten percent above market on a million dollar house is a hundred thousand in the bank. Three said they'd be happy to move into one of the new townhouses if it enabled them to stay in the community, saving them a quarter million by moving into a smaller house. It took us a year to assemble the plans and put it all together."

"What about the planners? Did they allow the increased density?"

"Not at first—it wasn't zoned for it, but only because nobody had ever proposed such an idea. But the Official Community Plan called for higher density in built-up areas to meet the demand for housing while protecting the farmland from development, so we had a sense that they'd be supportive. It helped that I was a planner: I knew the rules and could talk the lingo. They ended up creating a Sustainable Suburban Village Zone, which allowed us to build the townhouses. The planners were all for it. They call us their PIMBYs, because we were saying 'Please In My Back Yard.'"

"What happened to the people whose homes were taken down while they were waiting to move into one of the new townhouses?"

"We didn't take them down. We sold them and had them taken away. There's a good market for them, since it's cheaper than building from scratch. You saw the photo of one being cut in half. But to answer your question, we paid to accommodate them for the eight months it took to build; that was built into the financing."

"So now you own the whole development?"

"We sold twenty units and kept four, including the café and the store. They bring us a useful income that we use for other purposes."

"Such as…?"

"Every year we canvass the street for the best ideas, and then we take a vote.

Last year, we had $36,000 to play with. A third goes to a charity we choose each year, a third to support sustainable development here in Rose Crescent and a third to a global cause. That's the formula we agreed on. Right now we're supporting two young people with scholarships and we're helping an older couple get the support they need for their autistic son who lives with them. Last year we paid Joshua—he's one of our neighbors—to work in southern Ethiopia for a year, helping the Solar Electric Light Fund set up a solar microgrid in a series of villages, providing power for all their needs, including their health center and irrigation for their fields. He Skypes us regularly on how they're doing; it's really encouraging."[891]

Wow. I know I've been using the word a lot in response to things I saw and heard, but this was amazing, and there was probably a lot more that Jenny hadn't told me.

"You said your zoning was for a sustainable suburban village. Did that come with any special requirements?"

"Yes. There were lots of conditions, mainly to prevent the quick-buck developers from exploiting the new zoning. There's a long sustainability checklist, and you need to score sufficient points to qualify for rezoning. It keeps the greed-minded developers at bay and rewards those who come with a shift in their thinking.

"We won points for the support we received from the Rose Crescent Community Association, and points for the street conversion, the food growing, the community market, the affordable rental units, the community café, the store, and the green features in the townhouses. It's a great system, because it sets a high benchmark and encourages others to get involved in community development. But you know what I like best about it? We've actually become the village I always wanted, where everyone knows your name and people's doors are open when you need them. It's doing wonders for the local children. Life would have been a lot easier for my parents if my brother and I had grown up like this."

"Did you have any push-back from people on the surrounding streets?"

"Yes, for a while. But it's usually the noisy ones who think they speak for everyone. When we did a house-to-house survey we found that seventy percent of the neighbors liked what we were doing as long as it didn't increase the traffic or bring any crime. The new townhouses are all car-free by covenant. There's plenty of public transit, bike-routes and carsharing, so you really don't need to own a car. We've been able to sell our car, which saves us several thousand a year. There are great local bike routes, it's easy to get to Vancouver on the SkyTrain, and we're only five minutes by bike from open farmland. Beats living in Vancouver any day."

"It's great for our Zara, too," Joel added. He had been quietly weeding a row of carrots and onions while we talked. "She's got half a dozen little tykes whose homes she can show up at without us worrying. If someone else's kid shows up we just phone the parents to say they're here."

"It's wonderful for children to grow up in a community like this," I said. I had fond memories of playing with other children in the villages where we lived in East Africa.

"Are you a planner with Vancouver, or one of the local municipalities?" I asked Jenny.

"Vancouver. My colleagues are always making jokes at my expense, since my day job is about squeezing more people into Vancouver and then I take off after work for life in the suburbs."

"Has there been much take-up on the new suburban village zoning?"

"Not as much as we'd hoped for: maybe ten or fifteen so far in the region. It's Jayashri, our consultant, who's doing all the legwork. She's set up a Sustainable Suburbs website and she invites people to contact her. For every ten groups that approach her, one is willing to go all the way, and she has investors willing to lend the money. I think it's the fact that she's *not* a developer that wins people's confidence. It'll take time. It's always an upheaval for the folks whose houses are sold. But I've had plenty of people tell me they'd happily sell for ten percent above market, so I think it's more to do with the points needed for rezoning. The Block Party Day that happens across the province on the August long weekend is encouraging more streets to self-organize. That's the key. It trumps everything else when it comes to change."

There it was again, that phrase: self-organize. "As a planner, do you encourage self-organization in Vancouver's neighborhoods?" I asked.

"Absolutely. We run training programs on everything from how to organize a block party to how to green your street, and we have an active Green Neighborhood program. It's a positive driver for house prices, for better or worse, since people want to live on a street with a strong sense of community. It makes them feel safe, as well as being more sociable."

"The annual Block Party Day: how long has that been happening?"

"About ten years. It's become an annual event. We've joined up with the next block over and we take it in turn to organize. It helps extend our sense of neighborhood."

"This is a crazy thought, but could there be such a thing as an ecosystem of suburban neighborhoods?"

"It's a neat idea," Jenny replied, taking a sip of beer. "In a true ecosystem all the parts relate to each other. They're interwoven. But in most subdivisions it's as if they are deliberately *not* relating to each other. They're like pods attached by an umbilical cord—the road—to the city. There's no connection. They were supposed to make people happy but in reality they make people sick, fat, angry and depressed because they cut people off from the connections they need to thrive, which are to people and to nature. There's only a few that stand out as positive examples, like Village Homes in Davis, California. They really got it right there."[892]

Jenny was full of passion; it was clear that her work fulfilled her.

"We're building a social organism," she continued. "If—no, *when*—when the other blocks organize in a similar way we'll have a structure that allows us to connect and communicate. We'll be able to look at the food we're growing, the energy we're using and the businesses people are running and find ways to support each other. I'd love that. We'd be a real community."

"She sure gets going once you get her started!" Joel said, sitting up from his weeding. "If someone could invent an energy device to attach to her brain we wouldn't need solar power!"

"Ah, but you love it, you've got to admit," Jenny responded with a smile.

"She's my sweet honey-bee," Joel said. "She brings sweetness to me and sweetness to the world. But man, she sure does talk a lot!"

"And, like you *don't* talk? Whenever we're on the bus, it's 'Hi mate, how's it going?' and 'Hi mate, lovely day, init!' Joel's the ultimate pollinator, connecting with everyone."

"Well, that's just our way. It's Australia's gift to the world." Then turning to Jenny, he said, "Aren't you glad I pollinated with you?"

Jenny wrapped her arms around Joel from behind in a romantic embrace, and then Joel moved quickly to flip her over onto the ground. I had to put my beer down, I was laughing so much. Then he was straddling her singing 'Pollinate, pollinate, my sweetest little Jenny-mate,' and they rolled over each other laughing.

It was hilarious, especially when their daughter Zara came running from the house and tumbled on top of them, followed by their small white puppy.

After the rumpus, when Jenny had composed herself, I asked her about any exciting things she was working on as a planner.

"Well, the new Community Farm Zones are creating a lot of interest," she replied, brushing down her jeans and topping up my beer. "They allow for the development of a community farm provided the neighbors agree; they serve the neighborhood with a weekly market; and they encourage children and people with disabilities to get involved. We're also seeing some interesting projects on the run-down industrial lands. There's one on the Fraser River where a cluster of electronics businesses are collaborating in an industrial ecology network, working together on research and development with a materials recovery pool to minimize their wastes. And there have been exciting new developments like the Agora in the new Urban Farming District.

"But the thing that's generating the most interest," she continued, "is subjective planning, or intimacy planning, as some people call it."

"What's that? It sounds different."

"It's not really; the best planners have been doing it forever. They just never gave it a name. It's about fixing up a neighborhood to maximize the convo-zones and eliminate the dead-zones, or zombie-zones, as some people call them."

"Zombie-zones!" little Zara shrieked, running around the garden. "Zombie-zones!"

"Joel, can you look after her while I finish talking to Patrick? So as I was saying, it's about fixing places in the city that have no charm, that attract graffiti and crime and feel intimidating, especially to women. We invite volunteers from the local Village Assembly to walk an area with us to identify the zombie-zones and we encourage them to come up with ideas to 'charm' them and make them more intimate. Sometimes it's a simple matter of planting a tree; sometimes it requires reconfiguring a street layout or asking a store owner to create a window full of interest instead of a blank concrete wall. We rough up three or four ideas and broadcast them on the CityTV channel, asking people to vote for their favorites. It's about increasing the level of happiness in the city, which is seventy-five percent interpersonal and only twenty-five percent personal. It's about encouraging social intimacy and connection, creating places where people want to linger and hang out. Have you come across the Vancouver writer, Charles Montgomery? He wrote a great book, *Happy City*, which set us on this path."[893]

I said no, I hadn't, but I recalled Laura showing me the mosaic at the Tree Frog Café and talking about black holes in the city, and urban acupuncture. Must be the same thing. I pictured myself sitting at the café with her, sharing a kiss... a lingering kiss.

"On a larger scale," Jenny continued, oblivious to my little romance, "we're just coming out of a period of change. The Regional Growth Strategy used to focus on the land around the transit stations with high-rises of twenty, thirty, sometimes even forty storeys. It was fine on paper, but people were forced to buy condos in the high-rises because they couldn't afford anything else, and survey after survey showed that they really didn't like them, especially if they had children. And nor did their neighbors down on the street below. The sky-folks wanted to be closer to the street while the ground-folks grieved for the loss of sunshine, and felt drowned out.[894]

"The new approach concentrates on developing four-storey, wood-framed, low-rise buildings on the main urban corridors. Wood's a very sustainable resource when it's harvested ecologically, and it's a great store for carbon, while high-rises are generally made from concrete, glass and steel, none of which are very sustainable. By running the buses and bike lanes down the arteries and planting more trees you get a more European ambience, like Barcelona or Copenhagen.[895] And the rooftop greenhouses are creating a whole new urban scene; the sky-homes are all the rage with the Changers, the artists and the neo-boes."

"The neo-boes?"

"Neo-Bohemians. They're the ultimate Changers. They're totally into doing things for themselves, growing their own food, stitching their own clothes and making their own furniture. They're really well organized, with cells around the city that work together to push for progress on whatever issue is hot. They have their own culture-jams and community feasts, as well as being tightly knit into their neighborhoods."

The Changers. I'd forgotten about them—people born in the early decades of the century. "Are you and Joel Changers?"

"You make us sound like a tribe of aliens," Joel said with a laugh.

"I'm sorry," I said, flustered. "It's not a term I'm used to."

"No worry, mate. We're supposed to belong to Generation Y, but it's more fun identifying with the Changers, who want to get things done."

"It sounds like you and Jenny belong with the Changers," I said. "Look, I hate to leave—I'm really enjoying talking with you, but I need to be getting on soon. What time is it?"

"Five-thirty."

"Before I go, can I ask one last question?"

"Fire away," Jenny said. "It's Sunday and we've no plans apart from putting our feet up, having dinner and watching *Bank Run*. It's a detective series set in a tax haven. It's got scalding hot sex: they do it in two versions, one family friendly and one strictly for adults. We wait until Zara's sound asleep. So anyway, what's your question?"

Now I was distracted by the scalding hot sex, but I dragged my mind back to my question. "How do you relate your changes here to the wider world, and the larger threats to humanity and nature?"

"Whoa, that's a biggy! Do we have any more beer? Joel, why don't you answer that while I fix us another drink?"

"Jenny doesn't like talking about that kind of thing," Joel said, when she had gone into the house. "Freaks her out... the thought that here we are in our own little paradise while the rest of the planet's going ass-first down the tube."

"And you? What do you think?"

"Well. I think something pretty big's gonna hafta happen if we're not gonna to come a gutser. My granddad used to work the railways back when I was a brat and sometimes he'd let me ride the caboose with him on one of those puffer trains they ran for the limeys. It feels like that today: we're the caboose, all green and spiffy, but the old puffer's still dragging us towards the cliff. Something's gonna hafta happen, like it has here and in Vancouver. But I've no idea what. The greenies have got things underway, but globally they're still being outpaced by the wallies who make so much bloody money from the old ways, stupid fools. And there's still way too many nations burning the old fossil fuels and trashing their forests and farmlands. It's like we're still living with the bloody slave-trade before it was finally abolished."

"I can see you've got Joel going," Jenny said, returning with a new tray of drinks. "I trust you can follow his Aussie way of speaking?"

"No problem, mate! What kind of work are you involved with, Joel, if you don't mind my asking?"

"I'm a consulting engineer on hydrological matters: dams, wells, solar desalination plants. Gives me a sense of what's happening globally. I've got various projects in China, Africa and the Middle East."

"I love what you've done here," I said, "but what's it going to take for the rest of the world to understand that sustainable living is so much better for everyone, and not worse, as some people fear?"

"What's it going to take?" Joel responded. "I dunno. We're always looking for new ways to get people inspired about a more durable, resilient future. It's like a rockslide, or an earthquake. It's impossible to predict when something will trigger it, but once it does, there's no looking back. It's where syntropy meets complexity. Syntropy supplies the forward pressure, complexity sends it down a billion different pathways, but it's serendipity that determines the breakthrough, and even the biggest Watsons with their petabytes of memory can't get the better of serendipity."[896]

I wanted to talk all night, but it was almost six and I'd promised Dezzy I'd be home for her family dinner and I'd no idea how I was going to get there.

"Local brewed amber ale or summer fruit punch?" Jenny asked, distracting me from my problem. "The punch is a hundred percent local, harvested right here on the street last summer, no alcohol. And I've put together some cheese and nibblies."

"I'd better have the punch, since I'm cycling. I'd love to learn more about your work, Joel, but I've got to get back to Vancouver for my dinner party. I'm heading for a street a dozen or so blocks south of City Hall. What's the best way to get there from here?"

"I'd say use your rental bike to Langley Town Centre, then take the SkyTrain to Columbia and change to the Millennium Line: the new extension will take you right to City Hall. Alternatively, you could take the new Interurban rail to New Westminster and pick up the Millennium line. Either way, it should take you about an hour."[897]

"Thanks. And thanks for all the time you've given me! You've got a lovely home and a delightful neighborhood."

"Well, we do our best, mate," Joel replied. "Mind how you go on that bike, now. And watch for those chickens!"

34
The Dinner Party

MY BIKE-RIDE TO the SkyTrain was easy and the terminal was new, surrounded by a busy urban village and a pedestrian plaza where people were hanging out in open-air restaurants. Near the entrance I saw what would be the last of the *Change The World* posters that I had been enjoying, this one with a photograph of the American football coach, Vince Lombardi, and these words of his:[898]

> *People who work together will win, whether it be against complex football defenses, or the problems of modern society.*

They were really encouraging, these posters. Then as I climbed the stairs to the platform, I heard a squeal. Three people came down the other side on a shiny green plastic slide, landing at the bottom on a cushioned air-bubble and brushing themselves down while laughing their heads off. A happy smile burst inside my heart.[899]

The ride to Vancouver was mellow, the evening sun shining in from the west. I didn't want to strike up a conversation so I contented myself with gazing out the window as the SkyTrain followed the Fraser Highway to Surrey and crossed the river to Vancouver. The landscape was mostly low-density sprawl. The highway had good separated bike lanes, and there were areas where the right-of-way along the SkyTrain line had been used to install solar arrays. Many homes were growing food in their yards, and most had solar systems on their roofs. At each station, the density increased and sky-homes proliferated on the rooftops.

In no time at all I was back in Vancouver, walking up the hill to Dezzy's place. Cyclists came and went in a peaceful dance. I was close to the heart of the city, yet on this Sunday evening it was permeated with such a gentle quiet. With so many people walking and cycling, the electric cars and buses and all the abundant greenery, there was no doubt that a transformation had taken place.

Passing a roundabout, I saw a curved cob bench decorated with painted wildflowers so I sat down to take in the moment. My time here was running out fast—too fast. I wanted to breathe it all in.

I loved this future. I wanted to stay here. People here were *happy*, in spite of the continuing global worries. They had seen that change was possible, and they had hope. The sound of Jewish Klezmer music came from an upper window, all bent notes and vitality. A sign in the window read *Peace in Palestine*.

A small voice in my mind said: *This is their world, not yours. Love and let go.* And with that thought I picked myself up and walked the last few blocks to Dezzy's home.

"Come in! You're just in time!" Dezzy said. "We're about to sit down to dinner. How was Joey's Farm? Did you have a good time?"

"It was great! I didn't want to leave!"

"Hi Patrick!" Lucas said, coming out of the living room. "Word has it that you've become romantically enchanted with a certain someone over at Joey's Farm."

"You look surprised—or should that be embarrassed?" Dezzy chuckled.

"Don't worry," Lucas said. "We didn't drone you. Laura's a good friend. She gave me a call when she got back from the Farm, and we were chatting. And Dezzy told me you had a meeting with Li Wei-Ping, who Laura works with, so we put two and two together. Seems like you two put two and two together too!"

"And tootoo to you too," I replied with an irritated laugh. "Now you've got me wondering how much more you know!"

"Hi Patrick! How's our travelling philosopher?" Leo greeted me as he came out of the living room, followed by Betska. "I hear you've been taking a page out of Pico's book."

"Give him a break, Leo," Betska replied, "and stop dragging Pico into every conversation. How about a hug? It's good to see you, Patrick."

Her embrace was lovely and unhurried, and it allowed me to catch my breath. I liked being welcomed home like this, as if to a family of my own. A family of my own, not just me and Daria....

"Welcome home, Patrick," Aliya said, coming out of the living room. She was wearing an even more exotic dress than the night I had first met her: a long black silk evening gown that shimmered with stripes and colors, matching the braids in her hair.

"You look stunning," I said. "How do you do it?"

A large black man came into the hallway, greeting Dezzy with an affectionate hug. Then Jake came screaming down the stairs and threw himself into his father's arms.

"Daddy!" he shouted exultantly. "You're home! You're home!"

"So how have you been, young man?" Thaba asked his son, lifting him up onto his shoulders.

"Patrick, this is Thaba Mabaleka, my ex-husband. He's joining us for dinner," Dezzy said. "I'm hoping he might help us with some of those questions you were asking, if Jake will allow it."

"Pleased to meet you," Thaba said, in a deep, sonorous voice. Then putting his son down, he said, "Jake, I'll come up and see you in a moment and maybe read you a bedtime story."

Jake yelled, "Whoopee!" and rushed back upstairs.

"Come on in, everyone," Dezzy said. "Aliya, can you offer Patrick a drink and see if anyone needs a top-up? Soluna will be joining us as soon as she's put her children to bed."

"So tell me about this work you're doing," Thaba said, as we moved into the living room, where the table had been set for dinner with lavender placemats and a vase of summer flowers—red roses and purple irises.

"I'm trying to pin down what motivated people to make such a big effort to change their world," I answered. "Was it their mind-space? Or perhaps their soul-space?"

"Soul-space: now there's a concept I can enjoy," he chuckled. "I'm a physicist at the University of Washington in Seattle and ten years ago we could never have spoken about something as immeasurable as soul-space. But following Satyanendra Mukherjee's breakthrough work on syntropy there's been a lot of talk about that kind of thing. I have several students working on related themes, carving it up for their Masters and PhDs. I even have one who's researching the physics of intention and agency in a choice-restricted matrix. The physics of intention—can you believe it? There's a wonderfully rich debate taking place in science these days about the nature of reality, and what it includes."[900]

Thaba was a big man, with an even bigger presence. He had a warm smile and tight, curly black hair. He was wearing a colorful loose African shirt with the image of a springbok in red, green and black on the front. On his wrist he wore several beaded bracelets. It was easy to feel a bit drab next to him.

"Are any of your students looking at the role intention plays in the evolution of civilization?" I asked.

"Well," he said, leaning back on the sofa and crossing his legs, "that would be a pretty big topic. You're talking the psychobiology of entire civilizations. I haven't heard of any, but consciousness research is very fashionable these days, so it wouldn't surprise me if someone was."

"Are we ready to eat?" Dezzy called out. "Soluna's arrived. Thaba, could you read Jake his story so that we can get started?"

"How was your trip to Joey's Farm?" Leo asked, as we took our places at the table.

"Amazing! I had no idea they'd be using horses, or that they'd have such a strong sense of community. Have you been there?"

"No, but it's on my list. Working at the supermarket ties up most of my time, and I need what's left for my reading. From Socrates to Syntropy, remember? The philosophy course I want to teach in China, if I can find a college to accept me."

Dezzy had been busy in the kitchen with Lucas and the result was a creamy onion soup, followed by a salad picked fresh from the garden and a broad bean and zucchini rice pilaf served with hemp and sunflower seeds, yoghurt and mint, topped with nasturtium flowers, served with a pleasant white wine from the Cowichan Valley on Vancouver Island. My taste buds still remember the summer flavors as I write this today. For the wall-art Dezzy had chosen a stunning piece that showed a human emerging from an egg, emerging from a cluster of atoms, emerging from a supernova explosion. I *liked* this new digital art revolution.

As well as the food and the art, Dezzy had prepared a sumptuous mental menu, which was the reason for Thaba's and Soluna's presence.

Soluna was a biologist from UBC. She was a small woman with long brown hair who arrived riding one of the standing mobility devices I had seen on Friday when I was exploring the city. I learned that she had been paralyzed following a snowboarding accident some years ago and the device allowed her to move around vertically and sit when needed, as she did for dinner. She was a long-time friend of Dezzy's, and how she coped with having children as well, I never did learn.[901]

"Friends," Dezzy said, when she had served the soup and Thaba had returned, "I have been wanting to throw a dinner party like this for years, ever since I started hearing about syntropy theory. And then our new friend Patrick came knocking on my door, asking all sorts of penetrating questions, and it struck me that now would be a good time, particularly since Patrick has to leave for Portland later this evening for the next leg of his journey. What I am hoping is that we can get a better understanding of what syntropy is, and what it means for us all."

Yay! I thought to myself. Finally!

"There has been so much talk about syntropy," she continued, "but I doubt there's any of us—apart from Leo, I suspect—who could give a clear explanation. So I invited my good friend and ex-husband Thaba to join us. I thought, if we're going to understand syntropy, who better to tell us about it?

"It's not my intention to turn this into a seminar; it's just a dinner party with friends, but unlike some of the theories physics has presented us with, this one seems different. As I understand it, it erases the distinction between the inner and the outer world, and if that's true we all need to be better informed. I'm also hoping it might help Patrick, as he tries to puzzle out what lies behind the changes we've been able to achieve here in Vancouver."

I felt both thrilled and daunted. Would I be able to follow the discussion without making a fool of myself? I had a degree in environmental science, but when it came to physics I felt like a shrimp in an ocean of highly evolved sea-life.

"Thaba, would you be willing to get things started?" Dezzy asked.

"I feel a bit self-conscious in Soluna's presence," he replied. "But I could start by talking about the way we see things in my physics faculty. I'll try to use plain English."

"As long as I can follow along, I'll be happy," Betska said.

"Me too," Aliya said. "I've got a hunch that syntropy is a lot more important than I've understood so far. This pilaf is really delicious. Thanks, Dezzy—and Lucas!"

"So, syntropy," Thaba started. "Where to begin? Let's start with a toast to our host, Dezzy, who has put together such a wonderful meal for us."

"And to Lucas!" Dezzy said. "He did most of the cooking."

"To Dezzy and Lucas!" Betska exclaimed as we all raised our glasses.

"So," Thaba began, "If I'm going to do syntropy justice I need to go back to the beginning of modern science in the 16th and 17th centuries, when Copernicus, Galileo, Kepler and Newton showed what good results you could get when you marry experimentation with detailed observation and measurement. Everyone knew that when you dropped an apple it fell to the ground. But no-one had made the effort to measure its rate of fall. So when Newton finally buried himself in the numbers and did some serious head-scratching, out popped the theory of gravity."[902]

"Was he really sitting under an apple tree?" Betska asked.

"Who knows? The point I want to make is that when science got started, external reality was seen as something solid and real, unlike thoughts and feelings in the realm of consciousness. Those were left to the priests, and considered their realm of expertise. Them and their inquisitions. We need to remember that there was a time when a discussion like this could have gotten us tortured, and burned alive at the stake.

"Believe it or not, the separation continued for centuries. Science was about the material world 'out there,' even when it delved into the working of the brain. Matters of the mind and soul were left to the psychotherapists, priests and shamans. The objective material world was on one side of reality; the subjective world of consciousness on the other. And as a scientist, woe betide you if you crossed the line. That could put your career at risk. Science required measurability and solid data, not the soft subjective stuff that goes on in the realm of consciousness.

"Using this model of reality, things proceeded smoothly for almost four centuries. Science was able to unravel the secrets of chemistry, electromagnetism, the human body and much more, bringing unparalleled progress. But then quantum theory arrived, pushing the conscious observer onto the scene as a critical factor in the determination as to whether a quantum-scale entity would express itself as a wave or a particle. That was a problem, and even the leading quantum physicists said if you thought you understood quantum theory, it was proof that you didn't. It was easier to concentrate on crunching the numbers, which gave absolute proof of the validity of the quantum model, than try to resolve the philosophical quandary at the heart of quantum physics."

"And there was me thinking I was dumb because I could never understand it," Betska said.

"You're not the only one," Thaba replied. "Even Einstein had difficulties with it. He went to his grave rejecting the uncertainty principle, and the notion that something might exist without any causal explanation, as it appears to do in the quantum paradox. Can you pass me some more of that wonderful pilaf? There's nothing uncertain about that."

After taking a few moments to savor his food, Thaba continued.

"Meanwhile, there were other problems relating to the separation between matter and consciousness. Take free will, for instance, or agency, as we prefer to call it these days. We take it for granted that we have free will, and we use it to make things happen, like this lovely dinner party. In pure physics, however, there is no such thing as free will—or rather, there didn't used to be. Everything was causally set in motion at the time of the Big Bang, when the first particles began bumping into each other. When we were in the laboratory wearing our white lab coats, we inhabited a world where free will and agency did not exist. But the moment we took our coats off and went home to our families it magically reappeared, for I can assure you, it's impossible to be a parent without the assumption of free will. If we took the idea that there was no agency seriously, everything would grind to an immediate halt."[903]

"So which Thaba is speaking to us now?" Leo asked. "The Thaba who wears a lab coat, or the Thaba who's the father of Jake?"

"Both, to answer your question: and there's the paradox, and we scientists hate a paradox, since it means we haven't got our models right. The standard model of physics had been wedded to bottom-up causality, the classic billiard balls. No free will—just A causes B, causes C, starting with the Big Bang, all the way to Z. Do you know what the famous biochemist Francis Crick wrote in his book *The Astonishing Hypothesis*? The same Francis Crick who shared the Nobel Prize for discovering the double-helix molecular structure of DNA? Have you got your Li-fi on, Dezzy? And is it okay if I displace your lovely wall-art?"

Dezzy nodded. Thaba spoke a few words to his device and threw Crick's words onto the wall:

> *The astonishing hypothesis is that you, your joys and your sorrows, your memories and your ambitions, your sense of personal identity and free will, are in fact no more than the behavior of a vast assembly of nerve cells and their associated molecules. As Lewis Carroll's Alice might have phrased it: You're nothing but a pack of neurons.* [904]

"That's pretty gloomy stuff—not much room for free will there. But what has been fascinating in recent years has been the way top-down causality has emerged as a serious player, recognizing the role of free will and agency at every level of conscious existence."[905]

This was fun. I was struggling to keep up, but so far, so good.

"But free will and causality are not the only problem," Thaba continued. "The standard model of particle physics also said there was no direction or purpose in the Universe: it was all just random chance, even though the journey of existence, from the origin of the Universe to the evolution of life, screams otherwise.

"Normally, this would not be a problem, since when we're wearing our lab coats we don't concern ourselves with philosophical matters such as whether civilization is advancing or not."

"But that makes no sense at all," Aliya said. "Surely, you don't believe that?"

"No, I don't, which is why I said 'normally.' But I have colleagues who do. I have one who likes to point to the awesome immensity of the Universe to remind us how utterly insignificant Earth is. If you scale the Universe down to the size of the Earth, he likes to say, the Earth would be 1/180th the size of an atom. And he's right. And if you line ten million atoms up side by side, you get one millimetre. Within this complete insignificance among all insignificances, he says, do we really think that anything we do or think actually matters?"[906]

"But that's horrible," Aliya said. "It goes against everything I believe."

"We've got to suck it up," Lucas said, taking a sip of wine. "Pretending it's not true won't make it go away. I prefer to turn it around and think how amazing it is that within this vast immensity, how great it still feels to be alive and to know that I can make a difference."

"You two have put your finger on it exactly," Thaba said. "That's the core of the problem, right there: the separation between the external reality of this vast, seemingly impersonal Universe in which there is apparently neither purpose nor free will, and the rich reality of the internal world where we experience purpose and free will… and love. There's the paradox, and scientists either love or hate a paradox, because it means something needs to change."

"So what's the solution?" Betska asked. "Do we go on ignoring our insignificance in the vastness of the cosmos, or does syntropy offer us a new way of seeing things? If it does, I hope this dinner party lasts a long time. Leo, can you pass the wine? Or better, can you give us all a top-up?"

"I think I should turn it over to Soluna at this point. I hope I've laid some useful groundwork."

"You have indeed," Soluna said. "There are several more things that the standard model of physics can't explain, but we don't need to go into them now. I come to this as an evolutionary biologist, and one of our tasks is to explain how life evolved, and how living things that started out as a few basic molecules that chose to hook up with each other in a sea of hot mud ended up as humans discussing these things around the dinner table.

"I've been working in the field for over twenty years, and I still remember something one of my professors told me when I was an undergraduate at Oxford. As long ago as 1963, he said, the same Francis Crick who you just mentioned told the maverick scientist Rupert Sheldrake in a student seminar that there were two major unresolved problems in biology: development and consciousness.[907]

"By development, he meant the mystery of why it was that molecules adhered to each other and became so much more than their parts, over and over, until there were humans, with our capacity to ponder the vastness of the universe. It's not sufficient to assume that it happened by random mutations and the instinct of a gene to replicate. Something else must have been at work. But what?

"And by consciousness, even though Crick was a materialist who took it for granted that consciousness had purely physical roots, he meant both the 'soft problem' of how consciousness is supported in the brain, and what the philosopher David Chalmers called 'the Hard Problem' of consciousness, with capital letters: the undeniable reality of our felt experience. Crick spent the last twenty-five years of his life working with the German scientist Cristof Koch on neuroscience research, trying to pin down the nature and origins of consciousness and how the brain produces it. He died never having solved the problem, but Koch became extremely ambivalent about the claims of pure materialism; it's almost as if he sensed that consciousness did indeed operate in a universal pan-psychic realm, but he didn't have the evidence to come right out and say so.[908]

"As an evolutionary biologist, I live surrounded by the wonder of evolution. Every day we get a better understanding of why Darwin was right, and how everything that exists on the tree of life has evolved from the same common origins. When people say, 'We are one,' it really is true."

"You're making me feel a lot happier now," Aliya chimed in.

"Well, I've hardly started, Aliya! I'm hoping you might feel even happier by the time we're finished with the evening."

Aliya smiled and snuggled up to Lucas, taking his arm.

"Darwin was fundamentally right about evolution," Soluna continued, "but we've added many new understandings since his time. Natural selection is an important factor, but it's by no means the only one. Bringing consciousness into play and giving it a role in evolution is huge.

"As a biologist, I deal a lot with animals and plants. But first, do you all agree that you are conscious?" We chuckled, and she continued. "So do you believe that the other people around this dinner table are conscious?" We laughed again. "Okay, how about cats and dogs? Do you think they're conscious?"

"Of course they are," Lucas said. "So is every creature."

"Okay. But remember, it's not that long ago that scientists used to perform vivisection on monkeys and dogs without anesthesia, claiming they had no feeling. René Descartes, the famous French 17th century philosopher, used to perform live vivisection on dogs, putting his hand inside their living bodies, because his philosophy told him that animals were only machines and could not possibly feel pain."[909]

"Eugh!" Aliya responded, putting down her food. "That's horrible. How could he do that?"

"Exactly. I'm really glad that we've stopped doing vivisection at UBC. Our kitchen chefs still boil lobsters alive, however, and they still tear the limbs off living crabs. I've complained, and I've shown them the evidence that crabs and lobsters feel pain, but so far, to no effect. But let me get back to consciousness. How about elephants?"[910]

"Absolutely," Betska said. "They are probably more conscious than humans. I feel so ashamed at how we have treated them over the years."

"All agreed? So elephants are conscious. What about worms? Are they conscious? The tiny one-millimetre-long worm c. *elegans* has 18,000 genes and more than 300 neurons. So is it conscious?"[911]

"I'd have to think about that," Lucas said. Then after a brief pause, "and having given it due consideration, I conclude that, yes, it is."

"Anyone else?"

No-one spoke, but people were slowly nodding their heads to say that yes, it was probable that worms were conscious.

"My friend Sophie told me once that she has experienced the consciousness of a mosquito," Aliya said. "She was meditating, and a mozzy started to bother her. But instead of brushing it away she put out an inner request to understand the mind of a mosquito. She found herself transported to a very strange place, which she had difficulty in putting into words, but she was pretty sure that's what it was: the mind of a mosquito. And ever since that day, she says, she has never been bitten by one."

"That's so trippy!" Leo said, laughing. "She should teach a course to show us how to do it."

"That's fascinating," Soluna said. "A very elderly friend called Andrew Watson told me a similar thing about ants: that once, he was meditating on a beach in South Africa and the ants were bothering him. So he drew a circle in the sand around him and told the ants not to bother him for an hour. And sure enough, they didn't. But after a while he felt a bite and he instinctively slapped his leg, killing an ant. What he saw next totally amazed him. Two ants had taken the body of the dead ant and they were holding it up to him. He looked at his watch, and guess what? It was exactly an hour since they had made the agreement."[912]

"That's incredible!" Aliya exclaimed. "So the ants were totally conscious of what they had agreed to?"

"Seemingly so, so let's take this a step further. What about bacteria? The largest bacteria have as many as 7,000 genes, compared to a human's 30,000. Do they experience some kind of proto-consciousness?"

Silence.

"Anyone vote for bacteria being conscious? They can communicate with each other, pass electrical current to each other and respond to light. They can breed, like you and I. Personally, I think it highly likely that bacteria are conscious, which they experience in whatever way their bacterial biology makes possible. And what are bacteria made from? From organic cells that are in turn made from a host of organic molecules. So could it be that even the molecules are conscious, in a very elementary way? And if they are, what about the atoms they are made from? I expect you can see where I'm going with this."

"Before we go any further," Dezzy asked, "how do you define consciousness? If molecules and bacteria have consciousness, is that the same consciousness that you and I experience? And by the way, would anyone like dessert? We've got pear purée with fresh cream, and quince and walnut ice cream. The quinces are probably conscious, but I'm not so sure about the spoon."

"Ha!" Thaba responded with a deep laugh. "That might be because the molecules that make up the spoon were not consulted before someone came along and mashed them together, so their self-organization never came into play, migrating their consciousness to the higher level. And the dessert sounds delicious. Can I have some of both?" There was a pause while Dezzy served dessert, and then Soluna continued.

"When people talk about consciousness they use the same word to mean three very different things. Some people mean self-consciousness, which is clearly nonsense, since children are not self-conscious when they are babies, but they are obviously conscious.

"Some people mean the content of consciousness: the taste of this pear purée, the sound of our voices, the feel of the chairs we're sitting on. These can all be correlated to neuronal activity in a specific area of the brain. No-one questions the role of the brain in generating the content of consciousness, and deciding which of our gazillion daily perceptions will emerge into consciousness and which will not. There's a mass of scientific research going on to investigate the nature of those correlations.[913]

"When I use the term, however, I mean consciousness beyond content. I mean pure consciousness, the fundamental experience of being that remains when you quieten every sense and silence every thought. You have to be a very serious Buddhist or something similar to experience consciousness in this raw form, without intrusion, but when people do they speak of something very profound. They speak of overpowering light. They speak of becoming part of a Universe filled with compassion. We hold a meditation group in our biology department every Friday afternoon, and we talk about these things afterwards."

"Are you following this, Patrick?" Dezzy asked me. "You've been very quiet."

"I'm hanging on every word," I assured her. It was true. I was transfixed by what Thaba and Soluna were saying, and where this might be going.

"So to recap," Soluna continued, "before syntropy theory we had an unresolved problem with consciousness, and another unresolved problem with the process of development. We also had problems with the standard model of physics that Thaba referred to, concerning free will and purpose. And since biology is ultimately underpinned by the standard model of physics, these problems concerned us biologists as well. If a dog has consciousness, does it have free will? My cat Molly certainly seems to: I see it in her eyes.

"And that's how things stood when the research into telepathy in identical twins was published, using fMRI chambers with pairs of twins to demonstrate with a high level of certainty that one third of identical twins are telepathic under specific circumstances. One black swan: that's all it takes to prove that not all swans are white. The premise I find the most plausible is that all beings are telepathic among their kin, but in humans it rarely surfaces into waking consciousness, enabling us to deny it, just as people denied other major scientific breakthroughs before they slowly accepted the evidence. Evolution pushed our conscious experience of telepathy down into an unconscious part of the brain because we needed the conscious brain-space for language—a problem other species don't have.

"If it had been demonstrated just once that a pair of twins was reliably telepathic, that should have been enough. But since the history of psychic research has been so controversial, they repeated the experiment many times in different ways. By the time they were done there was no denying it any longer: something associated with consciousness was either travelling across space-time without any known means of doing so, or the mind is not restricted in space-time. Just because we experience it as such, does not mean it is."

This was the same research that Pelly had spoken about. It was reassuring that there was consistency in what I was hearing.[914]

"The twins research put the trans-dimensional nature of consciousness firmly on the table," Soluna continued. "Researchers all over the world began to focus on different theories. It was no longer sufficient to propose that consciousness originated in a specific nerve centre in the brain, or that it was an emergent property associated with the interconnection of the brain's neurons. If thoughts, feelings and biological responses could travel across space and arrive intact in another being, there was clearly something much more advanced going on. Thirty years ago, when the Nobel prizewinner Brian Josephson from the Cavendish Laboratory in Cambridge suggested that quantum theory might help us to understand the nature of telepathy, his views were met with total disdain by other mainstream scientists and labeled 'utter rubbish.' Not any longer!" Soluna leaned back to let it all sink in.[915]

"This is fascinating!" exclaimed Dezzy, and the others nodded, still processing silently.

"Let's have a break for coffee," said Dezzy, ever the hostess. "Lucas makes a very good spiked chili and chocolate blend, and I've got iced peppermint coffee and hot Senegalese coffee, though it's not for the faint of heart."

"Don't go anywhere near it!" Leo cried out. "That stuff's lethal unless you've got asbestos lips."

"I wouldn't say that," Soluna said. "I got quite a taste for it when I lived in Mexico. It's certainly an acquired taste, though."

"So, where were we?" Dezzy asked, when everyone had their coffee.

Soluna put her mug down. "We have been obliged to accept that consciousness is more than an emergent feature of the encapsulated brain," she said. "So let's get straight to it. The assumption I have embraced, in company with many of my colleagues, is that consciousness is omnipresent in the Universe, similar to space-time. I find it the only theory that makes sense, when you consider all the variables. Dualism makes no sense at all, since the dualities would need to be connected. This means you've either got to be a materialist monist, believing that the whole Universe is ultimately material, or a mystical monist, believing that it's ultimately made from consciousness. For me, assuming the omnipresence of consciousness is the only way in which the realm of mind can interact with the realm of matter without breaking the laws of physics. Without it, there's no way for mind to trigger the neurons to provide the content we enjoy in our conscious experience."

"I'm fine with this, but how do your colleagues at the university react when you talk this way?" Betska asked.

"When Mukherjee's paper on syntropy was published sometime around 2020, it met with a lot of scorn. But slowly, people are coming round to it. I'm actually in very good company, which is helpful on days when I question it all. Do you know who Max Planck was? He was the founder of quantum theory in the early 20th century. Take a look at what he said."

Soluna threw Max Planck's words onto the wall:

> I regard consciousness as fundamental. I regard matter as derivative from consciousness. We cannot get behind consciousness. Everything that we talk about, everything that we regard as existing, postulates consciousness. [916]

"And he wasn't the only one. Here's his colleague, the Austrian Wolfgang Pauli, another quantum theory pioneer:

> It is my personal opinion that the science of the future reality will be neither 'psychic' nor 'physical', but somehow both and somehow neither. [917]

"There's also the British scientist, Sir Arthur Eddington, whose book *The Expanding Universe* made a big impression in the 1930s. He had this to say:

> The universe is of the nature of a thought or sensation in a universal Mind. To put the conclusion crudely – the stuff of the world is mind-stuff. As is often the way with crude statements, I shall have to explain that by "mind" I do not exactly mean mind and by "stuff" I do not at all mean stuff. Still that is about as near as we can get to the idea in a simple phrase. [918]

"This has huge implications for developmental biology," Soluna continued. "It opens the door to the idea that the evolution of species is an intelligent learning process in nature, as the biologist Elisabet Sahtouris believes. Every creature and perhaps even every cell operates with the same fundamental tools of consciousness that we humans experience: awareness, agency, goal-seeking intention and effort, informed by the sensory input of information and organized by memory and intelligence. By agency, I mean the experience of being conscious, which brings the ability to act and respond. It's a fundamental precondition for free will, which we can choose to exercise or not. In a nutshell, consciousness provides a perceptual organizational matrix that enables the experiencer, whether hookworm or human, to use organized information to apply effort to engage in intentional action. Even proteins rearrange themselves when they're under stress."[919]

There was total attention around the dinner table as Soluna spoke. This was so different from the biology I had learned during my home-schooling years. A hookworm, a conscious intentional being? Back in my time, a description like that would have been criticized as anthropomorphizing, distracting from the objective analysis of a hookworm's life. This was huge, I began to realize. If mainstream science was embracing the omnipresence of subjective experience there would no longer be any barriers between science and spirituality.

"Can we measure consciousness the way we measure matter, time and space?" Dezzy asked.

"We're making progress on ways to measure its existence biologically in terms of correlated brain activity, but to measure consciousness as an absolute, a fifth dimension, equivalent to time and space— for that we may need an entirely new breed of math, going right back to zero; perhaps some new kind of non-differential ultracalculus that can measure the continuity of flow in analog reality without breaking it up into digital pieces. It may or may not be an inherent problem. Who knows? Perhaps one day there'll be a breakthrough that will allow us to measure raw consciousness. Maybe the very reason why quantum uncertainty exists is because there is agency and choice at the most nano-level of existence."[920]

"If consciousness is omnipresent in the Universe," Leo asked, "what about its interaction with things like gravity, space-time and electromagnetism? That's something I've always been curious about."

"It's something we're all curious about," Soluna replied. "Do you have any insights, Thaba?"

"There's some exploratory work being done around the potential coaxial nature of fields of consciousness and electromagnetic fields," he replied. "When it comes to gravity, which as Einstein taught us is the warping of space-time by mass, I know of research that's looking into quantum entanglement at the moment of the Big Bang, communication between entangled atoms, and whether syntropy might be an expression of the same mutual attraction of like-for-like in consciousness that gravity expresses in matter. But maybe we've laid enough groundwork. What do you think, Soluna?"[921]

"Yes," she replied. "I would just like to recap the shortcomings of the standard model of physics: the peas under the mattress that make for an uncomfortable sleep and drive scientists to seek a new model. As well as the known shortcomings, such as its inability to explain gravity or dark matter and its inability to explain the fixed universal constants, we've got the problems with free will, consciousness, development and intentionality.

"In a stable, peaceful world there might not be an urgency to solve these problems. After all, philosophers have been trying to understand them at least since the Greek philosopher Thales of Miletus, who lived around 600 BC.[922] But in a world in such turmoil, where we face such enormous threats to our existence, the questions become extremely important. If we're about to blow it, it would be good if we at least knew what it is we're about to blow. Who among us has never asked those big, fundamental questions—the 'Who are we, what are we doing, where are we going?' kind of question?"

"Count me in," I said. "I sometimes feel as if they're the *only* questions I'm asking."

"Me too," Aliya said. "Some Muslims say the Koran contains all the knowledge we need, but I don't accept that. So yes, what *are* we doing? What *is* our purpose in the Universe?"

"Syntropy doesn't answer all those questions," Thaba said, "but it's a big step forward. It's being proposed as a fifth fundamental interaction, alongside gravity, electromagnetism, weak interaction and strong interaction. It's the first time science has been able to consider a possible Theory of Everything that includes the subjective realities of consciousness, intention and life alongside the objective realities of matter, energy and space-time.

"Syntropy has been around as an idea for almost a hundred years, but the version we're talking about is Satyanendra Mukherjee's, which he published during the OMEGA Days. Let me see if I can find his First Law of Syntropy. Thaba picked up his device, said 'Search, Mukherjee, syntropy, first law,' and projected Mukherjee's words onto the wall:

> *Acting through consciousness, syntropy motivates individual units of being to self-organize cooperatively within their empathic reach to achieve greater organizational power, range, competence, integrity and freedom for their common good.*

"That sounds rather grand," Thaba continued, "but when we understand that 'units of being' embraces everything from a particle to a human we can see how radical it is. The fundamental premise of syntropy theory is that the Universe contains an omnipresent unifying force that causes all units of being to seek greater self-organization for their mutual benefit, using consciousness as the agency for motivation, intention and change. So the premise that consciousness is an omnipresent reality is very much entwined with syntropy theory. There are those who argue that syntropy can exist without bringing consciousness into the picture, just as gravity appears to operate without consciousness, but this would imply that syntropy was simply a means of delivering a pre-determined reality, which most of us intuitively reject. How are we doing? Are we making sense so far?"

"What does Mukherjee mean by the phrase 'within their empathic reach'?" Betska asked.

"That's a very important question," Thaba replied. "As humans, we experience compassion when we feel empathy for someone who is suffering, for a creature that has been hurt or a child who is crying. But if we really want to understand empathy we need to consider its reach, which includes its boundaries. History is full of humans who had empathy for their fellow tribe-members but not for other humans, and not for most creatures in the animal realm, who we have treated abominably. Empathic reach is the limiting condition that denotes these boundaries."

"Like Hitler?" Leo asked. "He had empathy for his fellow Nazis, but not for the Jews and communists, the gypsies and the gays. And the Nazis were very good organizers, too."

"Precisely," Thaba replied. "So now we come to the interesting part. We are all part of nature. We have all evolved through the same combined intelligent learning processes of syntropic self-organization, cooperative symbiosis, mutation and natural selection.

"Many of syntropy's critics fail to understand Mukherjee's point about empathic reach, and its gradual extension. When Hitler organized to lead Germany into war against the rest of Europe, the German people's drive to attack was immediately matched by the instinct of the British and their allies to defeat them. The empathic reach of the Nazis, who simply wanted to impose their will, was narrower than the empathic reach of the Allies, who were defending the sacred principles of truth, justice and freedom, so ultimately, the Germans didn't have the inner resources to win. Their higher cause of a thousand-year Reich was less motivational than the Allies' higher cause. There were many other factors at play, of course, such as who had control over the world's oil supply, who had the best code-breaking capacities, and the military oomph that the Americans provided when they entered the war, but we should never underestimate the power of the motivational factor.

"Following Hitler's defeat the Allies went on to form the United Nations to try to prevent such a war from ever happening again. The world's nations had tried to self-organize after World War One with the League of Nations, but the League didn't have the teeth or the willpower to do anything, so nations continued to invade each other and seize territory during the 1920s and 1930s. The United Nations, on the other hand, is still with us, and for all its shortcomings it still represents our highest impulse for global self-organization and the common good."[923]

"So let me get this right," Betska said. "Is syntropy then, in effect, a guarantee of ultimate happiness? Are we destined to self-organize ourselves into some kind of cosmic bliss?"

"That's a really big question," Thaba replied. "The way I see it, syntropy is an invitation, which we are free to accept or reject. It's a choice. I'm sure you know the words of Dr. Martin Luther King, who said 'the arc of the moral universe is long, but it bends towards justice.' I'm sure my countryman the great Nelson Mandela—Madiba—saw things the same way. Where else would he have found such courage and determination during his years of solitary confinement? That's how I see syntropy working among humans. We are the ones who bend the moral arc of the Universe towards justice—or who fail to.

"No-one is suggesting that entropy is not also a powerful force. There are plenty of negative social, economic and political conditions that feed entropy. Maybe all social and political activism is a struggle between entropy and syntropy: between entropic forces that generate cynicism, despair and defeat and syntropic forces that inspire hope, determination and courage.

"It's right there in the Pre-amble of the Constitution of the United States," Thaba continued: "'We the People of the United States, in order to form a more perfect Union....' A more perfect union. That's what syntropy is all about. And so we need syntropic politics, syntropic economics and syntropic families, as well as syntropic science."

There was a deep, concentrated silence around the table.

"Mukherjee has suggested that since there's a fundamental unity to all existence, as long as we accept the invitation, the pull of syntropy will gradually cause the boundaries of empathy to expand until they embrace all living beings, and all existence. His thinking is very similar to that of the celebrated French Catholic priest and scientist, Teilhard de Chardin, who saw evolution as a co-creation of consciousness and material complexity, which would culminate in the Omega Point, the mystical apex of all creation. Teilhard saw things the same way as Schrödinger: he believed that we live in a pan-psychic Universe. Let me see if I can find the quote...." Thaba spoke some words to his device and projected Teilhard's words onto the wall:

> We are logically forced to assume the existence in rudimentary form...
> of some sort of psyche in every corpuscle, even in those (the mega mol-
> ecules and below) whose complexity is of such a low or modest order
> as to render it (the psyche) imperceptible.[924]

"That's pretty trippy!" Leo said. "And to think that we are part of all this—that this is our heritage! It certainly beats feeling defeated because of the miserableness of human existence."

Thaba continued. "Mukherjee is fond of quoting Albert Einstein—you probably know the quote. You'll have to excuse the sexist pronouns; he was writing in the mid-20th century. Here, let me pull it up...."

> A human being is part of a whole, called by us the Universe, a part lim-
> ited in time and space. He experiences himself, his thoughts and feel-
> ings, as something separated from the rest—a kind of optical delusion
> of his consciousness. This delusion is a kind of prison for us, restricting
> us to our personal desires and to affection for a few persons nearest

us. Our task must be to free ourselves from this prison by widening our circles of compassion to embrace all living creatures and the whole of nature in its beauty. [925]

"I love that quote!" Aliya said. "I would love to believe that we will ultimately be drawn together into one compassionate family, embracing all living creatures and the whole of nature."

"Maybe we will," Thaba replied. "But we've got to remember, it's a choice: it was syntropy too that inspired the Nazis to believe they were the Master Race, who would rule the world for a thousand years. They self-organized too, but their empathic reach was limited to the Aryan people. If you embark on a conflict using empathy that only embraces a limited circle, you will ultimately lose when you confront the self-organizing power of circles with larger empathic reach. This is the dichotomy that led Mukherjee to formulate his Second Law of Syntropy:

In the long run, due to the deep fundamental unity of the Universe, any unit of being that extends its empathy beyond its familiar reach will discover affinity with other units of being. Over time, the syntropic impulse will result in ever-widening circles of empathy, until they embrace the entire Universe.

"The entire Universe?" Dezzy queried. "But that's incredible! You were just telling us how absolutely tiny and insignificant Earth was compared to the size of the Universe."

"That's true," Thaba replied. "It certainly stretches the imagination. But you've also got to realize that your human body has a thousand times more atoms in it than there are stars in the Universe, and somehow or other they have self-organized themselves to create you and me."

"More atoms in my body than there are stars in the Universe?" Aliya chimed in, her eyebrows raised high. "That's amazing. I had no idea."

"Yes: ten to the power of twenty-seven compared to ten to the power of twenty-four for stars in the Universe," Thaba replied.[926] "Many people use the expression 'God' or 'The Great Creator' when they contemplate such an enormous mystery. I have many friends who use the term 'God' to express the sacred unity of all that exists and the process of creation in all its wonder, both subjectively and objectively. I relate to syntropy in a more immediate way, since it provides a useful explanation for the symbiotic impulse towards mutual aid and cooperation, and the self-organizing tendency among atoms and molecules. It may even be the frustrating 'X factor' that has dogged complexity theorists for so long."[927]

"My father used to say God was a G.O.D.—a General Omnipresent Diaphany," I interjected, happy to be able to bring him into the conversation.

"It sounds like you had a very thoughtful father," Thaba replied. "That's an interesting definition, especially if we equate the word 'diaphany' with a field of consciousness."

"But where does this force of syntropy come from?" Aliya asked. "And how do we know it's real, and not an imagined fantasy? I *want* to believe, but I don't want to be taken in by an idea just because it's warm and fuzzy. I've seen enough self-organization by warring Sunnis and Shiites to last me a lifetime."

"Aliya, this may quite possibly be the most important question of all," Thaba replied. "It's one of those questions that make me feel I might go to my grave without having resolved it. Maybe death itself will be the doorway to understanding, when we finally lose the flood of daily detail that prevents us from experiencing pure consciousness.

"Speaking as a scientist, however, we might as well ask where gravity and space-time come from, or magnetism. We don't know the answers to those questions either. For all that we do know, we are still very limited in our knowledge compared to the immensity of what we don't know. We have only been seeking answers in a scientifically rigorous manner for a few hundred years. Imagine a civilization that has been at it for forty thousand years, or four hundred thousand years. Imagine how much more they will have had time to learn. If we can get through the current global crisis and learn to live together as a family of nations, maybe there will be a golden age of tranquility on the other side. After all, the Sun will be good for more than a billion years, which gives us a long time to enjoy the fruits of consciousness and harmony with nature. Dezzy—what on Earth did you put in my coffee? I don't normally speak like this."[928]

"It's wonderful," Aliya said. "Please don't stop!"

"I don't normally think of myself as religious," Thaba replied, "but I can enjoy my imagination being blown wide open as much as anyone. Give me Mahler's *Second Symphony* or some Hugh Masekela jazz any day. As a scientist, however, I prefer to leave the question marks hanging rather than bundling them up and calling them 'God.' I find that it serves to keep me curious. But forgive me: what was your question?"

"I asked where the force of syntropy comes from," Aliya said.

"Right. We know that consciousness is real, and we believe that it may permeate all existence. We know that units of existence have self-organized cooperatively throughout evolution for their own benefit to create greater capacity and reach; and we know that the self-organizing impulse operates in physics, chemistry and biology as well as among humans. So the hypothesis is that there's a deeper universal force at work, a fifth fundamental interaction, which Mukherjee calls syntropy. He didn't invent the term; he simply brought it into the mainstream. The concept was dreamt up by Luigi Fantappiè, an Italian, nearly a century ago. He was a well-regarded mathematician, a colleague of the physicist Enrico Fermi. He was working on an aspect of quantum theory concerning the anticipated potentials of a wave equation when he had this sudden insight that there was a new category of phenomena that he termed 'syntropic' that were totally different from entropic phenomena, which obey the principle of classical causation and the second law of thermodynamics—the law of entropy.

"If I can borrow your device, Soluna, I'll show you the page from his journal where he related his discovery." Thaba spoke the relevant words to the device, swiveled his chair to face the wall and projected Fantappiè's words:

> *I have no doubts about the date when I discovered the law of syntropy. It was in the days just before Christmas 1941, when, as a consequence of conversations with two colleagues, a physicist and a biologist, I was suddenly projected in a new panorama, which radically changed the vision of science and of the Universe which I had inherited from my teachers, and which I had always considered the strong and certain ground on which to base my scientific investigations.*
>
> *Suddenly I saw the possibility of interpreting a wide range of solutions (the anticipated potentials) of the wave equation that can be considered the fundamental law of the Universe. These solutions had been always rejected as 'impossible,' but suddenly they appeared 'possible,' and they explained a new category of phenomena that I later named 'syntropic,' totally different from the entropic ones, of the mechanical, physical and chemical laws, which obey only the principle of classical causation and the law of entropy.*
>
> *Syntropic phenomena, which are instead represented by those strange solutions of the 'anticipated potentials,' should obey two opposite principles of finality (moved by a final cause placed in the future, and not by a cause which is placed in the past): differentiation and non-causability in a laboratory. This last characteristic explained why this type of phenomena had never been reproduced in a laboratory, and its finalistic properties justified the refusal among scientists, who accepted without any doubt the assumption that finalism is a 'metaphysical' principle, outside Science and Nature. This assumption obstructed the way to a calm investigation of the real existence of this second type of phenomena; an investigation which I accepted to carry out, even though I felt as if I were falling into an abyss, with incredible consequences and conclusions.*
>
> *It suddenly seemed as if the sky were falling apart, or at least the certainties on which mechanical science had based its assumptions. It appeared to me clear that these 'syntropic,' finalistic phenomena that lead to differentiation and could not be reproduced in a laboratory, were real, and existed in nature, as I could recognize them in living systems. The properties of this new law opened consequences which were just incredible, and which could deeply change the biological, medical, psychological and social sciences.* [929]

"Syntropic phenomena obey opposite principles of finality, Fantappiè said, being moved by a final cause placed in the future, not in the past. That needs a lot of thinking about. A final cause, set in the future."[930] Thaba paused, as if pondering the thought himself.

"That's hard for me to wrap my mind around," Betska said.

"At first blush, it certainly seems so," Thaba replied. "When we observe the material world it seems clear that causation flows from the past to the present. When we observe the world of consciousness, however, which we can do any time we're awake, we define the goals we want to achieve through intentions set in the future and we use agency, effort and free will to move towards them. Causation flows from an anticipated future, back to the present. That's how Dezzy organized this lovely dinner party; that's how we achieve everything in life apart from routine, unconscious habits."

"This is getting beyond me," Aliya said. "If atoms have some kind of rudimentary consciousness, and if, in the world of consciousness, causation flows from the future to the present, does this mean that even atoms experience agency and causation this way? What does that mean for the nature of time?"

"A lot of things in physics appear far-fetched," Thaba replied. "When you contemplate the immensity of the Universe and the mystery of our origins it's hard not to blow a fuse. So far, we have no means of knowing if atoms experience agency and causation. But we know that atoms are drawn to each other, and we know that they self-organize to form molecules and ultimately to form elephants and humans.

"Self-organization occurs in every realm of existence.[931] If you remove the assumption of consciousness, it becomes very difficult to explain. Who or what is doing the self-organizing? People like the polymath Stu Kauffman talk about sets of molecules that are collectively autocatalytic, emerging spontaneously from their previous level of order.[932] Biologists talk about organisms having plasticity, and an ability to self-organize that emerges internally without being caused by any external factor. The South African mathematician and cosmologist George Ellis, who taught me when I was a student at the University of Capetown, was very clear that the Universe was not entirely a bottom-up creation, as most physicists believed, and that there are multiple levels of what he calls top-down causation, without going so far as to attribute them to consciousness and agency.[933]

"Back in the 1970s, the Hungarian biologist Albert Szent-Gyorgyi used the same term 'syntropy' to describe the way living systems evolve into forms of organization that are more complex and harmonic, in contrast to 'entropy,' which leads to the disintegration of all types of organization. He defined it as the 'innate drive in living matter to perfect itself.' Earlier in the century, the British philosopher and mathematician A.N. Whitehead spoke about the primacy of process; and the South African thinker and political leader Jan Smuts, one of my countrymen, spoke about holism, which he defined as the tendency in nature to form wholes that are greater than the sum of their parts, through creative evolution. It's very similar to syntropy. Einstein thought very highly of Smuts' concept; he wrote that it would be the most influential concept in directing human thinking over the next millennium, alongside relativity.

"Even the legendary biologist Richard Dawkins spoke about selfish genes as if they had purpose and intention, with the ability to mold matter and create form. When pressed, he said he didn't actually mean that, but he often spoke as if he did.[934] Kauffman believes there is a ceaseless creativity in the Universe, which comes from existence always being poised on the edge of chaos, where there is maximum choice. He has never suggested that organisms are conscious, however, or that it is the experience of agency experienced within consciousness that enables an organism to self-organize, the way we do. That's the leap Mukherjee made when he integrated syntropy with consciousness. Elisabet Sahtouris, the famous evolutionary biologist, believes that the Universe itself is consciousness, creating living systems within itself, and that all living systems are therefore conscious, intelligent and able to learn. Mukherjee built on the work of Sahtouris and many others, pulling it all together."[935]

"I'm still stuck on the implication for the nature of time," Aliya said. "I can understand an intention being set in the future; that's imaginary. But you seem to be saying that Fantappiè claimed that all living systems respond to a cause set in the future."

"Fantappiè did not relate syntropy to consciousness either," Thaba replied. "Modern scientific research into consciousness did not begin in earnest until the late 1980s. He just had the intuition about syntropy, as did many others, including Szent-Gyorgyi, and Whitehead, who used the term 'creativity' where Fantappiè used 'syntropy.'[936] They're not the same, but they're very similar. Since the early years of this century, Fantappiè's work has been championed by an Italian couple, Ulisse Di Corpo and Antonella Vannini. They publish a journal and organize conferences that bring scientists and philosophers together to explore the theory of syntropy.[937]

"Fantappiè had to frame the concept of syntropy within the classical quantum paradigm he was familiar with, not the new psi-quantum paradigm, which includes the reality of consciousness. In classical quantum physics, time has no inherent direction: it can go both forward and backward. There is also no free will, so if something has a cause set in the future it doesn't matter, since there's no choice about the way things work out. It's not a way of thinking I embrace any more, but it's the way most physicists used to think, myself included.

"When we embrace the psi-quantum paradigm, consciousness takes center-place, and subjective agency with its potential to act arises as an active response to observation and change. That causes us to think about time very differently. You referred to imagination. In the old paradigm, imagination belonged to the realm of the mind, which either co-existed dualistically alongside the material realm or was totally secondary to that realm, as a subset of brain activity. The psi-quantum paradigm opens up the relationship between consciousness and time. There have always been anecdotes about precognition; about people, for instance, who find themselves thinking about someone they haven't met for years, and suddenly he or she is right there on the street."

"That happened to me just recently," Dezzy said. "I was having a coffee in a café on 4th Avenue and I started thinking about an old school friend I'd known in Montreal. When I got home, there was an email from her. It was really weird."

"We know this kind of precognition happens; there's very solid evidence for it," Thaba continued.[938] "Until recently, however, we didn't have a clue how to understand it, so it was easier to ignore it or deny that it happened. In this new way of seeing the world, consciousness is an omnipresent dimension that may pre-empt time, making time a secondary phenomenon. If that's the case, then a glimpse into the future becomes possible, and so does the conscious creation of the future by intention, accompanied by effort. I'm not sure if this answers your question, Aliya, but it's the best I can do for now."[939]

"Maybe this is a good time to open it up for discussion," Dezzie said.

"This is fascinating," Betska said. "Your mother would have loved it, Leo. I'm wondering whether it speaks to Jung's idea of a collective unconscious, and the idea that we swim in an ocean that contains deep unconscious currents of memory and experience which occasionally surface into consciousness."

"I'm not a psychologist," Soluna said, "but when I worked in Mexico I had friends who were Mayan, and they certainly thought that way. I'm beginning to think that we should require our future physics students to spend a year in an ashram or a monastery or with an indigenous tribe before they join us, to give them familiarity with the different realms of consciousness. My snowboarding accident did wonders for me in that department. It made me sit still and go within, opening new doors of perception. Did I tell you, by the way, Dezzy, that I'm on the waiting list for a stem cell nerve repair operation?"

"Does that mean you'll be able to walk again?" Dezzy responded with excitement. "That would be incredible!"

"My specialist has warned me not to raise my hopes, since the science is still quite new. [940] But we'll see what happens. Sometime in the next two or three months, he said. I have also been using a form of electro-biotherapy called functional electrical stimulation. I wear an electrode cap that picks up my brain signals whenever I think about walking or standing, and it responds by activating the nerves in my leg muscles. I've been doing it for about a year now, and it has enabled me to walk about five metres. So combined with the stem cell repair, I'm feeling hopeful, in spite of what my doctor says."[941]

"That's amazing," Dezzy replied. "Sometimes I think that the entire progress of humanity has scientific progress at its core."

"Has anyone found a way to test syntropy theory, to see if it's false?" Leo asked.

"It's not as easy as measuring the rate of fall of an apple to test the theory of gravity," Thaba replied. "You can do a simple thought experiment in which you remove consciousness and see what happens: everything grinds to an immediate halt. What we're looking for is evidence of an omnipresent field of influence, similar to gravity, which shapes the way units of existence operate, driving or pulling them to greater self-organization and complexity. We can observe it happening in any realm we choose to study, from anthropology to economics and from physics to biology, but no-one has been able to locate the source of the influence, or test what would happen if you removed it. We face the same quandary with gravity. We know what it does, and we can measure its effect down to the nanometre, but nobody has been able to explain how it integrates with the other fundamental interactions. It's a mystery. Gravity, which comes from the interaction of mass with space-time, must have a fundamental entanglement with the other dimensions of existence, but we have no idea how it combines with the fundamental syntropic drive within matter and space-time towards unification."

At that moment the lights in the house flickered for four or five seconds, then returned to normal.

"Are we about to have a power cut?" Soluna asked. "Or is that an answer to Aliya's questions?"

"No," Dezzy laughed. "That's our daughter Gabriela in Montreal. She does it every night when she's about to go to sleep. It's our way of saying goodnight. She knows that if I'm home I'll respond by doing… this." Dezzy reached for her device and pressed some buttons. "There: I've just sent her a goodnight kiss."

"That's so cute!" Soluna said. "Are you using the SoulTouch app I've been reading about?"

"Yes. Gabriela has coded our home's password into the app, so all she has to do is touch it and the lights dim."

"That's so sweet," Betska said. "You must show me how it works."

At this point, I thought I'd better jump in before the opportunity was gone. "During the last few days," I asked, "I've heard several people refer to syntropy as an important factor in motivating people to work for a better world. How does that work?"

"It's to do with the motivational power of the stories we tell ourselves," Soluna said, turning to face me. "The stories about who we are, what we're doing and where we're going: the big questions we spoke about earlier. I have a colleague in the history department, Frances Wellsmore, who is researching what she calls 'ultimate storytelling': the foundational framing stories which humans have used throughout history to answer the huge, imponderable questions. She is fascinated by syntropy as a new ultimate story, in addition to its value as a scientific hypothesis. Every culture needs an ultimate story, she says. The need is deeply embedded into our psyche. It's probably got to do with the mystery of death, which is so absolute, and makes us wonder what it's all about.

"For thousands of years, she says, our palaeolithic forebears told themselves a story about how their ancestors enjoyed the happy hunting grounds in the spirit world after they died. Through their shamans, they discovered portals to a world filled with magic, which integrated them with nature and the great beyond. She calls it Frame One in the history of ultimate storytelling.

"When we settled down and started farming we created Frame Two. Our needs turned to the sky, for good rain and a safe harvest, so our stories grew to include the gods and spirits of the sky, the earth and the trees, who governed our lives. As empires grew, however, we became conscious of the enormous diversity of gods, and how little sense they made, so we created Frame Three, in which there was just one God, divine and omnipotent, who ruled over everything. If you obeyed God's commands, the story said, when you died you'd join God in Heaven. Misbehave, on the other hand, and you'd go to Hell. That was very handy for keeping social control in a complex society.

"But then science arose with its powerful ability to explain the world, and it shattered many gods, new and old. In their place humans created two new stories. Frame Four told of the incredible progress that could be achieved if we discarded kings and bishops, ignorance and superstition, and embraced in their place science and reason, exploration, enterprise and commerce. It brought us the Age of Enlightenment, inspired by philosophers like Voltaire, Locke and Rousseau and geniuses like Benjamin Franklin, and it continued to inspire until Europe collapsed into the brutality of the Great War in 1914.

"Frame Five ran alongside it during the 19th century and well into the 20th century. This was the story of socialism, which promised peace and the universal brotherhood of man if we would cast off the shackles of capitalism, which condemned so many to be prisoners of poverty, low wages and the bourgeoisie. When the Soviet Union finally collapsed, the hope of socialism died with it. God was dead, and the optimistic faith in progress that the Enlightenment brought had long since been chased away by the villainies of the 20th century. There are strands of socialism that are alive and well, such as our healthcare system here in Canada, Citizen's Income, and the rediscovery of public banking, but as a stand-alone story it has lost its pull. With its death, we were left with no new stories at all: only the old religious stories. There was a vacuum, which people tried to fill with shopping, alcohol, sex, drugs and fundamentalist religions, whether Christian, Hindu, Jewish or Muslim.

"Then came the assault on nature, with global warming, the pollution of rivers and oceans, the extinction of so many species, the destruction of forests and all the rest. So a new story emerged which Frances calls Frame Six. It speaks of humans as aliens in our own land, transgressors against the beauty of nature, destroyers of everything good. In its darkest expression, it says that it might be better if we allowed ourselves to go extinct and left the Earth for nature to recover."

OMG. This was the story of my generation, back in my time.

"Hollywood picked up on the theme and packaged it into a host of dark movies about apocalyptic plagues and disasters," Soluna continued. "The looming catastrophe of global warming hung over the world like a doom-laden cloud, making people feel deeply worried about the future and driving others into full-on denial. Fundamentalist religions made a comeback, with their simpler stories. It's quite remarkable, when you dig into religious predictions. When it comes to the long-term future, Christianity, Islam, Hinduism and Buddhism all prophesy apocalypse and disaster.

"Among the world's major religions," Soluna continued, "only the Jewish faith has a positive vision of the long-term future. Jewish belief has always been tied to their covenant with God, who would deliver the Jews from bondage and bring the ultimate return of the Messiah to Jerusalem, leading to a Paradise on Earth. It has always puzzled me why the other major religions revert to fatalism in their eschatology when they look into the future, as if they have never escaped the ancient Sumerian belief in the endlessly repeating wheel of birth and death. Only the Jews developed a positive vision of the future and a progressive sense of time, thanks to their covenant with God. It was such a tragedy that in the years after World War II and the Holocaust, they believed that they needed to keep this paradise to themselves in Israel, to the exclusion of the Palestinians from whom they took the land. I'm so glad that they seem to be finally making progress, after so many years of conflict and suffering."[942]

"I may be only Jewish on my father's side," Betska said, "but I'm very proud of my heritage. I much prefer that we don't all have to die in order to experience paradise."

"It has been a long time since anyone believed that scientists could deliver a Garden of Eden," Soluna said. "For many years, people saw us as being responsible for toxic chemicals, genetic manipulation and new weapons of war. In recent years science has been quite useless when it came to providing a story. Our miserable attempt said that all existence was material, life had happened only by chance, and there was no inherent meaning, purpose or direction in the Universe. Subjective reality was an illusion, and there was no such thing as choice or free will; but what the heck, wasn't the Universe amazing? The Earth was insignificant in the measure of the Universe, and everything was ultimately going to collapse, since the second law of thermodynamics stated that entropy and disorder would always increase. It was unrelentingly pessimistic. No wonder people felt hopeless and preferred to go shopping."

"Whoa. You're getting me depressed!" Lucas said. He had been sitting quietly during the discussion so far. "We never thought about any of these things during OMEGA Days, when we were putting everything on the line."

"No?" Soluna asked.

"No. We weren't thinking that humanity was some kind of plague, or that it might be better if we died off, leaving Earth to the bears and the earthworms. We simply had a determination to make a difference. It's true, we didn't have a larger story to frame our beliefs; we didn't feel we needed one. My engagement didn't come from a story in my head. It came from my gut, my anger at the abuses that were going on against people all over the world, and against nature."

"So you didn't have a deeper story that motivated you to act?" Soluna asked.

"No. Some of us felt motivated by a personal sense of spiritual purpose. Some joined because they could see we were having more fun, and it was better to change the world than complain about it. Personally, I don't have a clue about physics, philosophy, or the things Leo goes on about. I just feel that whatever's happening in the Universe, and whatever life is really about, it's just friggin' amazing to be alive, to be part of it, and to feel it in my body. I felt really happy when I was engaged in making a difference, compared to moping around, feeling that I couldn't contribute anything."

"Bravo!" Thaba exclaimed. "You are my kind of man."

"So let me modify what I just said," Soluna replied. "The kind of instinctive rebellion that you describe has happened throughout history. It simply needs enough people to feel a strong enough sense of injustice, and a feeling that 'this is wrong: we deserve something better.' Life itself provides the motive and the determination. But to sustain a movement so that it becomes more than a rebellion: that needs a deep, compelling story, and a vision that inspires. Negative energy slays hope, surrendering the field to entropy. Positive energy inspires hope, inviting syntropy to flourish.

"Regarding the OMEGA Days," she continued, "it was the power of the commitment to make Vancouver the greenest city in the world, combined with the belief that it was possible and the determination that it was necessary that provided the deeper, more lasting inspiration. It was the vision of the greenest city itself that provided us with the fuel to do what we did."

It was then that the light bulb clicked on in my mind. It was so simple; it had been staring me in the face all the time. I had wanted to know what inspired people to make Vancouver the greenest city in the world. It was *the vision itself* that inspired them. It had sufficient power, without any need for the understanding of syntropy that I was gaining. I felt a smile light up inside me.

"So let me relate this to syntropy," Soluna continued. "For everything we do in life we need both vision and intention. We set them as markers in the future, and we move our lives towards them. We do this for everything, from huge global campaigns to small dinner parties. But what is the story that inspires our intentions? We need a story that is *a positive attractor,* which will attract us to build a better future, giving us purpose and hope, reason to dream and reason to work. When a story tells of desolation, painting humanity as a transgressor against all that is good and beautiful, it's hard to have hope.

"My friend Frances Wellsmore believes that when people understand what syntropy theory is really saying, it will transform the entire way we think about our purpose, and our reason for being here on this planet. It will be like taking the power of the greenest city vision and multiplying it a thousandfold. Humanity has never known a story which carries such power, she says—one that embraces the scientific impulse, the spiritual impulse *and* the impulse for social and political change, and which also provides such a positive vision of the future."

My chance to jump in. "Do you think the new syntropy story will increase people's motivation to build a better world?"

"I think I've believed in something like this all my life," Betska responded. "I just didn't know it had a name. In my work as a therapist I have so often observed a deep resilience within the human spirit, however wounded someone might be. Humans have a deep unconscious drive to seek wholeness, and an internal capacity for healing. Where does it come from? I concluded that it was inherent in the human condition, and that deep down, the bottom is solid and can be trusted, if we are willing to surrender to it. But maybe it's also because I have Jewish roots, that I do in fact believe that one day we will restore the Earth to the Garden of Eden."

"What about you, Lucas?" I asked.

"I'm not a big one for philosophy," he replied. "I leave that to people like Leo. But what I'm hearing is that syntropy says all beings are related and it's natural for humans to want to come together instead of fighting. It's natural to want to live in harmony with nature instead of abusing her. It's natural to want to love instead of hate; to cooperate instead of compete. It's natural to feel drawn to a vision of unity and harmony instead of one of hatred and hostility."

When Lucas spoke, the room became quieter. He had a raw magnetism, which must have been very powerful when he was in the thick of the OMEGA Days.

"Dezzy said you could be pretty inspiring, Lucas. I can see why!" Soluna said.

"Lucas, can I clone you and bring you back to Seattle?" Thaba said, smiling. "We could do with energy like yours. Personally, I love big picture thinking, but you're right: most people get by quite happily without it. They just need to believe that their instinct to make the world a better place is on solid ground and not about to disappear down some post-modernist hole, destructuring the context of trans-dialectical vision through post-textual analysis, shredding the neo-cultural narrative to trigger a post-paradigmatic collapse. So yes, as the popular understanding of syntropy theory spreads, I believe it will accelerate positive social change."

"That was hilarious, Thaba!" Soluna said. "How ever do you come up with that stuff?"

"I have a post-modernism generator chip embedded in my brain. I find it very useful at dinner parties with my fellow academics."[943]

Leo laughed uproariously, and everyone chuckled.

"How about you, Aliya?" I asked.

"I find these ideas deeply inspiring," she replied. "It's more than a little bit amazing. It enables me to integrate my love of science with my love of God and my activism. Is it really true that syntropy has been operating since the very beginning of the Universe?"

"That's the theory," Thaba replied.

"And that it fits with both physics and biology?"

"Yes. The syntropy concept says that the impulse we experience to organize an activity or to plan a new venture is the same impulse that hummingbirds experience when they build a nest and the immune cells in our bodies experience when they heal a wound."

"That's so beautiful," Aliya said. "It gives me incredible hope. I've heard people talk about syntropy at the hospital, but I didn't understand it properly until now. Mind you, I'm still not sure I really do. It feels as if syntropy is expressing the creative will of Allah, peace be upon Him. It's telling me that the Universe Allah created has a moving, dynamic aspect in which we, who are part of the beauty of Allah's creations, seek a greater and more perfect union. Not with Allah Himself, Peace be upon Him, but with His creation."

"Don't the Sufis seek union with God directly?" Betska asked.

"Yes, but I'm not a Sufi; I was raised as a Sunni Muslim. I was taught that it's blasphemous to even suggest that a human could have union with something as great and unknowable as Allah. But I love the impulse towards greater unity that syntropy theory seems to express."

"What about you, Leo?" I asked.

"It's very powerful," he replied. "What matters for me is to strip it of any woo-woo factor and be able to present it with as much gravitas as we do the theory of gravity, if you'll excuse the pun. Less than half the human population responds to things that are intuitive and philosophical. If it's going to have an impact, it's got to be practical and grounded.

"I would go further," Leo continued. "If early Chinese, Greek and Islamic science is Phase One of science, and Copernicus to the present is Phase Two, then maybe syntropy is launching Phase Three. That's how fundamental the integration of the inner and the outer is, after so many centuries of separation."

"I agree," Soluna said. "It tells us that the Universe is biofriendly, as the physicist Paul Davies has claimed."[944]

"Can you explain in simple words the difference between syntropy and entropy, and how they relate to each other?" Leo asked.

"That's a big question," Thaba replied. "If we look at them separately, syntropy operates in the realm of consciousness, while entropy operates in the old-fashioned realm of matter. In the material worldview there is no free will, no purpose, and entropy's the only game in town. Heat has never been observed to pass from a colder to a warmer body. And when we measure events in the material world, the second law of thermodynamics, which states that the entropy of an isolated system will always increase, always holds.

"Strictly speaking, entropy only speaks about heat. It does not speak about organization, though many people have misunderstood the second law, thinking that it also says that disorganization in a system will always eventually increase.

"But now we know that the Universe is not solely material, and that consciousness and matter are intrinsically entangled. We also know from personal experience that disorganization does not always increase. Indeed, we have observed a tendency to self-organization throughout evolution that is clearly *negentropic*—it has negative entropy. Syntropy, operating through consciousness, appears to balance entropy, enabling the progress of evolution and civilization to occur. How they integrate in the long run is still a mystery, just as it's a mystery how the Universe came to have such a low state of entropy at the time of the Big Bang, when it all kicked off. Does time flow with entropy, with syntropy, or with both—or is time strictly a secondary experience? It's a big unanswered question."

Silence around the dinner table.

"How about you, Dezzy?" I asked. "What do you think?"

"I'm wondering what Derek might have thought if he was with us today. He would probably have wanted to make a movie about it, to reach the widest possible audience. Something that showed the tension between entropy and syntropy in the world that would make people realize that we *do* have a choice, we *can* influence what happens in the world."

"And what do you think, Patrick?" Soluna asked me.

I wasn't expecting that; my mind was still processing. "I'm still taking it all on board," I replied, playing for time. And then it came to me: "Would you say that on days when we doubt everything, syntropy offers us a deep confidence that the Universe wants us to succeed?"

"Maybe," Soluna replied. "But we can only succeed if enough humans get involved to make it so. The Universe does appear to be programmed to want to make it so, which is the good news, but the decision to proceed always rests with us. There have been many civilizations that collapsed because the hubris and self-entitlement of those who controlled things inhibited innovation and change and brought about their downfall."

"But they didn't have the story of syntropy to encourage them," I responded. "And their people were probably following one of monotheism's apocalyptic stories, which said the world was full of sin and evil and the only goodness lay in Heaven, after death."

"I see what you're saying: that the very fact that we understand syntropy theory makes the Universe a more hopeful place."

"Yes. Something like that."

"My worry is that we might be fooling ourselves," Betska said. "The human mind is at its most vulnerable when it really wants to believe something. It's one thing to believe that the human soul can find healing if it surrenders to a greater whole, whether one calls it God, Nature or The Universe; but it's quite another to believe that the entire Universe is set up that way. It would be truly amazing, if it is."

"Science is not a perfect art form," Thaba responded. "If you think of 'mystery' as a veil that covers all reality, we're still only lifting a tiny corner of the veil. The veil still hides almost all of the known Universe—and the entire unknown Universe. We lift the veil a tiny bit, and we tackle the puzzles we find. Sometimes we find a piece that makes sense of some loose edges. Sometimes we see a pattern. And sometimes we see a larger pattern, which obliges us to throw away our previous ideas.

"Syntropy is one of those larger patterns. It's totally possible that future scientists will find a new pattern that makes more sense, in which case they'll discard syntropy, or limit it to a special case. For now, however, it's making sense, and it's enabling us to pull a lot of pieces together. Our understanding of consciousness is still incredibly young; who knows where it will go when we integrate modern understandings from the West with ancient understandings from the East? To think that consciousness may be the fundamental substance that the Universe is made of—and we experience it every minute of every waking day. I find that totally mind-blowing. But there's an awful lot that's still taboo. Take death, for instance, and the fact that some people seem to have memories of a past life, backed by evidence that seems pretty solid...."

At that moment, a phone rang. Dezzy recognized the tone and looked around to find it. She pressed a button on her wristband and the phone made a stronger beeping sound, enabling her to find it down the back of the sofa. She listened for a while, saying a few words to whoever was calling, and then sat bolt upright, silencing us all and signaling to get our attention.

"It's Laura," she said. "It's about Derek. Lucas—quick, turn on the news! CBC, Channel 26! I'll put Laura on the speakerphone." Dezzy used her phone to bring the CBC News up on the wall-screen, with Laura in the corner of the screen.

My heart jumped a beat, excited to see Laura and eager to know what was happening. And all this with only fifteen minutes left before I had to leave.

"Holy shit—they've got Syd Brockle!!" Laura said, as clearly as if she was in the room.

TV reporter: "The Calgary police reported today that they have arrested the prominent Australian businessman and oil tycoon Syd Brockle on multiple charges of conspiracy to commit murder relating to the death of OMEGA Movement leader Derek Brooks, who was shot and killed during a demonstration in Vancouver many years ago. His death has remained an unsolved crime, so while the police are remaining tight-lipped, Brockle's arrest indicates that they have made a breakthrough in the case. We will follow the story closely as it unfolds."

We were glued to the screen, with Laura on speakerphone.

"Hey—it looks as if they've got someone in Vancouver too!" Lucas said.

"My god—that's our retired Chief Constable, Ray Robinson!" Dezzy said. "Holy shit!"

TV reporter: "Vancouver's recently retired Chief Constable, Ray Robinson, has also been arrested on charges of conspiracy to commit murder and conspiracy to obstruct justice. His lawyer says his client is completely innocent and if necessary he will fight the charges all the way to the Supreme Court."

"Stay on the line," Laura said. "My dad's ringing." Laura picked up her father's call and spoke to him. When the TV went on to a story about the coming Summer Olympics she switched to bring her father—Donald—onto the screen so that he could talk to us.

"This is pretty big news," Donald said. "How are you taking it, Dezzy? I expect you're pleased there's finally been an arrest after all these years." Dezzy couldn't speak, since she was sobbing and shaking, with Betska and Aliya comforting her.

"She'll be okay," Lucas said. "It's a pretty big shock. What happened?"

"Laura called me last night and told me about the conversation she'd had with your visitor, Patrick, and the connection to Syd Brockle that he'd picked up from someone he met on a farm. Liu Cheng suggested that we visit Syd's place up in the Kananaskis Country and knock on his door to see what would transpire. We arranged backup, just in case, and when Syd answered the door in his dressing gown he was all very charming, happy to invite us in. He's a very genial fellow, but he had this guy staying with him—Buffalo John, they call him. He used to work in private security, where he gained a reputation for being an ace marksman, a sniper. Buffalo John totally freaked when he learned who we were and he escaped out the back of the house and into the forest. But he won't be free for long—the police have tagged him in the leg so now he's got a GPS marker telling them his every move. The Calgary police have had Syd in the cells all day, but he's holding out for something big. He's giving the sense that there were far deeper forces at work."

By now Dezzy had recovered and was watching Donald talk, her face glued intensely to the screen.

"Syd's claiming it was never his intention that Buffalo John should kill Derek. That was all a misunderstanding, he claims, based on some intemperate words that might have been spoken. We know Syd was financing the Alphas, but we think Syd was acting for others higher up the food chain, but we'll need time to work that angle. And guess what else? It has become clear that Syd was tight with Vancouver's Deputy Chief Constable Ray Robinson, which is why there was so little progress on the file. The Vancouver police are arresting him as we speak."

"But there were two shooters," Dezzy said, now fully engaged. "Who was the second one?"

"They still don't know."

"So what was their motivation? Why did they do it?"

"Our first theory was that it was most likely connected to Derek's campaigning to close the tax havens. But now we're questioning that. Like I said, we think it goes deeper. It may be connected to Derek's campaigning to end the monopoly on the creation of credit that the private banks enjoyed."

"So the Vancouver Police Department was involved all the time?" Lucas asked.

"We don't know for sure. Liu Cheng's hoping Ray was acting alone. He's really pissed. He was Chief Constable at the time, so it all falls back to him. He's taking it very badly: he's probably going to make a statement tomorrow."

At that moment, just as I was about to announce that I had to leave in five minutes, Leo held up his device and exclaimed, "Hey! Look at this! There's something really big happening in Beijing. There's a mass occupation happening in Tiananmen Square; they're calling for a global uprising. There must be thousands of them! Tens of thousands!

"It's huge! Hashtag #EarthRiseUp, it says. It's happening!!"

"Can't it wait?" Dezzy said. "I've had fifteen years of frustrated grief and finally we've got the news I've been waiting for…."

"But Derek would love this!" Leo responded. "They're calling on people all over the world to join them! I've got to get involved, to get Vancouver on board!"

"Look, I'd best be getting on," Donald said, aware of the distraction and keen to wrap things up. "But before I go, Dezzy, I have a message for you from Liu Cheng. He is really distraught that his Deputy Police Chief seems to have been involved in the cover up. He wants to say how profoundly sorry he is, and he hopes that, some day, you might find it in your heart to forgive him for not figuring this out sooner. He's too cut up to talk right now."

"Please give him my love," Dezzy said. "And tell him I do forgive him. I know how much he took Derek's case to heart."

It was all happening. Laura was telling me how much her time with me had meant to her; Dezzy was sobbing in Aliya's arms, supported by Lucas; Leo was trying to get us to pay attention to the new global developments. "It's happening in Jakarta, too! And Mumbai! There's a call going out for RiseUp gatherings in every city, every town. This is it! This is it!"

Betska, meanwhile, was cradling her cup of tea, smiling contentedly. "It's all coming round," she said to me quietly. "It's all coming round. I'm so glad I'm here to see this day."

And I had to leave.

I went around thanking everyone and saying goodbye, and then I quietly slipped out the door.

Outside, the night was peaceful, the late evening sky a glorious glow of red above the rooftops. I crossed the road and sat down on the bench in the cob shelter, taking in my last few moments in the future… this wonderful future.

 Epilogue
My Return

IT HAS BEEN almost nine months since I left the Vancouver of the future and returned to our apartment off Commercial Drive, in the Vancouver of the present.

When I realized that I was no longer in the future, it was a profound shock. I went around for days in a state of grief, as if the people I had become friends with had died. I missed them so much, and I could not reconcile myself to the thought that I would never see them again, except in my memory. So then I began to write it all down.

Do I have any final words? If I do, they relate to my reason for undertaking my journey, which was to visit a future in which Vancouver had become one of the greenest cities in the world, and to learn what motivated people to make it so.

During my time in the future I kept hearing snippets about this new concept, syntropy, which was finally explained during that wonderful final dinner party. Was this the secret that had motivated them? If syntropy *is* real, the idea that there is a fundamental force in the Universe akin to gravity that drives all existence to self-organize for the ultimate benefit of all beings… well, that would be amazing.

But when I reflect on the people I met, and the way they talked, syntropy didn't really come into it, not as a conscious motivator or a new ultimate story, to use Soluna's words.

What motivated them was the same muddle of thoughts and feelings that motivate so many of us today: the same mixture of hope, frustration, compassion and anger. What they *did* have was the determination to make the world a better place. They weren't wasting time being cynical, or feeling hopeless about the future. So in this sense, they were acting as syntropy predicts we should, and in doing so they were shaping a powerful new story about who we are, and our future on this planet.

They didn't need to understand syntropy as a scientific theory. If it turns out to be true, it would indeed be an incredible new story. But we don't have the luxury of waiting to see if a scientific breakthrough will validate it. We just need to be alive, to feel compassion in our hearts and inspiration in our minds, and to roll up our sleeves and get to work.

ENDNOTES

JOURNEY TO THE FUTURE contains over 940 endnotes, almost all of which have web-links referencing the original material on which the novel is based. You can find the endnotes on the book's website, along with many other resources.

www.journeytothefuture.ca

ACKNOWLEDGEMENTS

MANY PEOPLE HAVE helped me write this book, and I'm very grateful for the time they took to review it, from short sections to the entire book. My thanks, therefore, to Alison Rustand, Amber Freer, Amelita Kucher, Aviad Sar Shalom, Bob Willard, Brenda Sawada, Brian Gould, Brian Pinch, Cathy Orlando, Don Salmon, Dr. Amanda Dauncey, Dr. Mary-Wynne Ashford, Dr. Roland Guenther, Duncan Sutherland, Dr. Elisabet Sahtouris, Felix Kramer, Gail Leondar-Wright, Gene Miller, Goksenin Sen, Graeme Taylor, Helene Ross, Herb Barbolet, Jack Barker, Jamie Kneen, Jane Dauncey, Jasmin Gerwein, JC Scott, Jeremy Finkleman, Jill Doucette, John Stonier, Judy Graves, Julie Carter, Katherine Palmer Gordon, Kerry Dawson, Kim Baird, Liz Courtney, Lodoe Lodoe, Mahla Shapiro, Martin Golder, Meredith Bingham, Mia Nissen, Michael Clague, Michael Nation, Miriam Kennett, Nicholas Shaxson, Nick Wilde, Nicole Chaland, Nicole Moen, Pat McMahon, Patrick Robertson, Pemba Doma Indup, Pete Russell, Professor Olaf Schuiling, Renate Sitch, Ric Cool, Rupert Sheldrake, Sam Carana, Sanjara Omoniyi, Sharon Quigley, Stephen Rees, Stephen Salter, Sylvia Olsen, Tim Roberts, Tom Hackney, Warren Bell and Wayne Madden. My appreciation in no way links them to the content, for which I bear sole responsibility.

I ALSO WANT to thank Cathy Reed, who edited an earlier version of the book, and Bruce Batchelor from the Agio Publishing House in Victoria, who guided me through the final editing and publishing stages, gave thoughtful and supportive advice, and helped the book become a reality. And thanks also to Agio's Marsha Batchelor for creating a great cover.

AND FINALLY I want to thank my partner, Carolyn Herriot, for her encouragement during the years that it took me to write this, when I was often buried away inventing this possible future. I'm fortunate to have been able to have a vision, along with the ability to express it. I'm also fortunate to have a life partner who believes in me, and who trusts what I am up to. The *Gratitude Pledge* that Patrick joined in reciting in *Song of the Universe* says it all.

APPENDIX 1: BECOMING ENGAGED

AFTER READING THIS book, you may ask, "How can I become more engaged?"

It's a difficult question to answer, for if you take a thousand engaged people you will find a thousand different ways in which they are making a difference in the world. But rest assured: there is a way that is specifically yours, unique to your passion, interests and skills. So how to find it?

The first step is commitment: I recommend taking a piece of paper, the grander the better, and writing, "I want to make a difference in the world." Then fix it to your fridge or put it by your bed where you will see it regularly. Your statement of intent is to yourself, to the Universe, to your friends, to God, however you prefer to see it.

The second step is context. Are you looking to find a deeper voluntary engagement, to change career, to embark on a period of training, to start a social change initiative, or make a difference through your current line of work? All are relevant, and all can be meaningful.

The third step is exploration. Out of the hundred realms where change is needed, two or three will call to you with a stronger voice. Of the many topics covered in this book, which interest you the most? Choose three and write them down. Then choose just one and spend a month learning more about it. Read books and articles. Find non-profit organizations that focus on it, and explore what they do. Find local groups that are active. Make friends with the people involved and go to their meetings. Offer to volunteer with them, to see how it feels. Don't limit yourself to social media interactions: they can be shallow and sometimes misleading. Read real books, and meet real people. For inspiration, as well as my regular research reading, I like to read biographies of people I admire, immersing myself in their lives and drawing inspiration from what they achieved. My most recent was *William Wilberforce*, by William Hague (Harper Press, 2007).

It is important to approach your quest as systematically as you would buying a house or finding a place to rent. Don't hesitate to call people out of the blue and ask if they have time to meet you. Make a plan with daily and weekly tasks, and stick to it. If you have a buddy you can share what you're doing with, so much the better. One thing will lead to another. Once you know your overall direction, you only need to know what your first step is. After that, you can consider your second step. As you start walking, your journey will begin.

And remember: the more deeply you become involved in an area of change and the better you get to know it, the more you will be able to contribute. People will respect you for your knowledge, and you will build a network of friends and contacts who can help you get things done.

There are very few for whom this will be a solitary affair. For everyone else it will involve friends, fun and relationships. Whenever I have launched a new initiative I have invited friends who share my interest to join me and we have done it together. In this way, changing the world becomes part of your wider life, and you will build friendships with people who share your values, and your desire to build a better world.

For a list of organizations that are active for change in Vancouver, and maybe also for other communities, see www.journeytothefuture.ca

— *Guy Dauncey*

APPENDIX 2: THE OMEGA SOLUTIONS

The five OMEGA themes were developed by a team of people from around the world. The regional solutions for each theme were crafted locally in a participatory process. Here are the Vancouver OMEGA solutions as they were explained to Patrick.

O for Open Democracy

1. Make all voting proportional and electronic, with a $100 tax credit to encourage voting.
2. Lower the voting age to 16, enabling most young people to experience their first election while still at school.
3. Give every party a public subsidy based on the number of votes received at the last election, and impose a limit on campaign fundraising, with no donations more than $1,000.
4. Create more transparency and accountability among politicians, lobbyists and civil servants, with disclosure rules to end closed doors meetings and secret deals.
5. Encourage political parties and governments to support the process of cooperative self-organization, which is the foundation of our efforts to build a better world.

M for Meaningful Work

1. Help Each Other. Create Business Networks and Cooperative Support Networks.
2. Put Your Money Where Your Heart Is. Change the way we create, store, circulate and lend money so that it serves the wellbeing of people and nature.
3. Share the Work, Share the Wealth. Embrace measures to increase worker and community ownership, co-operatives, employee-shareholding, and fair pay.
4. All for One and One for All. Citizen's Income, community-based welfare, fair taxes and fair benefits, to reduce poverty and inequality.
5. Go Green, Now and Forever. Support Green Business Certification, and Benefit Corporations.

E for a New Economy

1. Embrace real world definitions of wealth and growth. Redefine wealth and growth to include their social, environmental and cultural dimensions, using new indicators and systems of analysis that recognize the goal of development over growth in consumption.

2. Embrace real world sustainable business. Reframe corporate law to require the pursuit of social and environmental benefit as well as profit.
3. Embrace real world sustainable banking, including publicly owned state and federal banks. Reframe corporate banking law to require the pursuit of social and environmental benefit as well as profit.
4. Embrace real world taxation and regulation. End tax avoidance; capture external costs through appropriate taxation; re-impose the necessary regulations with strict standards of transparency and public scrutiny; end corporate capture of the regulators.
5. Embrace real world global financial architecture. Reframe the charters and goals of the World Trade Organization and the Bretton Woods Institutions so that they support global social, environmental and cultural progress as well as economic progress.

G for a Green Future

1. Cherish and protect nature. Enact legislation to give nature a safe, secure home with legal rights, and a global statute to made ecocide a crime against peace, with a lawyer to represent nature in court. Require the restoration of habitat in all new developments.
2. Cherish and protect the atmosphere. Phase out all fossil fuels, and replace them with 100% renewable energy for electricity, heat and transport. Work to capture as much surplus carbon from the atmosphere as possible, using natural methods of farming, pasturing and forestry.
3. Cherish and protect the waters. Give every river, lake and stream the legal right to exist in a pure, uncontaminated condition; strict development codes on aquatic and waterside development. Enact a *Global Oceans Treaty* to create Marine Protected Areas over 40% of the global ocean. Embrace a community-based approach to fisheries and fish-stock preservation and restoration.
4. Cherish and protect the earth. Develop sustainability policies for the farmlands, grasslands and forests. Enact an *Organic Farms Transition Act*.
5. Cherish and protect the humans. Enact legislation to enshrine the right of every child to have uncontaminated blood, breathe uncontaminated air, drink uncontaminated water and eat uncontaminated food.

A for Affordable Living: Housing, Food, Childcare, Healthcare and Transportation

Patrick received a good explanation of Affordable Housing goals from Lucas. Some points for the other goals are included.

Affordable Housing

1. Gather the money. Enact a progressively increasing tax on real estate sales over a million dollars; a sales tax on properties bought through offshore companies; an annual levy on properties registered offshore for tax reasons; an escalating land tax on empty properties; and a special development levy on commercial property development. Use the income to build affordable housing and to buy out slum landlords and restore their properties.

2. Establish a Community Land Trust to buy land and build rental housing, housing co-ops and cohousing projects, financed by the income received above.

3. Create a Micro-Village Zoning that allows mini-estates of micro-homes to be built for and by homeless people on temporarily vacant land.

4. Encourage tenants renting apartments to form Tenants' Stewardship Councils. Create a Tenants' Charter, recognizing tenants' desire to live in a building that is comfortable and energy efficient, free of fumes and infestations, with space to grow food and store bicycles and recyclables, in return for a commitment to look after the property, abide by a code of respectful conduct, and agree to a set of conditions if someone can't pay the rent. Give Stewardship Councils first right of refusal to buy a building if it comes on the market to convert into a Housing Co-op, with the land being owned by a Community Land Trust.

5. Create a Farm Village Zoning that allows any farmer owning more than twenty hectares to be allowed to sell one hectare for development as a farm village, with conditions attached to ensure that the people living there farm the land.

Affordable Food

1. Create more Community Garden Plots, Urban Orchards and Farmers Markets.

2. Grow food and fruit trees and bushes on boulevards and other public green spaces.

3. Assist Food Banks to become Community Food Centres where people can learn to grow their own food, form community buying clubs,

participate in community kitchens, and receive food vouchers that can be used at Farmers Markets.

4. End the practice of supermarkets throwing out waste food.

Affordable Childcare
1. Ensure that there are enough subsidized quality childcare spaces for all who need it.
2. Introduce one year of paid maternity or paternity leave.
3. Break the cycle of disadvantage for the children of low-income parents.[945]

Affordable Healthcare
1. Maintain Canada's system of socialized healthcare.
2. Change from a disease-oriented to a health-oriented system of functional healthcare based on nutrition, prevention and community care.
3. Add Pharmacare, providing universal and equitable coverage for the cost of prescription drugs in conjunction with Medicare.[946]

Affordable Transportation
1. Redesign neighborhoods to make walking attractive and easy.
2. Build a comprehensive safe, separated cycling infrastructure.
3. Encourage all schools to teach safe cycling.
4. Invest in a major expansion of quality electric public transit.
5. Support the expansion of car-sharing and ride-sharing.

APPENDIX 3: BOOK CLUBS, HOUSE READINGS & STUDY GROUPS

BOOK CLUBS

As a member of a great men's book club, I'm reluctant to suggest any limits to discussion. With a book that covers so many topics, however, it might make sense to first ask which parts of the book people found the most interesting, and whether people found it a believable future. Then review the various themes, and choose which to focus on from the list below. I have also suggested some discussion questions.

One useful discussion may be the theme of the book as a whole: do you feel hopeful about the future? And if not, what would it take to allow you to feel hopeful? What are the important things we need to do to enable people to feel hopeful about the future, and become engaged in making a difference?

HOUSE READINGS

Here is one successful way of holding an evening with up to 40 people. Before your guests come, go through the book and select 15 varied passages that lend themselves to public reading, marking them with a pencil. Take 15 slips of paper, write a heading for each selection with the relevant page number, and put the slips in a bowl.

When people arrive, ask who would like to do a reading, and invite them to pick a slip at random. After your introduction, invite people to come up one at a time to introduce themselves, say something about what interests them in life, and then do their reading. This makes the evening about personal connection and networking, as well as the book. After each reading, or after all the readings, open the evening for discussion.

STUDY GROUPS

For a six-week study group you have two choices. In Choice A, everyone reads the whole book and then you vote on the topics you want to discuss, choosing one or two per session from the list below. In Choice B, you read a number of chapters each week and discuss them together. The author welcomes feedback on which approach works best.

QUESTIONS FOR BOOK CLUBS, HOUSE READINGS & STUDY CIRCLES

- Before you started reading *Journey to the Future*, how would you rate your hopefulness about the future of our world on a scale of 0-10, where 0 = completely hopeless and 10 = very hopeful?
- How do you rate yourself on the cynicism scale, where 0 = not cynical at all

and 10 = totally cynical? Did reading this book make you more or less cynical, and if so, why?

- Have there been any other changes in your outlook as a result of reading the book, and if so, what are they? If not, what would it take to enable you to feel more hopeful?
- What are the important things we need to do to enable people to feel more hopeful about the future, and to become more engaged in making a difference?
- Choosing any of the topics listed below, is the author being reasonably optimistic or hopelessly naïve? If you think he is being naïve, what other changes do you think will be needed to make a difference?
- How important is the 'ultimate story' that you use to answer the big questions in life in shaping your vision of the future? (Chapter 34, pages 396-397)

SUGGESTED TOPICS FOR BOOK CLUBS AND STUDY GROUPS

Hungry for hope	First Nations
Green buildings, solar	Open democracy
Building a neighborhood	Closing the tax havens
The OMEGA Days	Food and farming
Affordable housing	The end of fossil fuels
The new synthesis	The Second American Revolution
Healthcare	Education
Cycling and transportation	Telepathy, consciousness
Poverty, homelessness	Forestry
The climate solutions	University of the future
A new local economy	Sustainability in suburbs
The next financial crash	Science, syntropy
Israel/Palestine	Ultimate stories
Green business	Overall responses

APPENDIX 4: SOURCES

CLIMATE AND ENERGY

Allen, Paul. *Zero Carbon Britain: Rethinking the Future*. Centre for Alternative Technology, 2013.

Berners-Lee, Mike, Duncan Clark. *The Burning Question: We can't burn half the world's oil, coal and gas. So how do we quit?* Greystone, 2013.

Braasch, Gary. *Earth Under Fire: How Global Warming is Changing the World*. University of California Press, 2007.

Brown, Lester. Plan B 3.0. *Mobilizing to Save Civilization*. Norton, 2008.

Dauncey, Guy. *Stormy Weather: 101 Solutions to Global Climate Change*. New Society Publishers, 2000.

Dauncey, Guy. *The Climate Challenge: 101 Solutions to Global Warming*. New Society Publishers, 2009.

Dyer, Gwynne. *Climate Wars*, Random House, 2008.

Flannery, Tim. *Now or Never. Why We Need to Act Now to Achieve a Sustainable Future*. Harper Collins, 2009.

Flannery, Tim. *The Weather Makers: How We are Changing the Climate and What it Means for Life on Earth*. Harper Collins, 2005.

Freese, Barbara. *Coal: A Human Story*. Penguin, 2003.

Goodall, Chris. *Ten Technologies to Save the Planet*. Green Profile, 2008.

Goodstein, Eban. *Fighting for Love in the Century of Extinction*. University of Vermont Press, 2007.

Gore, Al. *An Inconvenient Truth: The Planetary Emergency of Global Warming and What We Can Do About It*. Rodale, 2006.

Hansen, James. *Storms of My Grandchildren: The Truth about the Coming Climate Catastrophe and Our Last Chance to Save Humanity*. Bloomsbury, 2009.

Hoggan, James. *Climate Cover-Up: The Crusade to Deny Global Warming*. Greystone, 2009.

Klein, Naomi. *This Change Everything: Capitalism vs the Climate*. Penguin Random House, 2014.

Leggett, Jeremy. *The Energy of Nations: Risk Blindness and the Road to Renaissance*. Earthscan, 2014.

Leggett, Jeremy. *The Winning of the Carbon War*. www.jeremyleggett.net, 2015.

Lovins, Amory. *Reinventing Fire: Bold Business Solutions for the New Energy Era*. Chelsea Green, 2011.

Lynas, Mark. *Six Degrees: Our Future on a Hotter Planet*. Fourth Estate, 2007.

Marsden, William. *Stupid to the Last Drop. How Alberta is Bringing Environmental Armageddon to Canada (and Doesn't Seem to Care)*. Vintage, 2007.

McKibben, Bill. *Oil and Honey: The Education of an Unlikely Activist*. Times Books, 2013.

Monbiot, George. *Heat: How to Stop the Planet from Burning*. Doubleday Canada, 2006.

Nikiforuk, Andrew. *Tar Sands: Dirty Oil and the Future of a Continent*. Greystone, 2008.

Oreskes, Naomi and Erik Conway. *The Collapse of Western Civilization: A View From the Future*. Columbia University Press, 2014.

Pearce, Fred. *The Last Generation: How Nature Will Take Her Revenge for Climate Change*. Key Porter, 2007.

Romm, Joseph. *Cool Companies*. Island Press, 1999.

Romm, Joseph. *Hell and High Water: Global Warming, the Solution and the Politics*. Morrow, 2007.

Romm, Joseph. The website *Climate Progress*: www.thinkprogress.org/climate

Scheer, Herman. *Energy Autonomy: The Economic, Social and Technological Case for Renewable Energy*. Earthscan, 2007.

Scheer, Herman. *The Solar Economy: Renewable Energy for a Sustainable Global Future*. Earthscan, 2002.

Simpson, Jeffrey, Mark Jaccard and Nic Rivers. *Hot Air: Meeting Canada's Climate Change Challenge*. Douglas Gibson, 2007.

Weaver, Andrew. *Keeping Our Cool: Canada in a Warming World*. Viking Canada, 2008.

ECONOMICS AND FINANCE

Alperovitz, Gar. *What Then Must We Do? Straight Talk about the Next American Revolution*. Chelsea Green, 2013.

Atkinson, Anthony. *Inequality: What Can Be Done?* Harvard University Press, 2015.

Botsman, Rachel and Roo Rogers. *What's Mine is Yours: The Rise of Collaborative Consumption*. Harper Collins, 2010.

Brown, Ellen. *The Public Bank Solution: From Austerity to Prosperity*. Third Millennium Press, 2013.

Chang, Ha-Joon. *23 Things They Don't Tell You About Capitalism*. Bloomsbury Press, 2010.

Chang, Ha-Joon. *Bad Samaritans: The Guilty Secrets of Rich Nations and the Threat to Global Prosperity.* Random House, 2008.

Eisenstein, Charles. *Sacred Economics: Money, Gift and Society in the Age of Transition*. Evolved Editions, 2011.

Freeland, Chrystia. *Plutocrats: The Rise of the New Global Super-Rich and the Fall of Everyone Else*. Doubleday Canada, 2012.

Graeber, David. *Debt: The First 5,000 Years*. Melville House, 2011.

Hartmann, Thom. *The Crash of 2016: The Plot to Destroy America—and What We Can Do to Stop It*. Twelve, 2013.

Hartmann, Thom. *Threshold: The Crisis of Western Culture*. Viking, 2009.

Hawken, Paul, Amory Lovins and Hunter Lovins. *Natural Capitalism: Creating the Next Industrial Revolution*. Little, Brown and Company, 1999.

Juniper, Tony. *What Has Nature Ever Done For Us? How Money Really Does Grow on Trees*. Profile, 2013.

Kennedy, Margrit. *Occupy Money: Creating an Economy Where Everybody Wins*. New Society Publishers, 2012.

Korten, David. *Agenda for a New Economy: From Phantom Wealth to Real Wealth*. Berrett-Koehler, 2009.

Korten, David. *The Post-Corporate World: Life After Capitalism*. Berrett-Koehler, 1999.

Lanchester, John. *Whoops! Why Everyone Owes Everyone and No One Can Pay*. Penguin, 2010.

Lewis, Michael and Pat Conaty. *The Resilience Imperative: Cooperative Transitions to a Steady-State Economy*. New Society Publishers, 2012.

Lietaer, Bernard and Jacqui Dunne. *Rethinking Money: How New Currencies Turn Scarcity into Prosperity*. Berrett-Koehler, 2013.

Luyendijk, Joris. *Swimming With Sharks: My Journey into the World of the Bankers*. Guardian Books, 2015.

MacLeod, Andrew. *A Better Place on Earth: The Search for Fairness in Super Unequal British Columbia*. Harbour Publishing, 2015.

Mason, Paul. *Meltdown: The End of the Age of Greed*. Verso, 2009

Mason, Paul. *Why It's Still Kicking Off Everywhere: The New Global Revolutions*. Verso, 2013.

Mazzucato, Mariana. *The Entrepreneurial State: Debunking Public vs Private Sector Myths*. Anthem Press, 2013.

McDonough, William and Michael Braungart. *The Upcycle: Beyond Sustainability—Designing for Abundance*. North Point Press, 2013.

Nasar, Sylvia. *Grand Pursuit: The Story of Economic Genius*. Simon & Schuster, 2011.

Perkins, John. *Hoodwinked: An Economic Hit Man Reveals Why the World Financial Markets Imploded—and What We Need to Do to Remake Them*. Broadway Books, 2009.

Pettifor, Ann. *Just Money: How Society Can Break the Despotic Power of Finance*. Commonwealth Publishing, 2014.

Ryan-Collins, John, Tony Greenham, Richard Werner and Andrew Jackson. *Where Does Money Come From? A Guide to the UK Monetary and Banking System*. New Economics Foundation, 2011.

Scott, Brett. *The Heretics Guide to Global Finance: Hacking the Future of Money*. Pluto Press, 2013.

Shaxson, Nicholas. *Treasure Islands: Tax Havens and the Men Who Stole the World*. The Bodley Head, 2011.

Shuman, Michael. *The Small-Mart Revolution: How Local Businesses Are Beating the Global Competition*. Berrett-Koehler, 2006.

Standing, Guy. *The Precariat: The New Dangerous Class*. Bloomsbury, 2011.

Stiglitz, Joseph. *Freefall: Free Markets and the Sinking of the Global Economy*. Allen Lane, 2010.

Taibbi, Matt. *The Divide: American Injustice in the Age of the Wealth Gap*. Spiegl & Grau, 2014.

Varoufakis, Yanis. *The Global Minotaur: America, Europe and the Future of the Global Economy*. Zed Books, London, 2015.

Wilkinson, Richard and Kate Pickett. *The Spirit Level: Why Greater Equality Makes Societies Stronger*. Bloomsbury Press, 2009.

Willard, Bob. *The New Sustainability Advantage: Seven Business Case Benefits of a Triple Bottom Line*. New Society Publishers, 2012.

PHYSICS, BIOLOGY AND CONSCIOUSNESS

Chalmers, David. *The Conscious Mind: In Search of a Fundamental Theory*. Oxford University Press, 1996.

Doidge, Norman. *The Brain That Changes Itself: Stories of Personal Triumph from the Frontiers of Brain Science*. Penguin, 2007.

Falk, Dan. *The Universe on a T-Shirt: The Quest for the Theory of Everything*. Viking Canada, 2002.

Kauffman, Stuart. *At Home in the Universe: The Search for Laws of Complexity*. Viking, 1995.

Kauffman, Stuart. *Reinventing the Sacred: A New View of Science, Reason, and Religion*. Basic Books, 2008.

Koch, Christof. *Consciousness: Confessions of a Romantic Reductionist*. MIT, 2012.

Lanza, Robert. *Biocentrism: How Life and Consciousness are the Keys to Understanding the True Nature of the Universe*. Benbella, 2009.

Laszlo, Ervin. *Science and the Re-enchantment of the Cosmos: The Rise of the Integral Vision of Reality*. Inner Traditions, 2006.

Lewin, Roger. *Complexity: Life at the Edge of Chaos*. Macmillan, 1992.

Nagel, Thomas. *Mind and Cosmos: Why the Materialist, Neo-Darwinian Conception of Nature is Almost Certainly False*. Oxford University Press, 2012.

Narby, Jeremy. *Intelligence in Nature: An Inquiry into Knowledge*. Jeremy Tarcher, 2005.

Pfeiffer, Trish and John Mack (editors). *Mind Before Matter: Visions of a New Science of Consciousness*. O Books, 2007.

Playfair, Guy Lyon. *Twin Telepathy: The Psychic Connection*. Vega, 2002.

Radin, Dean. *Entangled Minds: Extrasensory Experiences in a Quantum Reality*. Paraview, 2006.

Radin, Dean. *Supernormal: Science, Yoga, and the Evidence for Extraordinary Psychic Abilities*. Deepak Chopra Books, 2013.

Radin, Dean. *The Conscious Universe: The Scientific Truth of Psychic Phenomena.* Harper One, 1997.

Rees, Martin. *Just Six Numbers: The Deep Forces That Shape the Universe.* Weidenfeld and Nicholson, 1999.

Sahtouris, Elisabet. *EarthDance: Living Systems in Evolution.* iUniversity Press, 2000.

Samanta-Laughton, Manjir. *Punk Science: Inside the Mind of God.* Iff Books, 2006.

Sheldrake, Rupert. *Science Set Free: Ten Paths to New Discovery.* Deepak Chopra, 2013.

Stannard, Russell (editor). *God For the 21st Century.* Templeton Foundation, 2007.

Strogatz, Steven. *Sync: The Emerging Science of Spontaneous Order.* Penguin, 2003.

Tononi, Giulio. *PHI: A Voyage from the Brain to the Soul.* Pantheon, 2012.

Turok, Neil. *The Universe Within: From Quantum to Cosmos.* Anansi, 2012.

OTHER READING

Ali, Ayaan Hirsi. *Infidel: My Life.* Atria Books, 2008.

Ali, Ayaan Hirsi. *Heretic: Why Islam Needs a Reformation Now.* Alfred A. Knopf Canada, 2015.

Ansary, Tamim. *Destiny Disrupted: A History of the World Through Islamic Eyes.* Public Affairs Books, 2009.

Boggs, Grace Lee. *The Next American Revolution: Sustainable Activism for the Twenty-First Century.* University of California Press, 2011.

Boyd, David. *The Optimistic Environmentalist: Progressing Towards a Greener Future.* ECW Press, 2015

Brockman, John (editor). *This Will Change Everything: Ideas That Will Shape the Future.* Harper Perennial, 2010.

Cahill, Thomas. *The Gifts of the Jews: How a Tribe of Desert Nomads Changed the Way Everyone Thinks and Feels.* Talese Anchor Doubleday, 1998.

Clover, Charles. *The End of the Line: How Overfishing is Changing the World and What We Eat.* Ebury Press, 2005.

Davidson, Ian. *Voltaire in Exile.* Atlantic Books, 2004.

Durant, Will and Ariel Durant. *The Age of Napoleon.* Simon and Schuster, 1975.

Fernandez-Armesto, Felipe. *Civilization: Culture, Ambition and the Transformation of Nature.* Simon and Schuster, 2001.

Fernandez-Armesto, Felipe. *Millennium: A History of the Last Thousand Years.* Simon and Schuster, 1995.

Fernandez-Armesto, Felipe. *1492: The Year The World Began.* Harper One, 2009.

Fisher, Rachel, Heather Stretch, Robin Tunnicliffe. *All The Dirt - Reflections on Organic Farming.* Touchwood, 2012.

George, Ernie and Sabrina Dugan. *How I Survived Lejac Residential School.* E-book (Amazon & Kobo), 2013.

Gordon, Katherine Palmer. *We Are Born With the Songs Inside Us: Lives and Stories of First Nations People in British Columbia*. Harbour, 2013.

Haidt, Jonathan. *The Righteous Mind: Why Good People are Divided by Politics and Religion*. Vintage, 2012.

Heath, Chip and Dan Heath. *Switch: When Change is Hard*. Random House, 2010.

Hessel, Stéphane. *Time for Outrage: Indignez-vous!* Twelve, 2010.

Higgins, Polly. *Eradicating Ecocide*. Shepheard-Walwyn, 2010.

Hitchcock, Darcy. *Dragonfly's Question: Principles for 'The Good Life' after the Crash*. Lulu, 2008.

Holmes, Richard. *The Age of Wonder: How the Romantic Generation Discovered the Beauty and Terror of Science*. Harper Press, 2008.

Karmi, Ghada. *Married to Another Man: Israel's Dilemma in Palestine*. Pluto Press, 2007.

Kimbal, Kristin. *The Dirty Life: On Farming, Food, and Love*. Scribner, 2010.

Lappé, Frances Moore. *EcoMind: Changing the Way We Think to Create The World We Want*. Nation Books, 2011.

Lyons, Jonathan. *The House of Wisdom: How the Arabs Transformed Western Civilization*. Bloomsbury Press, 2009.

MacKinnon, J.B. *The Once And Future World: Nature As It Was, As It Is, As It Could Be*. Houghton Mifflin Harcourt, 2013.

Mitchell, Alanna. *Sea Sick: The Global Ocean in Crisis*. McClelland & Stewart, 2009.

Montgomery, Charles. *Happy City: Transforming Our Lives Through Urban Design*. Doubleday Canada, 2013.

Montgomery, David. *Dirt: The Erosion of Civilizations*. University of California Press, 2012.

Pagden, Anthony. *The Enlightenment, and Why It Still Matters*. Random House, 2013.

Pinker, Steven. *The Better Angels of Our Nature: Why Violence Has Declined*. Viking, 2011.

Porritt, Jonathon. *The World We Made: Alex McKay's Story from 2050*. Phaidon, 2013.

Rong, Jiang. *Wolf Totem*. Penguin Press, 2008.

Sahlberg, Pasi. *Finnish Lessons: What Can the World Learn From Educational Change in Finland?* Teachers College Press, 2011.

Saul, Nick and Andrea Curtis. *The Stop: How the Fight for Good Food Transformed a Community and Inspired a Movement*. Random House, Canada, 2013.

Schama, Simon. *Citizens: A Chronicle of the French Revolution*. Penguin, 1989.

Walsh, Joan. *What's The Matter With White People? Why We Long For a Golden Age That Never Was*. John Wiley, 2012.

INDEX